SECOND EDITION

Fuel Cells

Dynamic Modeling and Control with Power Electronics Applications

POWER ELECTRONICS AND APPLICATIONS SERIES

Muhammad H. Rashid, Series Editor
University of West Florida

PUBLISHED TITLES

Advanced DC/DC Converters
Fang Lin Luo and Hong Ye

Alternative Energy Systems: Design and Analysis with Induction Generators, Second Edition
M. Godoy Simões and Felix A. Farret

Complex Behavior of Switching Power Converters
Chi Kong Tse

DSP-Based Electromechanical Motion Control
Hamid A. Toliyat and Steven Campbell

Electric Energy: An Introduction, Third Edition
Mohamed A. El-Sharkawi

Electrical Machine Analysis Using Finite Elements
Nicola Bianchi

Electricity and Electronics for Renewable Energy Technology: An Introduction
Ahmad Hemami

Fuel Cells: Dynamic Modeling and Control with Power Electronics Applications, Second Edition
Bei Gou, Woonki Na, and Bill Diong

Modern Electric, Hybrid Electric, and Fuel Cell Vehicles: Fundamentals, Theory, and Design
Mehrdad Eshani, Yimin Gao, Sebastien E. Gay, and Ali Emadi

Modeling and Analysis with Induction Generators, Third Edition
M. Godoy Simões and Felix A. Farret

Uninterruptible Power Supplies and Active Filters
Ali Emadi, Abdolhosein Nasiri, and Stoyan B. Bekiarov

SECOND EDITION

Fuel Cells

Dynamic Modeling and Control with Power Electronics Applications

Bei Gou • Woonki Na • Bill Diong

CRC Press is an imprint of the
Taylor & Francis Group, an **informa** business

CRC Press
Taylor & Francis Group
6000 Broken Sound Parkway NW, Suite 300
Boca Raton, FL 33487-2742

First issued in paperback 2020

ISBN-13: 978-1-4987-3269-7 (hbk)
ISBN-13: 978-0-367-65591-4 (pbk)

Visit the Taylor & Francis Web site at
http://www.taylorandfrancis.com

and the CRC Press Web site at
http://www.crcpress.com

Contents

Preface

In the first edition of this book that was published in August 2008, the main focus was the modeling of PEM (proton exchange membrane) fuel cells and nonlinear control design using the feedback linearization control method. Since then, the authors have continued conducting research in academia and industry on fuel cells and its power electronics applications. In the second edition, additional power electronics applications as well as material on direct methanol fuel cells and fuel cell temperature model have been added based on the authors' research. In addition, an implementation of a digital signal processor controller-based power electronics system has been described in detail. Thus, we believe that the subject of this book will be of interest to faculty, students, consultants, manufacturers, researchers, and designers in the field of power electronics who will find a detailed discussion of fuel cell modeling, analysis, and nonlinear control throughout the book with simulation examples, test results, and power converter designs and their control algorithms. We assume that readers have a fundamental knowledge of control theory and fuel cell chemical reactions, electric circuits, and power electronics.

This book presents a comprehensive description of modeling and control of PEM fuel cells and its power electronics applications. It describes typical approaches and achievements in the modeling and control of PEM fuel cells. Both linear and nonlinear models and control designs are included. Also the power converter control designs for fuel cell power applications are explained including a linear controller, the sliding mode control technique, and predictive control.

For readers' convenience, this book is organized in a self-contained way to introduce, in sufficient detail, the essence of recent research achievements in the modeling and control design of PEM fuel cells and its power electronics applications including simulations and its digital signal processor-based implementation. Mathematical preliminaries in linear control and nonlinear control, induction machine d–q modeling, and space vector PWM (pulse width modulation) are provided in the Appendix of the book.

Chapters 1 and 2 start with a brief introduction to fuel cells and fuel cell power systems. The fundamentals of fuel cell systems and their components are provided in Chapter 2. These two chapters serve as the background and preparation for the following sections.

Chapter 3 presents the linear and nonlinear modeling of fuel cell dynamics. It serves as a preparation for the linear and nonlinear control designs covered in Chapter 4.

Chapter 4 discusses the typical approaches of linear and nonlinear modeling and control design methods for fuel cells. It also serves to compare the linear and nonlinear control designs and their effectiveness.

Chapter 5 presents the Simulink® implementation of fuel cells, which includes the modeling of PEM fuel cells and control designs. Simulation results of linear and nonlinear controllers are presented and compared.

Chapters 6 and 7 discuss the applications of fuel cells in vehicles and utility power systems and stand-alone systems. Details of typical models and control strategies are presented and discussed.

Chapter 8 discusses the modeling and analysis of hybrid renewable energy systems, which include the integration of fuel cells, wind power, and solar power. Details of configuration and control schemes are presented and studied.

Chapter 9 introduces a multiobjective optimization method for PEMFC by considering its cost and efficiency.

Chapter 10 discusses power electronics applications in fuel cells, in which several control methodologies for designing power converters in fuel cells have been described in detail.

Chapter 11 introduces a feedback temperature controller for PEMFC using its thermal transfer function analysis.

Chapter 12 discusses a digital signal processor-based control design and implementation for power converters that can be used in a fuel cell vehicle and fuel cell-based power systems.

We have intentionally preserved the generality in discussing the advanced technology in modeling and control of PEM fuel cells and the current power electronics applications of fuel cells in vehicles and utility power systems and stand-alone systems.

We take this opportunity to acknowledge California State University, Fresno, the University of Texas at Arlington (UTA) and Kennesaw State University (KSU), as well as the U.S. Department of Energy (DOE), for the support of this research on fuel cells and its power electronics applications. Our thanks also to Dr. Kai S. Yeung from the Department of Electrical Engineering at UTA; former graduate students, Yunzhi Chen and Dr. Zheng Hui from the same department at UTA; former graduate students, Lu-Ying Chiu, Eduardo Hernandez, and Wolf Carter from the University of Texas at El Paso (UTEP); former undergraduate students Kshitiz Singh, Diego Estrada, and Bao Nguyen from Texas Christian University and KSU; and Dr. Randall S. Gemmen from DOE. Many thanks to Pengyuan Chen, former graduate student from Fresno State, Dr. Tahyung Kim at the University of Michigan-Dearbon, and Dr. Taesik Park at Mokpo National University, South Korea for helping with the power electronics research in this book.

Additionally, we express our gratitude to Taylor & Francis Group for affording us the opportunity to publish this book.

Bei Gou
Woonki Na
Bill Diong

Authors

Bei Gou earned his PhD in electrical engineering at Texas A&M University in 2000. He was a power application engineer with ABB System Control at Santa Clara, California from 2000 to 2002, and worked as a senior analyst at the Independent System Operator-New England (ISO-NE) from 2002 to 2003. Dr. Gou joined the Department of Electrical Engineering at The University of Texas at Arlington as an assistant professor from 2003 to 2008. He is the founder of a start-up company—Smart Electric Grid, LLC. He was with the Department of Electrical and Computer Engineering as associate professor at North Dakota State University from 2011 to 2013. His current research areas include power system real-time monitoring, nonlinear control of fuel cells and power electronics theory and applications, blackout and cascading failures of power systems, phasor measurements and state estimation for power systems, and power system reliability. Dr. Gou has published about 90 journal and conference papers. He has authored two books, *Fuel Cells: Modeling, Control and Applications* (Taylor & Francis/CRC Press, Boca Raton, Florida, 2009) and *Monitoring and Optimization of Power Transmission* and Distribution Systems (VDM Publishing House Ltd., Germany, 2009).

Bill Diong is an associate professor in the Department of Electrical Engineering at Kennesaw State University (KSU), and also the coordinator of its MS Applied Engineering—Electrical concentration program. He earned his PhD in electrical engineering from the University of Illinois (Urbana-Champaign) in 1992, and gained valuable practical experience as a senior research engineer with Sundstrand Aerospace (now part of UTC [United Technologies Corporation] Aerospace Systems) before returning to academia. Prior to joining KSU in 2011, he had been an assistant professor at the University of Texas at El Paso where he was the Forrest and Henrietta Lewis professor of electrical engineering between 2000 and 2002. His research interests encompass advanced power and energy systems, including electric vehicles and electronic power converters, and other dynamic systems and their control; he has authored or coauthored more than a dozen journal publications and several dozen conference publications in these areas. Funding for his work has come from various organizations that include the National Science Foundation, the Ballistic Missile Defense Organization, and the U.S. Air Force. He is a senior member of the IEEE (Institute of Electrical and Electronics Engineers).

Woonki Na earned BS and MS degrees in electrical engineering from Kwangwoon University, Seoul, South Korea, in 1995 and 1997, respectively, and earned his PhD in electrical engineering from the University of Texas at

Arlington in May 2008. From 2008 to 2009, he worked with Caterpillar Inc., Peoria, Illinois and participated in several hybrid electric drives programs as senior engineer. He was an assistant professor with the Department of Electrical and Computer Engineering, Bradley University, Peoria, Illinois from 2010 to 2013. Currently he is an assistant professor with the Department of Electrical and Computer Engineering, California State University, Fresno since August 2013. His research and teaching interests include power electronics, battery management, and digital signal processor (DSP)-based control designs for hybrid electric vehicles and renewable/alternative energy applications.

1

Introduction

1.1 Past, Present, and Future of Fuel Cells

1.1.1 What Are Fuel Cells?

A fuel cell operates like a battery by converting the chemical energy from reactants into electricity, but it also differs from a battery in that as long as the fuel (such as hydrogen) and an oxidant (such as oxygen) is supplied, it will produce DC electricity (plus water and heat) continuously, as shown in Figure 1.1. In the 1960s, the first practical fuel cell was developed and then used in the US Gemini and Apollo programs for space applications. Since then, fuel cells have been increasingly applied in other areas although it remains a "new" technology since its commercialization is still a hot research topic today. As some of the fundamental obstacles are being overcome, fuel cells have become more feasible for a few applications and are gradually being developed and commercialized.

For example, in 1993, Ballard Power Systems demonstrated fuel cell-powered buses. Then almost all major automotive manufacturers developed fuel cell vehicle prototypes in the late 1990s and the early 2000s, which are undergoing tests in the United States, Japan, and Europe. For stationary power applications, more than 2500 fuel cell stationary power systems have already been installed globally at hospitals, office buildings, utility power plants, and so on. In 2005, Samsung Electronics also unveiled a prototype of fuel cells for portable power applications that can run a laptop for about 15 h. However, challenges remain to the commercialization of fuel cells still exist. The most significant problems are reducing their cost as well as improving their operating reliability.

The recent increasing impetus in developing and commercializing fuel cells are due to its several advantages. These include "clean" by-products (e.g., water when operated on pure hydrogen), which means it is "zero emission" with extremely low (if any) emission of oxides of nitrogen and sulfur. They also operate quietly, not having any moving parts, even when working with extra fuel processing and supply equipment. Furthermore, they have high-power density and high efficiency, typically more than 40% efficiency

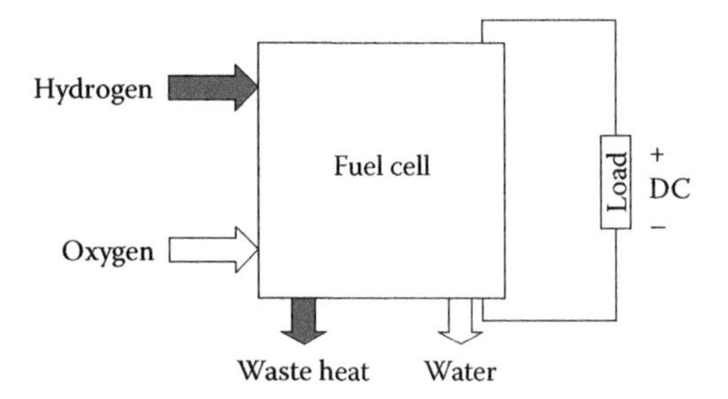

FIGURE 1.1
A fuel cell's inputs and outputs.

in electric power production, which is better than traditional combustion engine/generator sets, and the "waste" heat from a fuel cell can be used for heating purposes, thus increasing its overall efficiency. Finally, they can increase national energy security, since different types of fuel cells can operate on various conventional and alternative fuels such as hydrogen, ethanol, methanol, and natural gas, and hydrogen itself can be produced by harnessing a variety of renewable energy sources; such capability can help to reduce US dependence on foreign oil.

1.1.2 Types of Fuel Cells

Fuel cells are most commonly classified by the kind of electrolyte being used. These include proton exchange/polymer electrolyte membrane fuel cells (PEMFCs), direct methanol fuel cells (DMFCs), alkaline fuel cells (AFCs), phosphoric acid fuel cells (PAFCs), molten carbonate fuel cells (MCFCs), solid oxide fuel cells (SOFCs), zinc air fuel cells (ZAFCs), and photonic ceramic fuel cells (PCFCs), which vary widely in their required operating temperature. But this book will only focus on PEMFCs, which are low-operating-temperature fuel cells intended for use in mass-production fuel cell vehicles that are currently under development by the major auto manufacturers, as well as in offices and residences.

1.2 Typical Fuel Cell Power System Organization

The fundamental components of a fuel cell power system are a fuel cell (most commonly a stack or multilayer connection of fuel cells), a fuel and oxidant supply, an electrical load, and an electric power conditioner (see Figure 1.2). The fuel and oxidant supply along with the electric power conditioner are typically lumped together with the fuel reformer, the thermal management

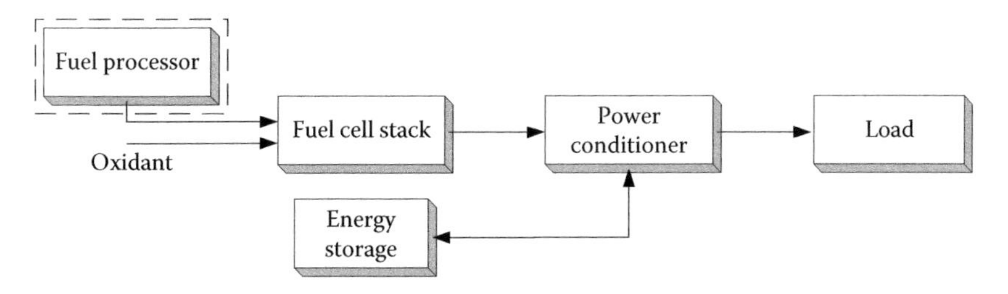

FIGURE 1.2
Block diagram of a fuel cell power system's interconnection.

subsystem, and the humidification subsystem, when these are present, under the term balance of plant (BOP).

1.3 Importance of Fuel Cell Dynamics

For the successful commercialization of fuel cell vehicles, their performance, reliability, durability, cost, fuel availability, and public acceptance should be considered [1]. The most important disadvantage of fuel cells now is their cost. However, the performance of the fuel cell systems during transients is another key factor. Therefore, during transients, to generate a reliable and efficient power response and to prevent membrane damage as well as detrimental degradation of the fuel cell stack voltage and oxygen depletion, it is necessary to design better control schemes to achieve optimal air and hydrogen inlet flow rates, i.e., a fuel cell control system that can perform air and hydrogen pressure regulation and heat–water management precisely based on the current drawn from the fuel cell [2,3].

This book essentially addresses the issue of fuel cells' slow transient response to load changes, which is important since the dynamic behavior of a fuel cell is integral to the overall stability and performance of the power system formed by the fuel supply, fuel cell stack, power conditioner, and electrical load. Normally fuel cells have transient (dynamic) responses that are much slower than the dynamic responses of the typical power conditioner and load to which they are attached. As such, the fuel cell's inability to change its electrical output (current) as quickly as the electrical load changes has significant implications on the overall power system design. In particular, some form of energy storage with a quick charge/discharge capability is needed to function as a firm power backup during electrical load increases if the fuel flow to the fuel cell is not being kept constant at its maximum level (which is wasteful and inefficient). The slower the fuel cell's response, the larger the amount of energy storage that is needed with the attendant increases in its size, weight, and cost; it also reduces the number of suitable energy-storage options (ultracapacitor [UC], flywheel, battery, etc.). Therefore, the fuel cell's dynamic response is of significant importance, particularly in mobile applications.

1.4 Organization of This Book

Chapter 2 describes PEMFCs and the fuel cell power system BOP components in more detail. Chapter 3 describes the modeling of a PEMFC's dynamic behavior as an initial step to prescribe controller designs to improve its transient behavior. It is followed by Chapter 4, which presents the understanding of the design of feedback controllers. Chapter 5 features the Simulink implementations of fuel cell models and controllers. Finally, Chapters 6 and 7 discuss two important applications of fuel cells where dynamic response is important to vehicles and to fixed-voltage hybrid power generation systems. Chapter 8 discusses hybrid renewable energy systems where the fuel cell is working with other alternative energy sources such as wind and solar energy. Chapter 9 presents a multiobjective optimization in terms of the cost and efficiency of PEMFC (proton exchange/polymer electrolyte membrane fuel cell). Chapter 10 discusses power electronics applications for fuel cells, especially the designing of linear and sliding mode control for a power factor correction converter and bi-directional converter. Chapter 10 describes predictive controllers for fuel cell vehicles. Chapter 11 deals with a temperature controller design for PEMFC by considering its thermal transfer function. Lastly, Chapter 12 presents the implementation of digital signal processor-based power electronics control for a sliding mode control in a DC/DC converter.

References

1. F. Barbir and T. Gomez, Efficiency and economics of PEM fuel cells, *International Journal of Hydrogen Energy*, 22(10/11), 1027–1037, 1997.
2. A.M. Borbely and J.G. Kreider, *Distributed Generation: The Power Paradigm for the New Millennium*, New York, CRC Press, 2001.
3. F. Barbir, *PEM Fuel Cells: Theory and Practice*, Elsevier Academic Press, Burlington, MA, USA, 2005.

2

Fundamentals of Fuel Cells

2.1 Introduction

The PEMFC, also called a solid polymer fuel cell, was first developed by General Electric in the United States in the 1960s for use by NASA (National Aeronautics and Space Administration) on their first manned space vehicles [1]. This type of fuel cell primarily depends on a special polymer membrane that is coated with highly dispersed catalyst particles. Hydrogen is fed to the membrane's anode side (possibly at a pressure greater than atmospheric pressure) where the catalyst causes the hydrogen atoms to release its electrons and become H^+ ions (protons)

$$2H_2 \rightarrow 4H^+ + 4e^- \tag{2.1}$$

as shown in Figure 2.1. The proton exchange membrane (PEM) only allows the H^+ ions to pass through it, whereas the electrons are collected and utilized as electricity by an outside electrical circuit (doing useful work) before they reach the cathode side. There, the electrons and the hydrogen ions diffusing through the membrane combine with the supplied oxygen (typically from air) to form water, which is a reaction that releases energy in the form of heat

$$4e^- + 4H^+ + O_2 \rightarrow 2H_2O \tag{2.2}$$

This water by-product must be removed to prevent the cell from being flooded and rendered inoperative (more details later). In addition, any unused hydrogen and oxygen (air) are exhausted from the cell anode and cathode outlets, respectively. For this reaction to proceed continuously, the electrons produced at the anode must flow through a circuit external to the fuel cell, and the protons must flow through the PEM as shown in Figure 2.1 [2].

The reaction in a single fuel cell produces an output voltage of around 0.7 V; for general applications, several individual cells are connected in series to form a fuel cell stack to produce the desired voltage additively. The required operating temperature for PEMFCs is 50–100°C, which enables a fast start-up

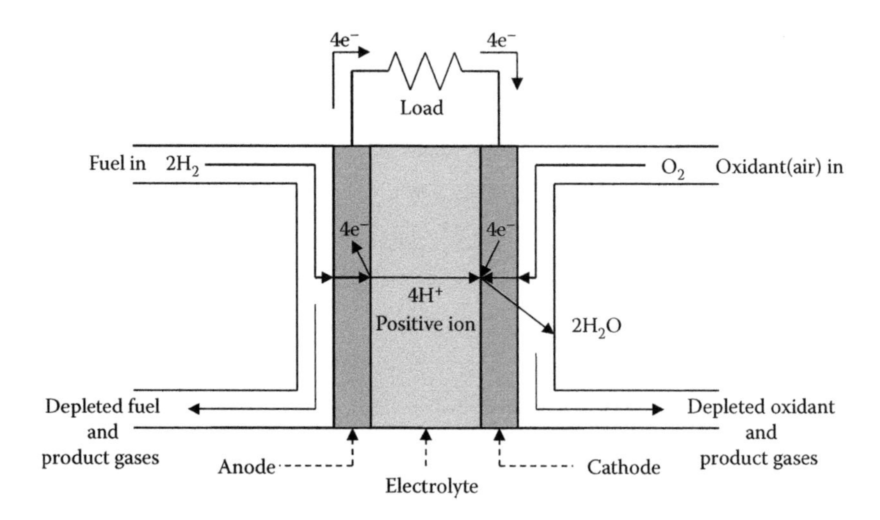

FIGURE 2.1
Electrochemical reaction in the PEMFC. (Adapted from W. Na, Dynamic modeling, control and optimization of PEM fuel cell system for automotive and power system applications, PhD thesis, the University of Texas at Arlington, May 2008.)

of operation. Thus, the PEMFC is particularly attractive for transportation applications as well as a small- or mid-size distributed electric power generator because it has a high-power density, a solid electrolyte, a long stack life, and low corrosion. Other advantages include its clean by-products (pure water when hydrogen is the fuel, which means "zero-emission"), high-energy efficiency of more than 40% typically in electric power production, and quiet operation [3]. Hence, PEMFCs have big application potential in power automobiles, aircraft (auxiliary power), homes and small offices, and portable electronics (as replacements for rechargeable batteries).

2.2 PEMFC Components

The main parts of a practical PEMFC are illustrated in Figure 2.2. The membrane electrode assembly (MEA) consists of the polymer membrane together with the electrodes and gas diffusion layers. Each electrode essentially consists of a layer of catalyst particles (usually platinum deposited on the surface of larger particles of carbon-support powder) and are affixed to either the membrane or the gas diffusion layer. The gas diffusion layer is made of a porous and electrically conductive material, such as carbon cloth, to enable the reactants to diffuse into and out of the MEA, and to collect the resulting current by providing electric contact between the electrode and the outside

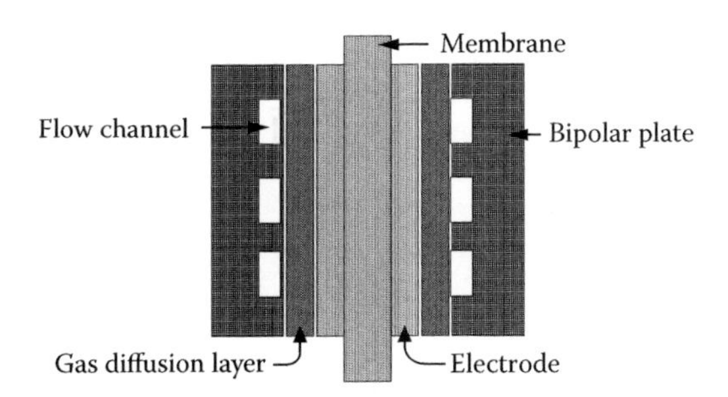

FIGURE 2.2
The main functional parts of a PEMFC.

bipolar plate. Furthermore, it allows the water formed at the cathode to exit to the gas channels.

The bipolar plates, also called flow field plates, distribute the reactant gas over the surface of the electrodes through flow channels on their surface: different channel geometries are available. They also collect the current and form the supporting structure of the fuel cell. For good electrical and thermal conductivity, plus physical strength and chemical stability, solid graphite is usually used as the material for these plates.

The composition and function of the various PEMFC parts are further discussed in the following sections.

2.2.1 Membrane

The electrolyte membrane is the key part in any PEMFC. A proton-conducting polymer is used as the electrolyte, thus giving rise to this type of fuel cell's name. The basic material used for the membrane is polyethylene, which has been modified by substituting fluorine for hydrogen to yield polytetrafluoroethylene. The bonds between the fluorine and the carbon make the membrane very durable and chemical resistant (inert). The basic electrolyte is then complemented with sulfonic acid; the HSO_3 group added is ionically bonded (see Figure 2.3). The result is the ability to attract H^+ ions into the electrolyte. This material, made by the DuPont Co. and sold under the trade name Nafion®, is significant in the development of PEMFCs.

The main properties of these polymer membranes are [3]

- They are resistant to chemical attacks
- They have very strong bonds, so that they can be made into very thin films
- They are acidic
- They can absorb a lot of water
- The H^+ ions they attract are well conducted through them if the membranes are adequately hydrated (but not flooded)

```
    F     F     F     F     F     F     F
    |     |     |     |     |     |     |
  — C  —  C  —  C  —  C  —  C  —  C  —  C —
    |     |     |     |     |     |     |
    F     F     F     O     F     F     F
                      |
                  F — C — F
                      |
                  F — C — F
                      |
                      O
                      |
                  F — C — F
                      |
                  F — C — F
                      |
                  O = S = O
                      |
                      O⁻  H⁺
```

FIGURE 2.3
Structure of a sulfonated fluoroethylene.

2.2.2 Membrane Electrode Assembly

The performance of the PEMFC is largely determined by the MEA, which is the central part of the fuel cell. The MEA, as illustrated in Figure 2.4, consists of the electrolyte membrane sandwiched between the anode and cathode electrodes. These electrodes include the catalyst particles and the gas diffusion layers.

Since the fuel oxidation and the oxygen reduction reactions are kinetically slow, a noble metal such as platinum or one of its alloys is used as the catalyst to increase the reaction rate: this catalyst is used at both the anode and the cathode. The platinum is formed into small particles that are spread onto the surfaces of larger particles of carbon support. The most common carbon

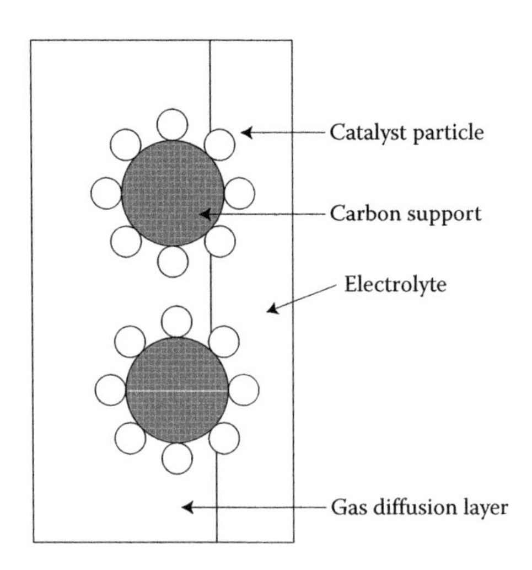

FIGURE 2.4
Membrane electrode assembly (MEA).

support used is a carbon-based powder XC72® (Cabot). The platinum spread into the carbon is highly divided to increase the surface area that is in contact with the reactants so as to maximize its catalytic effect. In the early days of PEMFC development, this catalyst was used at loadings of about 50 mg of platinum per cm^2 [4] but that has now been reduced to less than 1 mg of platinum per cm^2 [5], thus significantly lowering the PEMFC's overall cost.

The catalyst particles and carbon support are affixed to a porous gas diffusion layer made of an electrically conductive material such as carbon cloth or carbon paper. This carbon cloth or carbon paper diffuses the reactant gases to the surfaces of the catalyst particles, while diffusing the water produced at the cathode away from the electrolyte membrane. In addition, it also provides the electrical connection between the electrode and its corresponding current-conducting bipolar plate.

2.2.3 Bipolar Plates

The bipolar (also known as field flow) plates that form an important part of the weight and volume of a PEMFC are used to bring the reactant gases via machined-flow channels (grooves) to the MEA. They help to distribute the reactants into the surface of the electrodes. In addition, they collect the current produced by the electrochemical reaction. The bipolar plates require good electrical and thermal conductivity, good mechanical strength, and chemical stability. Graphite is the most common material used for these plates currently, although extensive research is being performed to develop new materials that can reduce the weight of the bipolar plate, thereby increasing the fuel cell's power density.

It is important to note that the geometry of the flow channels varies depending on the needs and design of each fuel cell. The specific geometry is also a significant factor in a PEMFC's operating performance. Figure 2.5 illustrates different geometries used for the bipolar plates.

2.2.4 Heating or Cooling Plates

These plates can be used to either heat the PEMFC or cool it to keep its temperature close to the one that yields an optimal operating performance. Heating plates typically rely on the use of electricity and ohmic (resistive) heating. Cooling plates are used when air cooling is insufficient; then, liquid, such as water, is actively circulated through these plates to cool the stack.

2.3 BOP Components

A practical fuel cell stack may get the hydrogen fuel from a pressurized tank through regulated valves or the hydrogen is indirectly obtained

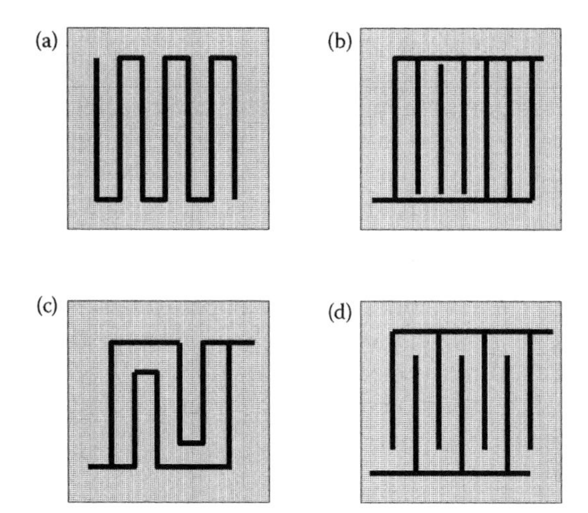

FIGURE 2.5
Different geometries of the flow channels: (a) serpentine; (b) parallel; (c) parallel serpentine; and (d) discontinuous.

from a hydrogen-rich fuel such as natural gas via a fuel processor called the reformer. The additional components, such as the above, that may be needed in operating a PEMFC stack are collectively known as the BOP. They help with the functions of fuel storage and/or processing, water management, thermal management, and power conditioning, to achieve the fuel cell system's design requirements. These are only briefly described below as they are not the main focus of the rest of this book.

2.3.1 Water Management

Without adequate water management, an imbalance will occur between water by-product and water removal from the fuel cell. It is critical to ensure that all parts of the cell are sufficiently hydrated, since adherence of the membrane to the electrode and also membrane lifetime will be adversely affected if dehydration occurs. Furthermore, high water content in the electrolyte membrane ensures high ionic conductivity, which improves the overall operating efficiency of the fuel cell. Since the electrochemical reaction produces heat, which increases evaporation, humidifiers are used to (pre-)humidify the incoming gases, particularly on the anode side. The humidifier may be as simple as a bubbler or else something more sophisticated such as a membrane humidifier or water evaporator [4].

Owing to the PEMFC's operation at less than 100°C and atmospheric pressure, water is produced at the cathode as a liquid. While sufficient hydration is important for optimal PEMFC operation as noted above, this water must not be allowed to flood that electrode because it would impede gas diffusion to that electrode and reduce the cell's operating performance.

2.3.2 Thermal Management

Most PEMFCs currently use cast carbon composite plates for current collection and distribution, gas distribution, and also thermal management. Active air cooling can be achieved through the use of fans. Liquid cooling requires the use of pumps to circulate fluid through the cooling plates of the stack. These fans and pumps are typically driven by electric motors.

2.3.3 Fuel Storage and Processing

Currently, the most common way to store hydrogen for use as PEMFC fuel is as a gas requiring pressurized cylinders or tanks. Then pressure-reducing regulators are also needed. Storing hydrogen in liquid form requires "only" adequate insulation, but this is a more inefficient way of storing and transporting hydrogen than as a gas.

An alternative to using hydrogen as the primary fuel is to use a hydrocarbon or alcohol compound as the source of hydrogen. But then, a fuel processor or reformer is needed to chemically convert that hydrocarbon or alcohol to a hydrogen-rich gas. Furthermore, because PEM and reformer catalysts are prone to inactivation by sulfur and CO (and also CO_2 to a lesser extent), other subsystems are also needed, to remove the sulfur from the primary fuel and/or the CO from the hydrogen-rich reformate. All of these various components add weight, volume, and expense to the overall system, and reduce its efficiency.

2.3.4 Power Conditioning

The power conditioner is an electronic system that is needed to convert the variable low DC voltage produced by a fuel cell into usable DC power (typically at a regulated higher voltage) or alternating current (AC) power, as appropriate to meet the operating requirements of the intended application. Various types of power converters, such as DC–DC converters and DC–AC inverters, may be employed in fuel cell power-conditioning systems.

For powering DC loads, typically a step-up DC–DC converter (to increase the voltage level) is employed. On the other hand, a switch-mode DC–AC inverter is typically used to convert a PEMFC's output DC voltage into regulated AC voltage at 60-Hz (or other) frequency. A filter at the output of such an inverter attenuates the switching-frequency harmonics and produces a high-quality sinusoidal voltage suitable for typical AC loads.

For many important applications, such as powering vehicle propulsion, a 5:1 or better peak-to-average power capability [4] is desired, e.g., compare the power needed for accelerating a vehicle as compared to cruising at constant speed. Since present-day fuel cells typically cannot change their power output as quickly as most of these load-demand changes, they are thus designed to satisfy average power requirements and then any additional amounts of power must be supplied from another energy source such as a battery or an UC. The power-conditioning unit must therefore also provide the means for interfacing to this energy-storage device and a control scheme is needed to properly coordinate its charging and discharging based on the fuel cell output and load-demand conditions.

References

1. *Fuel Cell Handbook* (7th edition), EG&G Technical Services, Inc., for U.S. Department of Energy, Office of Fossil Energy, National Energy Technology Laboratory, Morgantown, West Virginia, November 2004.
2. W. Na, Dynamic modeling, control and optimization of PEM fuel cell system for automotive and power system applications, PhD thesis, the University of Texas at Arlington, May 2008.
3. J. Larminie and A. Dicks, *Fuel Cell Systems Explained* (2nd edition), John Wiley & Sons, Chichester, UK, 2003.
4. G. Hoogers (editor), *Fuel Cell Technology Handbook*, CRC Press, Boca Raton, Florida, 2003.
5. H.A. Gasteiger, J.E. Panels, and S.G. Yan, Dependence of PEM fuel cell performance on catalyst loading, *Journal of Power Sources*, 127, 162–171, 2004.

3

Linear and Nonlinear Models of Fuel Cell Dynamics

3.1 Introduction

For preliminary fuel cell power system planning, stability analysis, control strategy synthesis, and evaluation, an appropriate dynamic model of a fuel cell system is desired. Since the existing control-oriented models [1–3] do not contain all water components, which are one of the important factors of fuel cell systems [4], it is difficult to design an accurate dynamic fuel cell model based on the existing models. The main motivation for developing dynamic models of PEMFCs is to facilitate the design of control strategies with objectives such as to ensure good load-following performance, to prevent fuel cell stack damage, and to prolong the stack life by controlling the anode and cathode gases pressures. In this chapter, various models of PEMFCs will be described. Using these models of PEMFCs, appropriate controllers of fuel cell systems can be designed as described in Chapter 5.

3.2 Nonlinear Models of PEM Fuel Cell Dynamics

3.2.1 Unified Model of Steady-State and Dynamic Voltage–Current Characteristics

The performance of a fuel cell can be expressed by the polarization curve, which describes the cell voltage–load current (*V–I*) characteristics of the fuel cell that are highly nonlinear [1–9]. Optimization of fuel cell operating points, design of the power-conditioning units, design of simulators for fuel cell stack systems, and design of system controllers depend on such characteristics [10]. Therefore, the modeling *V–I* characteristics of fuel cells is important.

It is observed that the known steady-state *V–I* characteristics computed from the formulated electrochemical modeling are divided into two components: one of which is named as the steady component including the thermodynamic potential and the ohmic overvoltage; the other is named as the transient component consisting of the activation and concentration overvoltages. Then, the former is fitted by a low-order least-squares polynomial and the latter is described by a high-order least-squares polynomial. The coefficients in the polynomials can be estimated by using the least-squares technique. The sum of these two components can be used to accurately model the steady-state *V–I* characteristics of PEM fuel cells. Furthermore, by introducing the first-order time delay to describe the dynamic response of PEM fuel cells, the developed mathematical modeling can also be used to accurately predict the dynamic *V–I* characteristics.

For PEM fuel cells, steady-state *V–I* characteristics of a fuel cell are determined by [1,4,5]

$$V_{cell} = E_N - V_a - V_c - V_{ohm} = V_{st} - V_{tr} \tag{3.1}$$

where

V_{cell} represents the output voltage of the fuel cell

E_N represents the reversible voltage of the fuel cell (also named as the thermodynamic potential)

V_a represents the voltage drop due to the activation of the anode and cathode (also named as the activation overvoltage)

V_c denotes the voltage drop resulting from the reduction in concentration of the reactants gases or from the transport of mass of oxygen and hydrogen (also named as the concentration overvoltage)

V_{ohm} denotes the ohmic voltage drop resulting from the resistance of the conduction of protons through the solid electrolyte and of the electrons through its path (also named as the ohmic overvoltage)

$V_{st} = E_N - V_{ohm}$ is the steady component of the cell voltage

$V_{tr} = V_a + V_c$ is the transient component of the cell voltage

The unified mathematical model of the steady-state and dynamic voltage–current characteristics of fuel cells is

$$\begin{aligned} v_{cell}(t) = {} & \sum_{k=0}^{2} p_k (I_{cell} - I_{st})^k - \sum_{k=0}^{5} q_k (I_{cell} - I_{tr})^k \\ & - \left[\sum_{k=0}^{5} q_k (I_{cell} - T_{tr})^k - (I_{cell} - I_{tr})^k \right] (1 - e^{-t/T_C}) \end{aligned} \tag{3.2}$$

where

$$V_{st}(I_{cell}) = \sum_{k=0}^{2} p_k (I_{cell} - I_{st})^k$$

$$I_{st} = \sum_{m=0}^{N_{st}-1} \frac{I_{cellm}}{N_{st}}$$

$$V_{tr}(I_{cell}) = \sum_{k=0}^{5} q_k (I_{cell} - I_{tr})^k$$

$$I_{tr} = \sum_{m=0}^{N_{tr}-1} \frac{I_{cellm}}{N_{tr}}$$

where I_{cell} is the cell current, p_k the coefficients determined by using the least-squares technique, I_{cellm} the known discrete current values, and N_{st} is the number of the given discrete current data for the steady component ($N_{st} \geq 3$).

Where q_k are the coefficients determined by using the least-squares technique, and N_{tr} is the number of the given discrete current data for the transient component ($N_{tr} \geq 6$).

3.2.2 Simulation Results

The Ballard Mark V PEM fuel cell is used to test the proposed model. Tables 3.1 and 3.2 show the computed coefficients in the steady and transient components polynomials, respectively. These coefficients are computed by using the least-squares technique, based on the data from the formulated electrochemical modeling and the parameters of the Ballard Mark V PEM fuel cell listed in Table 3.3 (Figures 3.1 and 3.2).

3.2.3 Nonlinear Model of PEM Fuel Cells for Control Applications

A PEM fuel cell consists of a polymer electrolyte membrane sandwiched between two electrodes (anode and cathode) in Figure 3.3. In the electrolyte,

TABLE 3.1

Steady Coefficients for the Ballard Mark V PEM Fuel Cell

k	0	1	2
p_k	0.110149E+01	−0.297366E−02	−0.183457E−04

TABLE 3.2

Transient Coefficients for the Ballard Mark V PEM Fuel Cell

k	0	1	2
q_k	0.516228E+00	0.298822E−02	0.265448E−04
k	3	4	5
q_k	−0.323092E−05	−0.603896E−07	0.349527E−08

TABLE 3.3

PEMFCs Ballard Mark V Voltage Parameters

Parameter	Value and Definition
N	Cell number: 35
V_o	Open-cell voltage: 1.032 V
R	Universal gas constant (J/mol K): 8.314 J/mol K
T	Temperature of the fuel cell (K): 353 K
F	Faraday constant (C/mol): 96,485 C/mol
α	Charge transfer coefficient: 0.5
M	Constant in the mass transfer voltage: $2/11 \times 10^{-5}$ V
N	Constant in the mass transfer voltage: 8×10^{-3} cm^2/mA
R_{ohm}	2.45×10^{-4} Ω cm^2
A_{fc}	Fuel cell active area: 232 cm^2
I_0	Exchange current density (A/cm^2)
I_n	Internal current density (A/cm^2)

Source: Adapted from J. Larminie and A. Dicks, *Fuel Cell Systems Explained,* Wiley, New York, 2002.

only ions can pass by, and electrons are not allowed to go through. So, the flow of electrons needs a path like an external circuit from the anode to the cathode to produce electricity because of the potential difference between the anode and cathode. The overall electrochemical reactions for a PEM fuel cell fed with hydrogen-containing anode gas and oxygen-containing cathode gas are as follows:

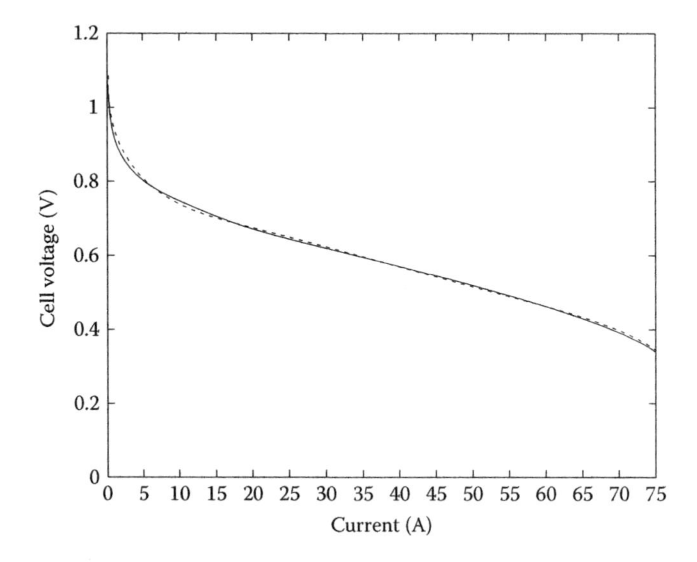

FIGURE 3.1
Computed and given steady-state *V–I* characteristics for the Ballard Mark V PEM fuel cell (dotted curve—computed values and solid curve—the given data).

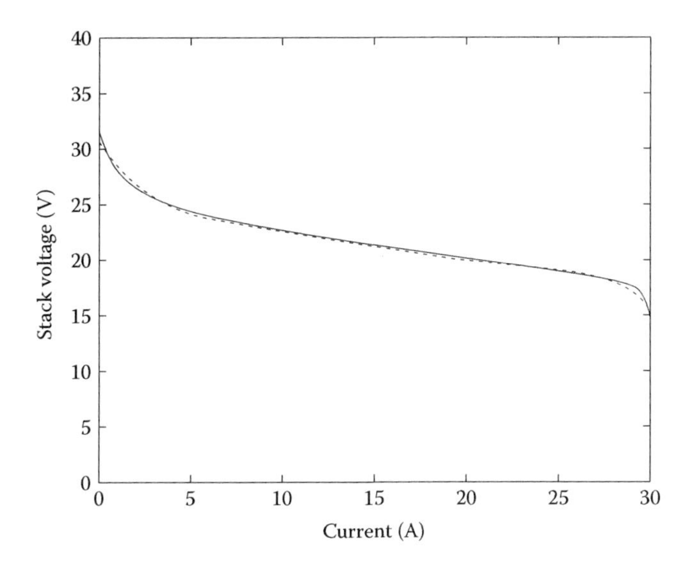

FIGURE 3.2
Polarization *V–I* curve (Ballard Mark V PEMFC at 70°C).

Anode: $2H_2 \leftrightarrow 4H^+ + 4e^-$

Cathode: $O_2 + 4H^+ + 4e^- \leftrightarrow 2H_2O$

Overall reaction: $2H_2 + O_2 \leftrightarrow 2H_2O$ + electricity + heat

In practice, a 5-kW fuel cell stack, such as a Ballard MK5-E PEMFC stack, uses a pressurized hydrogen tank at 10 atm and oxygen taken from

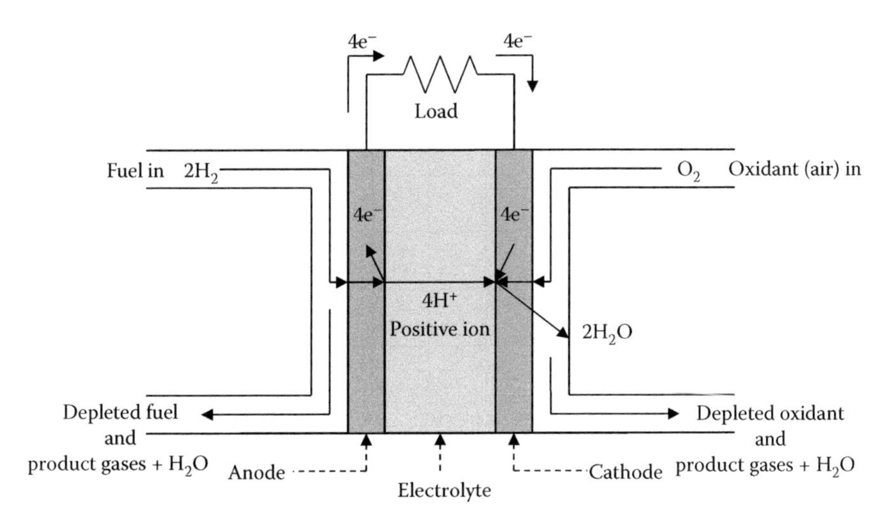

FIGURE 3.3
Schematic of fuel cell operation.

atmospheric air [11,12]. In case of using a reformer, on the anode side, a fuel processor called a reformer that generates hydrogen through reforming methane or other fuels such as natural gas, can be used instead of the pressurized hydrogen tank.

A pressure regulator and purging of the hydrogen component are also required. On the cathode side, an air supply system containing a compressor, an air filter, and an air flow controller are required to maintain the oxygen partial pressure [2,4,7,13]. On both sides, a humidifier is needed to prevent dehydration of the fuel cell membrane [2,4,7]. In addition, a heat exchanger, a water tank, a water separator, and a pump may be needed for water and heat management in the fuel cell systems [2,4,7].

To produce a higher voltage, multiple cells have to be connected in series. Typically, a single cell produces voltage between 0 and 1 V based on the polarization *I–V* curve, which expresses the relationship between the stack voltage and the load current [2,4,7]. Figure 3.4 shows that their relationship is nonlinear and mainly depends on current density, cell temperature, reactant partial pressure, and membrane humidity [2,4,7].

The output stack voltage V_{st} [4] is defined as a function of the stack current, reactant partial pressures, fuel cell temperature, and membrane humidity:

$$V_{st} = E - V_{activation} - V_{ohmic} - V_{concentration} \tag{3.3}$$

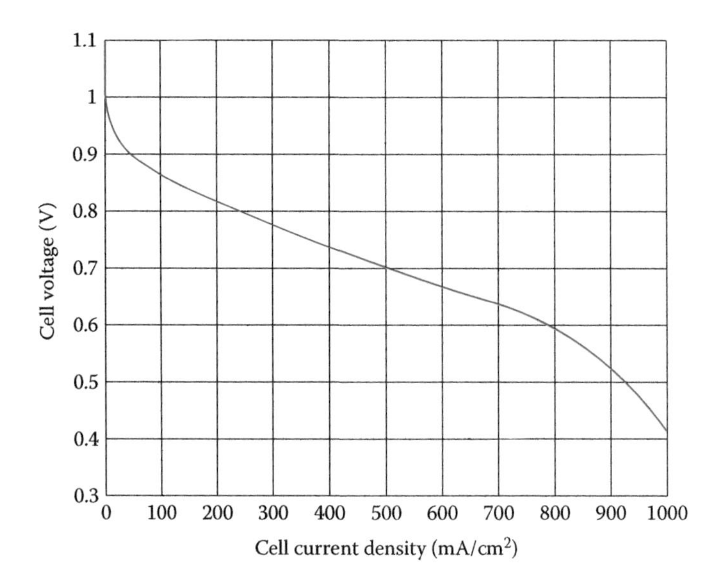

FIGURE 3.4
Polarization *V–I* curve (Ballard Mark V PEMFC at 70°C) [4].

In the above equation

$$E = N_o \cdot \left[V_o + \frac{R * T}{2F} \ln \left(\frac{P_{H_2} \sqrt{P_{O_2}}}{P_{H_2O_c}} \right) \right]$$

is the thermodynamic potential of the cell or reversible voltage based on Nernst equation [4], $V_{activation}$ the voltage loss due to the rate of reactions on the surface of the electrodes, V_{ohmic} the ohmic voltage drop from the resistances of proton flow in the electrolyte, and $V_{concentration}$ is the voltage loss from the reduction in concentration gases or the transport of mass of oxygen and hydrogen. Their equations are given as follows:

$$V_{activation} = N \cdot \frac{R * T}{2\alpha F} \cdot \ln \left(\frac{I_{fc} + I_n}{I_o} \right) \tag{3.4}$$

$$V_{ohm} = N \cdot I_{fc} \cdot R_{ohm} \tag{3.5}$$

$$V_{concentration} = N \cdot m \exp(n \cdot I_{fc}) \tag{3.6}$$

In Equation 3.3, P_{H_2}, P_{O_2}, and partial pressure of water vapor ($P_{H_2O_c}$) are the partial pressures of hydrogen, oxygen, and water, respectively. Subscript *c* means the water partial pressure, which is vented from the cathode side.

3.3 State-Space Dynamic Model of PEMFCs

To derive a simplified nonlinear dynamic PEMFC model, the following assumptions are made:

- Owing to slower dynamics of the stack temperature, the average stack temperature is assumed to be constant.
- The relative humidity can be well controlled to a little over 100%, and thereby, the liquid water always forms the stack. This liquid water is perfectly managed by the water tank and water separator and the water-flooding effects can be controlled.
- A continuous supply of reactants is fed to the fuel cell to allow operation at a sufficiently high flow rate.

- The mole fractions of inlet reactants are assumed to be constant in order to build the simplified dynamic PEMFC model. In other words, pure hydrogen (99.99%) is fed to the anode, and the air that is uniformly mixed with nitrogen and oxygen by a ratio of, say, 21:79, is supplied to the cathode.
- The full state has to be measured to utilize feedback linearization [14].

The ideal gas law and the mole conservation rule are employed. Each partial pressure of hydrogen, the water from the anode, and the oxygen, nitrogen, and water from the cathode are defined as state variables of the PEMFC. The relationship between inlet gases and outgases is described in Figure 3.5 [15].

The partial-pressure derivatives are given by the following equations.

Anode mole conservation:

$$\begin{aligned} \frac{dP_{H_2}}{dt} &= \frac{R*T}{V_A}(H_{2in} - H_{2used} - H_{2out}) \\ \frac{dP_{H_2O_A}}{dt} &= \frac{R*T}{V_A}(H_2O_{ain} - H_2O_{aout} - H_2O_{mbr} + H_2O_{back} - H_2O_{l,aout}) \end{aligned} \tag{3.7}$$

Cathode mole conservation:

$$\begin{aligned} \frac{dP_{O_2}}{dt} &= \frac{R*T}{V_c}(O_{2in} - O_{2used} - O_{2out}) \\ \frac{dP_{N_2}}{dt} &= \frac{R*T}{V_c}(N_{2in} - N_{2out}) \\ \frac{dP_{H_2O_c}}{dt} &= \frac{R*T}{V_c}(H_2O_{cin} + H_2O_{cproduced} - H_2O_{cout} + H_2O_{mbr} - H_2O_{mbr} - H_2O_{l,cout}) \end{aligned} \tag{3.8}$$

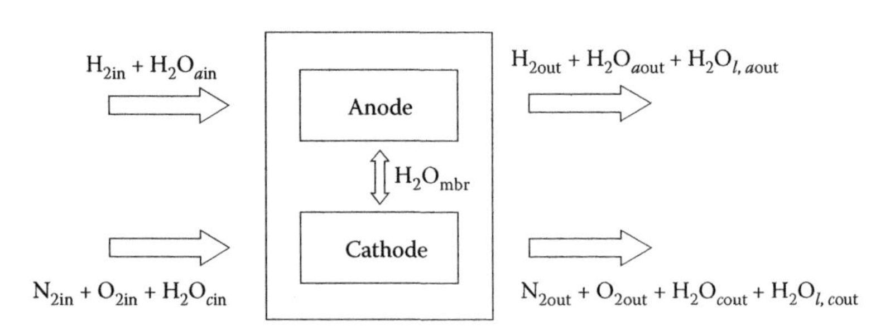

FIGURE 3.5
Gas flows of PEMFCs.

where H_{2in}, O_{2in}, N_{2in}, H_2O_{ain}, and H_2O_{cin} are the inlet flow rates of hydrogen, oxygen, nitrogen, the anode-side water, and the cathode-side water, respectively. In addition, H_{2out}, O_{2out}, N_{2out}, H_2O_{aout}, and H_2O_{cout} are the outlet flow rates of each gas. H_{2used}, O_{2used}, and $H_2O_{cproduced}$ are the usage and the production of the gases, respectively. In general, the membrane water inlet flow rate H_2O_{mbr} across the membrane is a function of the stack current and the membrane water content λ_m. By assuming that $\lambda_m = 14$ [7,16], H_2O_{mbr} is defined as a function of the current density only, and $H_2O_{mbr} = 1.2684\,(N \cdot A_{fc} \cdot I_{fc})/F$ [7,16], where A_{fc} (cm^2) is the fuel cell active area, N the number of the fuel cells, and I_{fc} is the fuel cell current density. Furthermore, in order to describe a more accurate dynamic model, the back diffusion of water from the cathode to the anode can be defined with $H_2O_{v,back} = \gamma \cdot H_2O_{v,mbr}$ [2]. The back-diffusion coefficient β is measured as to be 6×10^{-6} cm^2/s with the water content being $\lambda_m = 14$ [2]. The flow rates of liquid water leaving the anode and cathode are given by $H_2O_{l,aout}$ and $H_2O_{l,cout}$, which are dependent on the saturation state of each gas [7]. To estimate the liquid water, the maximum mass of vapor has to be calculated from the vapor saturation pressure as follows:

$$m_{v,\max a\,\mathrm{or}\,c} = \frac{p_{vs} V_{a\,\mathrm{or}\,c}}{R_v T_{st}} \tag{3.9}$$

The saturation pressure p_{vs} is calculated from an equation presented in Reference 17

$$\begin{aligned} \log_{10}(p_{vs}) = &-1.69 \times 10^{-10} T^4 + 3.85 \times 10^{-7} T^3 - 3.39 \times 10^{-4} T^2 \\ &+ 0.143T - 20.92 \end{aligned} \tag{3.10}$$

where the saturation pressure p_{vs} is in kPa and temperature T is in Kelvin. If the mass of water calculated in Equations 3.7 and 3.8 is greater than the maximum mass of vapor in Equation 3.9, the liquid water formation occurs simultaneously. The mass of liquid water and vapor is calculated as follows [7]:

Logic 1:

$$\text{if } m_{w,a\text{ or }c} \le m_{v,\max a\text{ or }c} \rightarrow m_{v,a\text{ or }c} = m_{w,a\text{ or }c},\ m_{l,a\text{ or }c} = 0$$

$$\text{if } m_{w,a\text{ or }c} > m_{v,\max a\text{ or }c} \rightarrow m_{v,a\text{ or }c} = m_{v,\max a\text{ or }c}$$

$$m_{l,a\text{ or }c} = m_{w,a\text{ or }c} - m_{v,\max a\text{ or }c}$$

Thereby, $\beta_{a\,\mathrm{or}\,c}$ can be used to estimate the liquid water formation in Equation 3.11. According to logic 1, if $m_{w,a\text{ or }c} \le m_{v,\max a\text{ or }c}$, then $\beta_{a\text{ or }c} = 0$; otherwise, $\beta_{a\text{ or }c} = 1$, and therefore $H_2O_{l,aout}$ and $H_2O_{l,cout}$ are defined by

$$H_2O_{l,a\,\mathrm{or}\,c\mathrm{out}} = \beta_{a\,\mathrm{or}\,c} \left| \frac{(p_{H_2O_{a\,\mathrm{or}\,c}} V_{a\,\mathrm{or}\,c})/(R_v T_{st}) - (p_{vs} V_{a\,\mathrm{or}\,c})/(R_v T_{st})}{M_{H_2O}} \right| \tag{3.11}$$

where M_{H_2O} is the water molar mass, 18.02 g/mol. All units of flow rates, usages, and the production of gases are defined in mol/s. However, because the liquid water is considered based on our assumption that each relative humidity stays over 100%, $\beta_{a \text{ or } c}$ will be 1, which means that $p_{H_2O_{a \text{ or } c}} > P_{vs}$ during the simulation. V_a and V_c are the anode and cathode volumes, respectively, and their units are m^3. According to the basic electrochemical relationships, the usage and production of the gases are functions of the cell current density I_{fc} [4], as follows:

$$H_{2\text{used}} = 2O_{2\text{used}} = H_2O_{c\text{produced}} = \frac{N \cdot A_{fc} \cdot I_{fc}}{2F} \tag{3.12}$$

For simplicity, let us define

$$\frac{N \cdot A_{fc}}{2F} = C_1 \quad \text{and} \quad 1.2684\frac{N \cdot A_{fc}}{F} = C_2$$

Thus, in Equations 3.7 and 3.8, H_2O_{mbr} and $H_2O_{v,\text{back}}$ can be simplified with C_1 and C_2. With the measured inlet flow rates and the stack current, the outlet flow rates are given by the summation of the anode and cathode inlet flow rates, that is, $Anode_{\text{in}}$ and $Cath_{\text{in}}$, minus the usage and production of gases as well as the pressure fraction proposed by El-Sharkh et al. [13]. The $Anode_{\text{in}}$ is defined by $H_{2\text{in}} + H_2O_{a\text{in}}$, and the $Cath_{\text{in}}$ is defined by $O_{2\text{in}} + N_{2\text{in}} + H_2O_{c\text{in}}$.

The outlet flow rates on the anode side are

$$\begin{aligned} H_{2\text{out}} &= (H_{2\text{in}} - C_1 \cdot I_{fc})F_{H_2} \\ H_2O_{a\text{out}} &= (H_2O_{a\text{in}} - C_2 \cdot I_{fc} + \gamma \cdot C_2 \cdot I_{fc})F_{H_2O_a} \end{aligned} \tag{3.13}$$

and the outlet flow rates on the cathode side are

$$\begin{aligned} O_{2\text{out}} &= \left(O_{2\text{in}} - \frac{C_1}{2} I_{fc}\right)F_{O_2} \\ N_{2\text{out}} &= N_{2\text{in}} \cdot F_{N_2} \\ H_2O_{c\text{out}} &= (H_2O_{c\text{in}} + C_1 \cdot I_{fc} + C_2 \cdot I_{fc} - \gamma \cdot C_2 \cdot I_{fc})F_{H_2O_c} \end{aligned} \tag{3.14}$$

where F_{H_2}, $F_{H_2O_a}$, F_{O_2}, F_{N_2}, and $F_{H_2O_c}$ are the pressure fractions of gases inside the fuel cell, given as follows [18]:

$$F_{H_2} = \frac{P_{H_2}}{P_{H_2} + P_{H_2Oa}} \quad F_{O_2} = \frac{P_{O_2}}{P_{O_2} + P_{N_2} + P_{H_2O_c}} \quad F_{N_2} = \frac{P_{N_2}}{P_{O_2} + P_{N_2} + P_{H_2O_c}}$$
$$F_{H_2O_a} = \frac{P_{H_2O_a}}{P_{H_2} + P_{H_2O_a}} \quad F_{H_2O_c} = \frac{P_{H_2O_c}}{P_{O_2} + P_{N_2} + P_{H_2O_c}} \tag{3.15}$$

To analyze the transient behavior of fuel cells, we take into account the pressure fraction of each gas proposed by Chiu et al. [18]. In Reference 18, only the three pressure fractions F_{H_2}, F_{O_2}, and $F_{H_2O_c}$ are considered, but in our study, by considering all pressure fractions of gases, a more accurate dynamic fuel cell model is achieved and a better analysis of the transient behavior of fuel cells is possible than in previous studies [3,7,11,18]. The state equations (3.16) and (3.17) are obtained by substituting Equations 3.13 and 3.14 into Equations 3.7 and 3.8.

The new state equations on the anode side are

$$\frac{dP_{H_2}}{dt} = \frac{R*T}{V_a}\left[H_{2in} - C_1 \cdot I_{fc} - (H_{2in} - C_1 \cdot I_{fc})F_{H_2}\right]$$
$$\frac{dP_{H_2O_a}}{dt} = \frac{R*T}{V_a}\left[H_2O_{ain} - (H_2O_{ain} - C_2 \cdot I_{fc} + C_2 \cdot I_{fc})F_{H_2O_a} - C_2 \cdot I_{fc} + \gamma \cdot C_2 \cdot I_{fc}\right] \tag{3.16}$$

and the state equations on the cathode side are

$$\frac{dP_{O_2}}{dt} = \frac{R*T}{V_c}\left[O_{2in} - \frac{C_1}{2}I_{fc} - \left(O_{2in} - \frac{C_1}{2}I_{fc}\right)F_{O_2}\right]$$
$$\frac{dP_{N_2}}{dt} = \frac{R*T}{V_c}\left[N_{2in} - N_{2in} \cdot F_{N_2}\right] \tag{3.17}$$
$$\frac{dP_{H_2O_c}}{dt} = \frac{R*T}{V_c}[H_2O_{cin} + C_1 \cdot I_{st} - (H_2O_{cin} + C_1 \cdot I_{st} + C_2 \cdot I_{st} - C_2 \cdot I_{st})\, F_{H_2Oc} + C_2 \cdot I_{st} - \gamma \cdot C_2 \cdot I_{st}]$$

Because the initial mole fractions Y_{H_2}, Y_{O_2}, and Y_{N_2} are set to be 0.99, 0.21, and 0.79, respectively [3,7,11], the input values H_{2in}, O_{2in}, and N_{2in} are defined by the mole fractions, which are given as

$$H_{2in} = Y_{H_2} \cdot Anode_{in}$$
$$O_{2in} = Y_{O_2} \cdot Anode_{in} \tag{3.18}$$
$$N_{2in} = Y_{N_2} \cdot Cath_{in}$$

The water inlet flow rates on the anode and the cathode are expressed in terms of the relative humidity, saturation pressure, and total pressure on each side, as follows [2]:

$$\begin{aligned} H_2O_{ain} &= \frac{\varphi_a P_{vs}}{P_A - \varphi_a P_{vs}} \cdot Anode_{in} \\ H_2O_{cin} &= \frac{\varphi_c P_{vs}}{P_C - \varphi_c P_{vs}} \cdot Cath_{in} \end{aligned} \tag{3.19}$$

where φ_a and φ_c are the relative humidity on the anode and the cathode sides, respectively; $P_a = P_{H_2} + P_{H_2O_a}$ is the summation of partial pressures of the anode; and $P_c = P_{O_2} + P_{N_2} + P_{H_2O_c}$ is the summation of partial pressures of the cathode. P_{vs} is the saturation pressure, which can be found in the thermodynamics tables [19]. The relative humidity φ_a and φ_c are defined from the water injection input u_{a_h} for the anode, and u_{c_h} for the cathode. Furthermore, $Anode_{in}$ and $Cath_{in}$ are defined as the products of the input control variables u_a and u_c and the conversion factors k_a and k_c [11,18] on each side, which are translated from standard liters per minute (SLPM) to mol/s. In other words:

$$\begin{aligned} Anode_{in} &= u_a \cdot k_a \\ Cath_{in} &= u_c \cdot k_c \end{aligned} \tag{3.20}$$

where the conversion factors k_a and k_c are 0.065 mol/s, respectively. The hydrogen and the air stoichiometric ratios are assumed to be constant to keep the reactants flowing through the stack [19]. Hence, both of these reactants are able to be fed to the fuel cell continuously, and the fuel cell control system can be mainly dependent on the input control variables u_a and u_c. First, the anode gas pressure $P_a = P_{H_2} + P_{H_2O_a}$ and the cathode gas pressure $P_c = P_{N_2} + P_{O_2} + P_{H_2O}$ will be controlled by u_a and u_c, respectively, to avoid an unwanted pressure fluctuation and prevent MEA damage; thus, it can lead to prolong the fuel cell stack life [15]. In terms of control for the relative humidity on both sides, the first-order time-delay water injection inputs u_{a_h} and u_{c_h} will be applied because the water injection system has a very slow time constant τ_d of about 70 s [20]. Thus, in our dynamics model of PEMFC, the first-order time-delay model for the water injection is considered and the state equation from the relationship between the water injection input and relative humidity is derived as follows:

$$\varphi_a = \frac{1}{1+\tau_d s} u_{a\_h}; \quad \varphi_b = \frac{1}{1+\tau_d s} u_{b\_h} \tag{3.21}$$

where φ is the relative humidity. As seen in Equation 3.19, because each water input is a function of humidity, the water injection inputs also affect the pressure controls. So, we can establish a dynamic model of PEMFCs and the details of the control design will be described in Chapter 4.

3.4 Electrochemical Circuit Model of PEM Fuel Cells

3.4.1 Equivalent Circuit

Another important modeling method of PEM fuel cells is to use equivalent electrical circuits. The benefit of this kind of models is that the analysis, simulation, or study of PEM fuel cells can be simplified by using equivalent electrical circuits to replace PEM fuel cells. In this section, we will introduce a typical electrical circuit as proposed in Reference 21.

A mathematical approach is presented for building a dynamic model for a PEM fuel cell stack. The following assumptions are made [2,4,8,18,22]:

- One-dimensional treatment.
- Ideal and uniformly distributed gases.
- Constant pressures in the fuel cell gas flow channels.
- The fuel is humidified and the oxidant is humidified air. Assume that the effective anode water vapor pressure is 50% of the saturated vapor pressure while the effective cathode water pressure is 100%.
- The fuel cell works under 100°C and the reaction product is in the liquid phase.
- Thermodynamic properties are evaluated at the average stack temperature, temperature variations across the stack are neglected, and the overall specific heat capacity of the stack is assumed to be a constant.
- Parameters for individual cells can be lumped together to represent a fuel cell stack.

A schematic diagram of a PEM fuel cell and its internal voltage drops are shown in Figure 3.6. For details of the workings of PEM fuel cell, the reader is referred to References 18, 22, and 23.

After the effective partial pressures of H_2 and O_2 are studied, and the instantaneous change in the effective partial pressures of hydrogen and oxygen is also considered through the ideal gas equations as given in Reference 24, the fuel cell output voltage can be written as follows:

$$V_{\text{out}} = E - V_c - V_{act1} - V_{ohm} \quad (3.22)$$

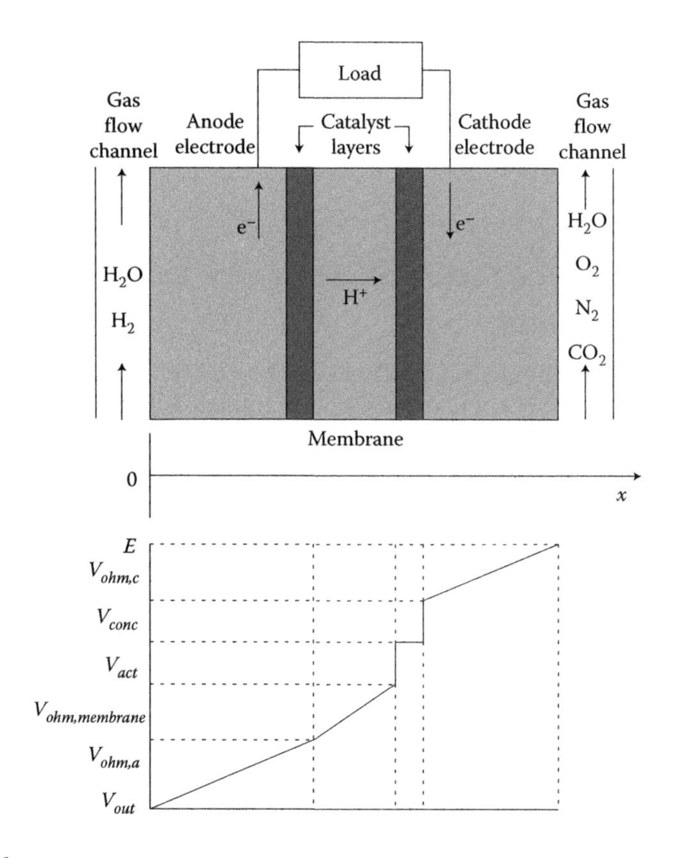

FIGURE 3.6
Schematic diagram of a PEM fuel cell and voltage drops across it.

where

- E is the reversible potential of each cell (in volts)
- V_c is the voltage across the capacitor
- $V_{act} = V_{act1} + V_{act2}$, where $V_{act1} = \eta_0 + a(T - 29)$ is the voltage drop affected only by the fuel cell internal temperature, while $V_{act2} = bT\ln(I)$ is both current and temperature dependent
- V_{ohm} is the overall ohmic voltage drop

According to the above voltage output equation, the equivalent circuit shown in Figure 3.7 is obtained

In the above circuit, C is the equivalent capacitor due to the double-layer charging effect.

3.4.2 Simulation Results

To validate the models built in Simulink and PSPICE, real input and output data shown in Figures 3.8 and 3.9 were measured on the 500-W SR-12 Avista

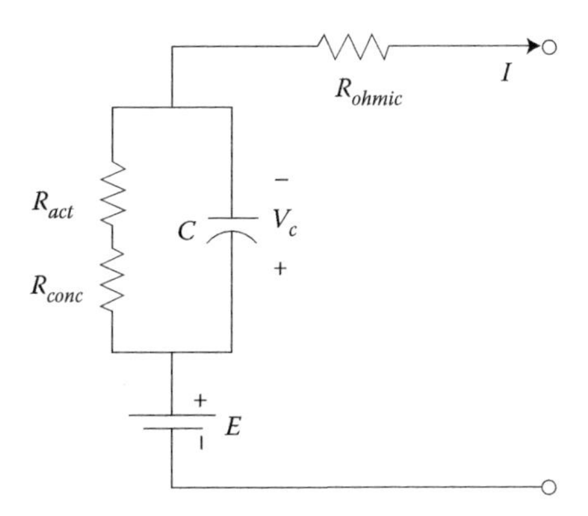

FIGURE 3.7
Equivalent circuit of a PEM fuel cell.

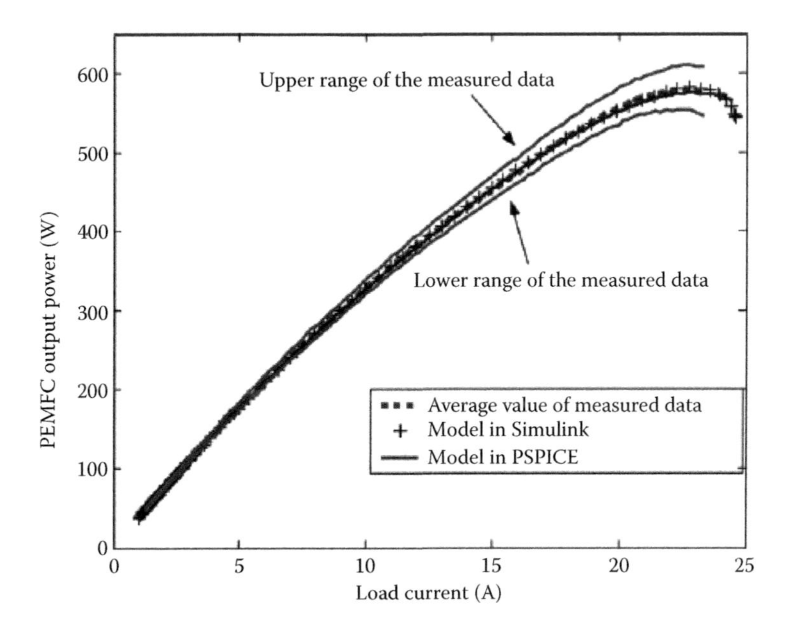

FIGURE 3.8
P–I characteristics of SR-12 and models.

Labs PEM fuel cell. The Chroma 63 112 programmable electronic load was used as a current load. Current signals were measured by LEM LA100-P current transducers; the output voltage was measured by an LEM LV25-P voltage transducer and the temperature was measured by a k-type thermocouple together with an analog connector. The current, voltage, and temperature data were all acquired by a 12-b Advantech data-acquisition card in a PC.

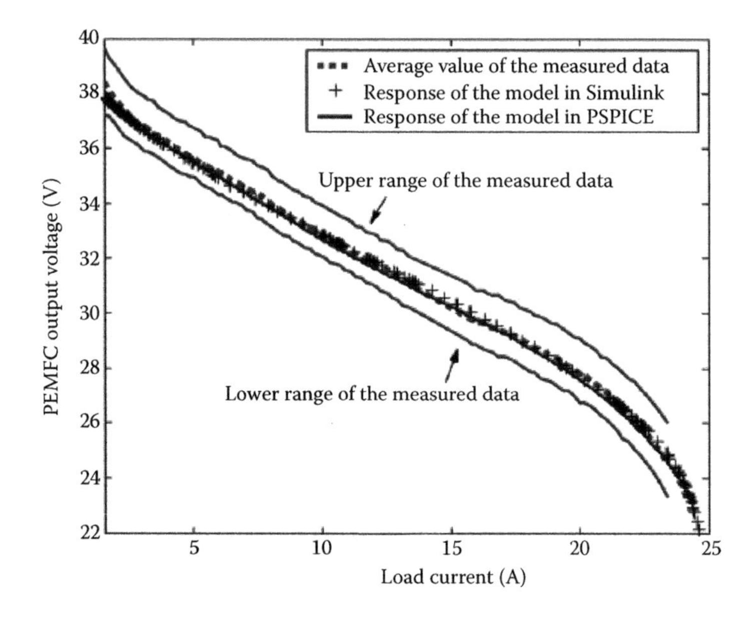

FIGURE 3.9
V–I characteristics of SR-12 and models.

3.5 Linear Model of PEM Fuel Cell Dynamics

We restrict our attention to models of PEMFC *dynamic* behavior, and so, we will not focus on models of their *static* behavior, such as those described in References 25–27. Of course, the static equations are obtainable as a limiting case from the dynamic equations, and the differences between the modeling performances of those derived from static equations can be compared, as in Reference 28. These steady-state models are able to solely simulate the cells' steady-state behavior, and cannot be used for describing and for model-based feedback control of transient conditions that are important for some specific applications. This is the case, especially, for vehicle applications, given the rapid changes of mechanical and electrical quantities.

Furthermore, our focus is on the fuel cell stack itself and thus, we will not be describing the behavior of the balance of plant (systems), which include the hydrogen and air supplies, the thermal management unit, humidifiers, etc.

Since the PEMFC's dynamic behavior is inherently nonlinear, any linear model of those dynamics is only an approximation of the original nonlinear equations. The different linear-approximating models studied thus far include the following.

3.5.1 Chiu's et al. Model

This was one of the earliest-proposed linear models of PEMFC dynamic behavior [18]. It is based on a set of nonlinear dynamic equations as the starting point. In order to linearize these equations, a small-perturbation method was used to model the fuel cell dynamics around particular operating points as an approximating linear system. Such a model is useful for transient response analysis and for control design by linear system techniques subject to the constraint of small perturbations.

The output voltage of the PEMFC is defined as in Reference 4, which gives the stack voltage equation as

$$E = N\left(E_0 + \frac{R^*T}{2F}\ln\left\{\frac{P_{H_2}(P_{O_2}/P_{std})^{1/2}}{P_{H_2O_c}}\right\} - L\right) \tag{3.23}$$

where E, N, E_0, T, and L denote the stack output voltage, number of cells in the stack, cell open-circuit voltage at standard pressure, operating temperature, and voltage losses, respectively. In addition, P_{H_2}, P_{O_2}, and $P_{H_2O_c}$ represent the partial pressure of each gas inside the cell. Also, R is the universal gas constant, F Faraday's constant, and P_{std} is the standard pressure. Note the implicit assumption in the equation that all the cells in a stack are identical.

The PEMFC's voltage losses L consist of the following:

- Activation losses—due to the slowness of the reactions taking place in the cell, which can be minimized by maximizing the catalyst contact area for reactions
- Internal current losses—due to the leakage of electrons passing through the membrane to the cathode side instead of being collected to be utilized, which has a significant effect on the open-circuit voltage
- Resistive losses—caused by current flow through the resistance of the whole electrical circuit including the membrane and various interconnections, with the biggest contributor being the membrane; effective water management to keep it hydrated reduces its ohmic loss
- Mass transport or concentration losses—caused by gas concentration changes at the surface of the electrodes

Hence, the voltage losses L can be expressed as

$$L = (i + i_n)r + a\ln\left(\frac{i + i_n}{i_o}\right) - b\ln\left(1 - \frac{i + i_n}{i_l}\right) \tag{3.24}$$

where i is the output current density, i_n the internal current density related to internal current losses, i_o the exchange current density related to activation losses, i_l the limiting current density related to concentration losses, r the area-specific resistance related to resistive losses, and a and b are constants.

3.5.1.1 Fuel Cell Small-Signal Model

From Equations 3.23 and 3.24, we can see that there are some nonlinear terms in the equations. In order to linearize the cell voltage equation, we use a small-perturbation method to model the fuel cell dynamics around particular operating points as an approximately linear system. Then, we can easily obtain the dynamic response of the cell's output voltage at these operating points for small input variable perturbations.

3.5.1.1.1 *State Equations*

First, we define the partial pressures of hydrogen, oxygen, and water (on the cathode side) as the three state variables of the system. Since the water management is a factor to affect the performance, we have to use humidifiers on both the anode and cathode side to control the humidity inside the cell. The consideration of water on the cathode side is more complicated than on the anode side because it includes not only the water supplied from the humidifiers, but also the by-product of the reaction. Figure 3.10 illustrates the various gas/vapor flows in and out of the cell.

On the basis of an ideal gas law $P^*V = n^*R^*T$, the partial pressure of each gas is proportional to the amount of the gas in the cell, which is equal to the gas inlet flow rate minus gas consumption and gas outlet flow rate. Thus, the state equations are

$$\frac{dP_{H_2}}{dt} = \frac{R^*T}{V_a}(H_{2in} - H_{2used} - H_{2out}) \tag{3.25}$$

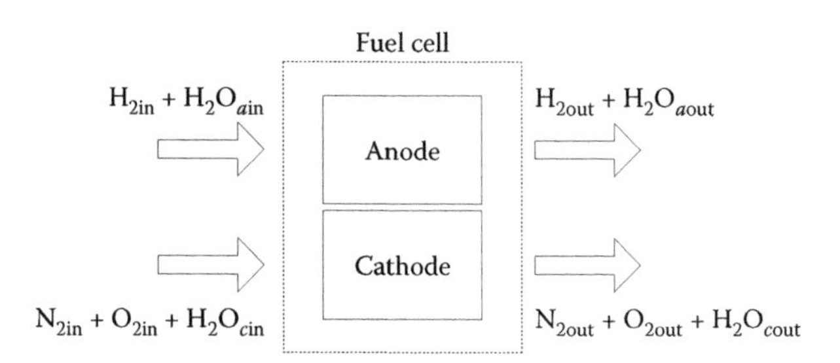

FIGURE 3.10
Illustration of gas flows of the PEMFC.

$$\frac{dP_{O_2}}{dt} = \frac{R^*T}{V_c}(O_{2in} - O_{2used} - O_{2out}) \tag{3.26}$$

$$\frac{dP_{H_2O_c}}{dt} = \frac{R^*T}{V_c}(H_2O_{cin} + H_2O_{cproduced} - H_2O_{cout}) \tag{3.27}$$

where H_{2in}, $O_{2in,}$ and H_2O_{cin} are inlet flow rates of hydrogen, oxygen, and water of cathode, respectively; H_{2out}, O_{2out}, and H_2O_{cout} are outlet flow rates of each gas. Furthermore, H_{2used}, O_{2used}, and $H_2O_{cproduced}$ represent usage and production of the gases, which are related to output current I by

$$H_{2used} = 2O_{2used} = H_2O_{cproduced} = 2K_r I = 2K_r A_c i \tag{3.28}$$

where $K_r = N/4F$, A_c is the cell active area, and i is the cell current density.

Since we can measure the inlet flow rates and output current, we can define the outlet flow rates by the equations

$$H_{2out} = (Anode_{in} - 2K_r A_c i)F_{H_2} \tag{3.29}$$

$$O_{2out} = (Cath_{in} - K_r A_c i)F_{O_2} \tag{3.30}$$

$$H_2O_{cout} = (Cath_{in} + 2K_r A_c i)F_{H_2O_c} \tag{3.31}$$

where $Anode_{in}$ and $Cath_{in}$ are the summations of anode inlet flows and cathode inlet flows, respectively, as defined by $Anode_{in} = H_{2in} + H_2O_{ain}$ and $Cath_{in} = N_{2in} + O_{2in} + H_2O_{cin}$, while F_{H_2}, F_{O_2}, and $F_{H_2O_c}$ are the pressure fractions of each gas inside the fuel cell. At this juncture, we point out the subtle but significant difference between the proposed model and the original U.S. Department of Energy (DoE) model [29] for defining the pressure fractions.

For the original DoE model,

$$F_{H_2} = \frac{P_{H_2}}{P_{op}} \tag{3.32}$$

$$F_{O_2} = \frac{P_{O_2}}{P_{op}} \tag{3.33}$$

$$F_{H_2Oc} = \frac{P_{H_2Oc}}{P_{op}} \tag{3.34}$$

but for the model proposed by Chiu et al. [18]

$$F_{H_2} = \frac{P_{H_2}}{P_{H_2} + P_{H_2Oa}} \tag{3.35}$$

$$F_{O_2} = \frac{P_{O_2}}{P_{N_2} + P_{O_2} + P_{H_2O_c}} \tag{3.36}$$

$$F_{H_2O_c} = \frac{P_{H_2O_c}}{P_{N_2} + P_{O_2} + P_{H_2O_c}} \tag{3.37}$$

In the original DoE model, it is assumed that the cell anode and cathode pressures remain constant and equal at P_{op}, which is the steady-state operating pressure. But because we are using the model to analyze the transient behavior of fuel cells, we have to account for the perturbation of each gas pressure as soon as we change some conditions. Thus, we use the summation of the gas partial pressures in Equations 3.35 through 3.37 instead of assuming a constant operating pressure.

In order to get the state-equation matrices, we substitute Equations 3.28 through 3.37 into Equations 3.25 through 3.27, and then differentiate both sides.

For example, we substitute Equations 3.29 and 3.32 into Equation 3.25, which becomes

$$\frac{dP_{H_2}}{dt} = \frac{R^*T}{V_A}\left[H_{2\text{in}} - 2K_r\,A_c\,i - (Anode_{\text{in}} - 2\,K_r\,A_c\,i)\frac{P_{H_2}}{P_{op}}\right]$$

Then, performing linearization (first-order approximation) about the given steady-state operating point yields

$$\frac{d\Delta P_{H_2}}{dt} = \frac{R^*T}{V_A}\left[-\left(\frac{H_{2\text{in}}}{Y_{H_2}} - 2K_rA_ci\right)\frac{1}{P_{op}}\Delta P_{H_2} + \left(1 - \frac{P_{H_2}}{Y_{H_2}P_{op}}\right)\Delta H_{2\text{in}} - 2K_rA_c\frac{P_{op} - P_{H_2}}{P_{op}}\Delta i\right] \tag{3.38}$$

where ΔP_{H_2}, $\Delta H_{2\text{in}}$, and Δi are the perturbations of each variable, and Y_{H_2} is the molar fraction of H_2 at the anode inlet defined by $H_{2\text{in}}/Anode_{\text{in}}$. Moreover, the state and input variables in the equation need to be replaced by their steady-state values at the chosen operating point.

Following this example, we obtained the remaining equations for the DoE and Chiu et al. models as

1. Original DoE model

$$\frac{d\Delta P_{O_2}}{dt} = \frac{R^*T}{V_c}$$

$$\left[-(Cath_{in} - K_r A_c\, i)\frac{1}{P_{op}}\Delta P_{O_2} - K_r A_c \frac{P_{op} - P_{O_2}}{P_{op}}\Delta i + \left(1 - \frac{P_{O_2}}{(Y_{O_2} + Y_{H_2Oc})P_{op}}\right)\Delta O_{2in} \right.$$

$$\left. - \frac{P_{O_2}}{(Y_{O_2} + Y_{H_2Oc})P_{op}}\Delta H_2O_{cin} \right] \tag{3.39}$$

$$\frac{d\Delta P_{H_2Oc}}{dt} = \frac{R^*T}{V_C}\left[-(Cath_{in} + 2K_r A_c\, i)\frac{1}{P_{op}}\Delta P_{H_2O_c} - \frac{P_{H_2O_c}}{(Y_{O_2} + Y_{H_2Oc})P_{op}}\Delta O_{2in} \right.$$

$$\left. + \left(1 - \frac{P_{H_2O_c}}{(Y_{O_2} + Y_{H_2Oc})P_{op}}\right)\Delta H_2O_{cin} + 2K_r A_c \frac{P_{op} - P_{H_2Oc}}{P_{op}}\Delta i \right] \tag{3.40}$$

where Y_{O_2} is the mole fraction of O_2 at the cathode inlet defined by $O_{2in}/Cath_{in}$, and $Y_{H_2O_c}$ is the mole fraction of water at the cathode inlet defined by $H_2O_{cin}/Cath_{in}$ so that $Cath_{in} = \dfrac{O_{2in} + H_2O_{cin}}{Y_{O_2} + Y_{H_2O_c}}$.

2. Chiu et al. model

$$\frac{d\Delta P_{H_2}}{dt} = \frac{R^*T}{V_A}\left[-(Anode_{in} - 2K_r A_c\, i)\frac{P_{H_2O_a}}{(P_{H_2} + P_{H_2Oa})^2}\Delta P_{H_2} \right.$$

$$\left. + \left(1 - \frac{P_{H_2}}{Y_{H_2}(P_{H_2} + P_{H_2Oa})}\right)\Delta H_{2in} - 2K_r A_c \frac{P_{H_2Oa}}{P_{H_2} + P_{H_2Oa}}\Delta i \right] \tag{3.41}$$

$$\frac{d\Delta P_{O_2}}{dt} = \frac{R^*T}{V_C}\left[-(Cath_{in} - K_r A_c\, i)\frac{P_{N_2} + P_{H_2Oc}}{P_{cathode}^2}\Delta P_{O_2} + (Cath_{in} - K_r A_c\, i)\frac{P_{O_2}}{P_{cathode}^2}\Delta P_{H_2O_c} \right.$$

$$+ \left(1 - \frac{P_{O_2}}{(Y_{O_2} + Y_{H_2O_c})P_{cathode}}\right)\Delta O_{2in} - \frac{P_{O_2}}{(Y_{O_2} + Y_{H_2O_c})P_{cathode}}\Delta H_2O_{Cin}$$

$$\left. - K_r A_c \frac{P_{N_2} + P_{H_2Oc}}{P_{cathode}}\Delta i \right] \tag{3.42}$$

$$\frac{d\Delta P_{\mathrm{H_2O}_c}}{dt} = \frac{R^*T}{V_C}\left[(Cath_{\mathrm{in}} + 2K_r A_c i)\frac{P_{\mathrm{H_2O}_c}}{P^2_{cathode}}\Delta P_{\mathrm{O_2}}\right.$$

$$-(Cath_{\mathrm{in}} + 2K_r A_c i)\frac{P_{\mathrm{N_2}} + P_{\mathrm{O_2}}}{P^2_{cathode}}\Delta P_{\mathrm{H_2O}_c} - \frac{P_{\mathrm{H_2O}_c}}{\left(Y_{\mathrm{O_2}} + Y_{\mathrm{H_2O}_c}\right)P_{cathode}}\Delta \mathrm{O}_{2\mathrm{in}}$$

$$\left.+\left(1 - \frac{P_{\mathrm{H_2O}_c}}{\left(Y_{\mathrm{O_2}} + Y_{\mathrm{H_2O}_c}\right)P_{cathode}}\right)\Delta \mathrm{H_2O}_{\mathrm{cin}} + 2K_r A_c \frac{P_{\mathrm{N_2}} + P_{\mathrm{O_2}}}{P_{cathode}}\Delta i\right] \quad (3.43)$$

where $P_{\mathrm{H_2O}_a}$ is the partial pressure of water vapor in the anode, $P_{\mathrm{N_2}}$ the partial pressure of the nitrogen in the cathode, and $P_{cathode} = P_{\mathrm{N_2}} + P_{\mathrm{O_2}} + P_{\mathrm{H_2O}_c}$.

For the fuel cell output, we substitute Equation 3.24 into Equation 3.23 and differentiate both sides. Then, we obtain a linear equation for the perturbation of the stack output voltage ΔE in response to the system state changes due to input perturbations at particular operating points as

$$\Delta E = N\left[\frac{R^*T}{2F}\frac{1}{P_{\mathrm{H_2}}}\Delta P_{\mathrm{H_2}} + \frac{R^*T}{4F}\frac{1}{P_{\mathrm{O_2}}}\Delta P_{\mathrm{O_2}} - \frac{R^*T}{2F}\frac{1}{P_{\mathrm{H_2O}_c}}\Delta P_{\mathrm{H_2Oc}}\right.$$

$$\left.-\left(r + \frac{a}{i + i_n} + \frac{b}{i_l - i - i_n}\right)\Delta i\right] \quad (3.44)$$

3.5.1.1.2 *Linear State-Space Model*

From Equations 3.38 through 3.44, we can form a linear small-signal state-space model of the hydrogen PEMFC described by

$$\Delta\dot{x} = \boldsymbol{A}\,\Delta x + \boldsymbol{B}\,\Delta u$$

$$\Delta y = \boldsymbol{C}\,\Delta x + \boldsymbol{D}\,\Delta u$$

$$\Delta x = [\Delta P_{\mathrm{H2}}\ \Delta P_{\mathrm{O2}}\ \Delta P_{\mathrm{H_2O}_c}]^T$$

$$\Delta u = [\Delta \mathrm{H}_{2\mathrm{in}}\ \Delta \mathrm{O}_{2\mathrm{in}}\ \Delta \mathrm{H_2O}_{\mathrm{Cin}}\ \Delta i]^T$$

$$\Delta y = \Delta E$$

where the three system states in Δx are the perturbations of the partial pressures of hydrogen, oxygen, and water vapor inside the cells, and the four inputs in Δu are the perturbations of the inlet flow rates of hydrogen, oxygen, and water vapor, and also the output current density, while the system output Δy is the perturbation of the fuel cell stack voltage.

For the DoE model, the matrices ***A, B, C,*** and ***D*** are

$$A = R^*T$$

$$\begin{bmatrix} -(Anode_{\text{in}} - 2K_r A_c i)\dfrac{1}{V_a P_{op}} & 0 & 0 \\ 0 & -(Cath_{\text{in}} - K_r A_c i)\dfrac{1}{V_c P_{op}} & 0 \\ 0 & 0 & -(Cath_{\text{in}} + 2K_r A_c i)\dfrac{1}{V_c P_{op}} \end{bmatrix} \tag{3.45}$$

$$B = R^*T$$

$$\begin{bmatrix} \dfrac{1}{V_a} - \dfrac{P_{H_2}}{V_a Y_{H_2} P_{op}} & 0 & 0 & -2K_r A_c \dfrac{P_{op} - P_{H_2}}{V_a P_{op}} \\ 0 & \dfrac{1}{V_c} - \dfrac{P_{O_2}}{V_c\left(Y_{O_2} + Y_{H_2O_c}\right)P_{op}} & -\dfrac{P_{O_2}}{V_c\left(Y_{O_2} + Y_{H_2O_c}\right)P_{op}} & -K_r A_c \dfrac{P_{op} - P_{O_2}}{V_c P_{op}} \\ 0 & -\dfrac{P_{H_2O_c}}{V_c\left(Y_{O_2} + Y_{H_2O_c}\right)P_{op}} & \dfrac{1}{V_c} - \dfrac{P_{H_2O_c}}{V_c\left(Y_{O_2} + Y_{H_2O_c}\right)P_{op}} & 2K_r A_c \dfrac{P_{op} - P_{H_2O_c}}{V_c P_{op}} \end{bmatrix}$$

$$C = N\frac{R^*T}{2F}\begin{bmatrix} \dfrac{1}{P_{H_2}} & \dfrac{1}{2P_{O_2}} & -\dfrac{1}{P_{H_2O_c}} \end{bmatrix}$$

$$D = N\left[0 \quad 0 \quad 0 \quad -\left(r + \frac{a}{i + i_n} + \frac{b}{i_l - i - i_n}\right)\right]$$

For the Chiu et al. model, the matrices ***A, B, C,*** and ***D*** are

$$A = R^*T$$

$$\begin{bmatrix} \begin{matrix} -(Anode_{\text{in}} - 2K_r A_c i) \\ \dfrac{P_{H_2}O_a}{V_a(P_{H_2} + P_{H_2}O_a)^2} \end{matrix} & 0 & 0 \\ 0 & -(Cath_{\text{in}} - K_r A_c i)\dfrac{P_{N_2} + P_{H_2O_c}}{V_c P_{cathode}^2} & (Cath_{\text{in}} - K_r A_c i)\dfrac{P_{O_2}}{V_c P_{cathode}^2} \\ 0 & -(Cath_{\text{in}} + 2K_r A_c i)\dfrac{P_{H_2O_c}}{V_c P_{cathode}^2} & -(Cath_{\text{in}} + 2K_r A_c i)\dfrac{P_{N_2} + P_{O_2}}{V_c P_{cathode}^2} \end{bmatrix}$$

$$B = R^*T \begin{bmatrix} \frac{1}{V_a} - \frac{P_{H_2}}{V_a Y_{H_2}\left(P_{H_2} + P_{H_2O_a}\right)} & 0 & 0 & -2K_r A_c \frac{P_{H_2Oa}}{V_a\left(P_{H_2} + P_{H_2Oa}\right)} \\ 0 & \frac{1}{V_c} - \frac{P_{O_2}}{V_c\left(Y_{O_2} + Y_{H_2O_c}\right)P_{cathode}} & -\frac{P_{O_2}}{V_c\left(Y_{O_2} + Y_{H_2O_c}\right)P_{cathode}} & -K_r A_c \frac{P_{N_2} + P_{H_2O_c}}{V_c P_{cathode}} \\ 0 & -\frac{P_{H_2O_c}}{V_c(Y_{O_2} + Y_{H_2O_c})P_{cathode}} & \frac{1}{V_c} - \frac{P_{H_2O_c}}{V_c\left(Y_{O_2} + Y_{H_2O_c}\right)P_{cathode}} & 2K_r A_c \frac{P_{N_2} + P_{O_2}}{V_c P_{cathode}} \end{bmatrix}$$

$$C = N\frac{R^*T}{2F}\begin{bmatrix} \frac{1}{P_{H_2}} & \frac{1}{2P_{O_2}} & -\frac{1}{P_{H_2O_c}} \end{bmatrix}$$

$$D = N\left[0 \quad 0 \quad 0 \quad -\left(r + \frac{a}{i + i_n} + \frac{b}{i_l - i - i_n}\right)\right] \tag{3.46}$$

3.5.1.2 Correspondence of Simulation and Test Results

MATLAB® was used to perform simulations of PEMFC dynamic response based on the small-signal linear state-space models described above. The values of the fuel cell models' main parameters, corresponding to a PEMFC stack (composed of four cells) that was tested at the DoE National Energy Technology Lab in Morgantown, WV, are listed in Table 3.4. From auxiliary tests, it was also estimated that the coefficient of Δi in Equation 3.44, yielding the instantaneous voltage change for a unit perturbation in current density, equaled 30.762 μΩ m². Furthermore, the state variables' steady-state values at the chosen operating point of 40 A load current were calculated from the input and output steady-state values, and then substituted into the matrices ***A***, ***B***, ***C***, and ***D***.

The following figures compare the responses between the test fuel cell stack, the original DoE model, and the Chiu et al. model. They are of the output voltage response to a step change of the load current with the input flow rates held constant. For the first test case with results shown in Figure 3.11, the stack was operating at a steady-state condition with $H_{2in} = 3664$ mL/min, $N_{2in} + O_{2in} = 11{,}548$ mL/min (assumed to be equally distributed

TABLE 3.4
Fuel Cell Parameters

Cell active area $A_c = 136.7$ cm²
Volume of anode $V_a = 6.495$ cm³
Volume of cathode $V_c = 12.96$ cm³
Number of cells $N = 4$
Operating pressure (for the original DoE model) $P_{op} = 101$ kPa

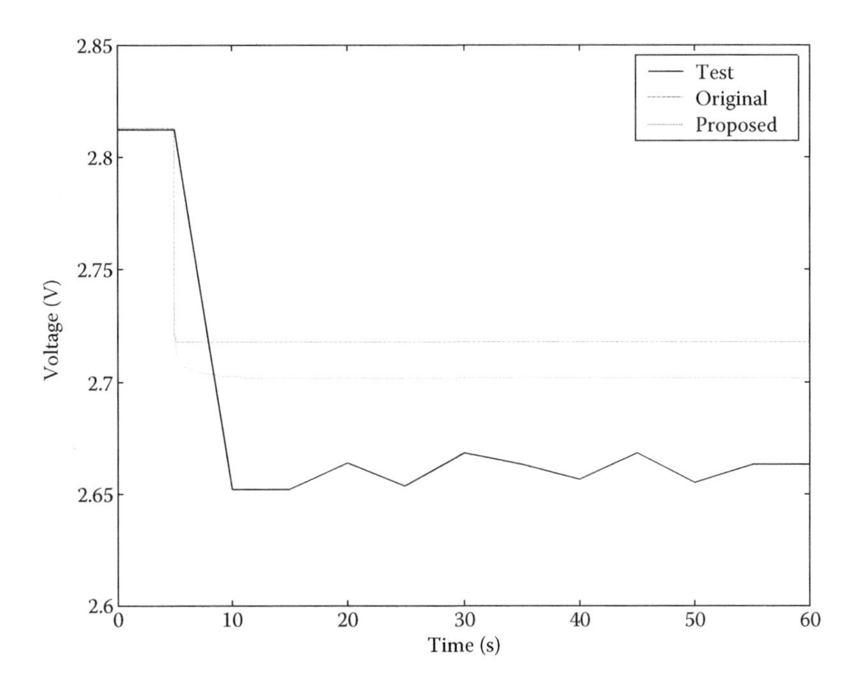

FIGURE 3.11
Comparison of simulated and test results.

between four cells), $T = 338.5$ K, and load current of 40 A, when the load current was abruptly increased to 50 A at the 5th second. The measured stack output voltage (with data-acquisition rate of 0.2 samples/s) shows a rapid decrease followed by a steady-state that appears to be a constant accompanied by some "noise." Simulations of this load current perturbation from the 40 A operating value were then performed using the DoE and the Chiu et al. small-signal dynamic models described by Equations 3.45 and 3.46, respectively. As shown in Figure 3.11, the Chiu et al. model predicts the transient response starting from the output voltage operating value of 2.813 V (the voltage drop at the instant of current change being given by $30.762^{*}\Delta i$) and also the steady-state response of the measured output voltage better than the DoE model. Note that the discrepancy in steady-state values is mainly due to the nonlinear functions found in the exact output voltage of Equation 3.23, i.e., the linear models used are only a fair approximation of the stack's exact behavior due to the fairly large perturbation involved in this test, as to be expected.

Operating condition:	$H_{2in} = 3664$ mL/min
	$N_{2in} + O_{2in} = 11{,}548$ mL/min
	T (operating or cell temperature) = 338.5 K
Perturbation:	I at 40 A then stepped up to 50 A at 5th second

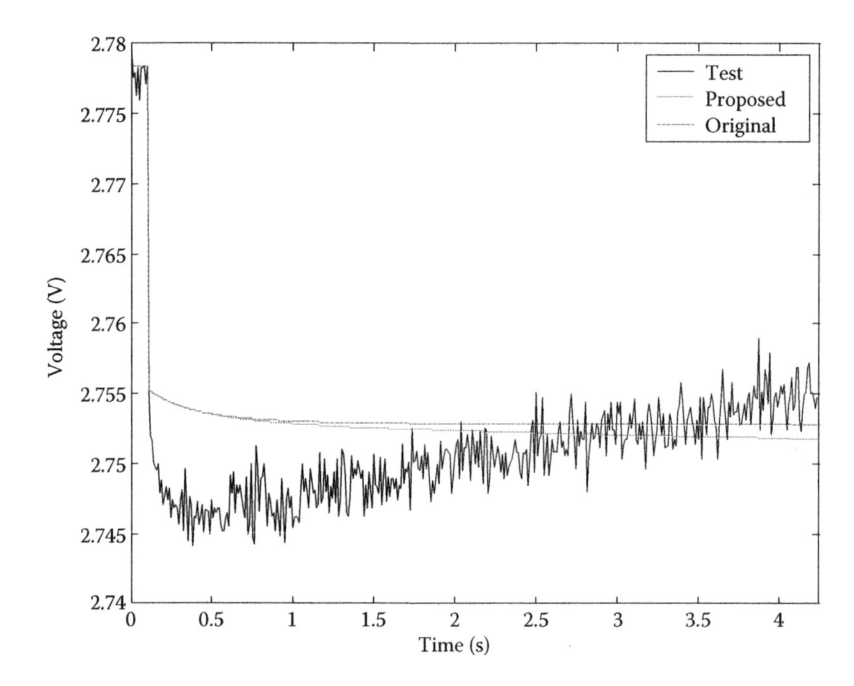

FIGURE 3.12
Another comparison of simulated and test results.

For the second test case with results shown in Figure 3.12, the stack was operating at a steady-state condition with $H_{2in} = 3000$ mL/min, $N_{2in} + O_{2in} = 10{,}000$ mL/min (assumed to be equally distributed between four cells), $T = 338.5$ K, and load current of 40 A, when the load current was abruptly increased to 42 A at 0.1 s. The measured stack output voltage (with data-acquisition rate of 100 samples/s) shows a rapid decrease followed by a gradual increase accompanied by some "noise." Simulations of this (smaller) load current perturbation from the 40 A operating value were then performed again using the DoE and the Chiu et al. small-signal dynamic models. As shown in Figure 3.12, the Chiu et al. model again predicts the transient response and the steady-state response of the output voltage better than the DoE model. However, their difference has been reduced as has the difference between them and the measured response, which was expected because of the smaller perturbation from the chosen operating point.

Operating condition:	$H_{2in} = 3000$ mL/min
	$N_{2in} + O_{2in} = 10{,}000$ mL/min
	T (operating or cell temperature) = 338.5 K
Perturbation:	I at 40 A then stepped up to 42 A at 0.1th second

Note that the test response showing the output voltage slowly increasing over time after the initial drop was also observed consistently at various operating conditions after a load increase; this was more noticeable for smaller perturbations (when a finer voltage scale was used) than for larger perturbations. But the phenomenon does not appear to represent an undershoot and subsequent recovery as they pertain to a linear dynamic system. It may be that the membrane conductivity changes as its level of hydration increases after a load change leading to the gradual increase in cell voltage. This will need to be addressed in future research.

3.5.2 Page's et al. Model

A different approach from Page et al. [30] was adopted toward obtaining a linear model of PEMFC dynamics in [13]. This approach is based on acquisition of test data from a single cell or a multiple-cell stack, followed by a least-squares estimation of the parameters (component values) for the proposed equivalent circuit model (Figure 3.13). But this is an input–output model where the equivalent circuit internal variables do not actually correspond to actual physical quantities.

3.5.3 University of South Alabama's Model

A partially linearized model of PEMFC dynamics is applied, where the partial pressures of hydrogen, oxygen, and water in the fuel cell were each modeled as having first-order dynamics [13,31]. But the PEMFC output voltage still nonlinearly depends on these partial pressures in their model.

3.5.4 Other Models

Reference 32 also adopts a small-signal approach toward modeling the dynamics of PEMFCs, but chooses to ignore the dynamics of all variables except temperature, as being the dominant (slowest evolving) system variable. In any case, this approach does not lead to a strictly linear model of PEMFC dynamics.

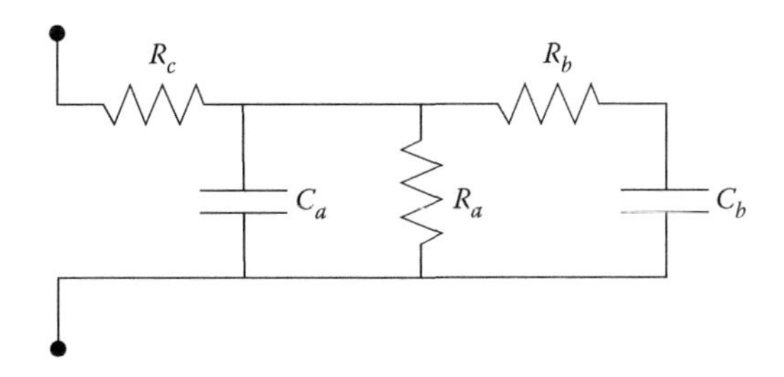

FIGURE 3.13
Page's et al. equivalent circuit model.

In Reference 33, the dynamic model presented (essentially amounting to a single equivalent capacitance) was not validated with respect to transient responses from actual PEMFCs, although the steady-state values predicted by the model were compared to the polarization curves of actual PEMFCs.

3.6 Parametric Sensitivity of PEMFC Output Response

There is significant motivation to identify which fuel cell parameters have greater impact on the cell's steady-state and transient responses so as to facilitate improved simulations and designs (internal modifications as opposed to external controller designs) of such cells. A sensitivity analysis was described in Reference 34 using a PEMFC electrochemical model and data from a 500 W PEMFC stack manufactured by BCS Technology [35]. The goal was to identify the relative importance of each parameter to the FC model simulation's accuracy. To investigate this, the parametric sensitivity of the FC electrochemical model was calculated using a multiparametric sensitivity analysis approach. To represent the FC stack electrochemical behavior, a first-order (one capacitor) electrical circuit was used to model the FC dynamic behavior. But the main focus was on FC *steady-state* response and evaluating the parameters effects on the stack polarization curve. To evaluate accuracy of the model's *dynamic* response, since the charge double-layer effect is responsible for a delay in the FC voltage change after a change in its current, the parameter used to describe this behavior is an equivalent capacitance; so, its effect on the model's step response (representing a real-world current interruption test) was studied. This capacitance does not influence the stack polarization curve because each point on this curve is obtained after the voltage has reached its steady-state value. As expected, the results of the transient response analysis were similar to that of a basic *RC* circuit.

The remainder of this section describes the results obtained in References 36 and 37 from a sensitivity analysis performed on an input–output transfer function (which is a linear systems concept) that is derived from the linear small-signal model of PEM fuel cell dynamics previously described in Section 3.6.1 [18]. These provide a greater insight into the issue of which physical parameters have the greater impacts on fuel cell dynamic response.

The transfer function being investigated represents the fuel cell's output voltage response (system output) to a perturbation of the load current (system input), which is equivalent to its output impedance. It is well known that this impedance has long been studied, usually through AC measurements termed impedance spectroscopy, as a means to characterize a cell's physical processes, electrical properties, and transient response [38–40]. But in this book, we describe a slightly different use of PEMFC output impedance.

Sensitivity is mathematically defined as the partial derivative of a function with respect to one of its parameters divided by the ratio of the function to that parameter. Here, it gauges the effect of a unit change in a given fuel cell parameter on the cell's input–output transfer function (for small load current changes). The sensitivity of the PEMFC's dynamic response can be evaluated for the following parameters: cell active area, parameter associated with cell activation losses (the slope of Tafel line), parameter associated with cell concentration losses, cell-limiting current density (corresponding to concentration losses), cell internal current density (corresponding to internal current losses), cell exchange current density (corresponding to activation losses), number of cells in the stack, cell area-specific resistance (corresponding to resistive losses), stack operating temperature (actually an operating condition parameter), cell anode volume, and cell cathode volume. Several plots will be presented below to illustrate and compare these sensitivity functions.

3.6.1 Fuel Cell Dynamic Response and Sensitivity Analysis

Now, note that the fuel cell's output impedance represents its output voltage response to a perturbation of the load current, which is the system transfer function of main practical interest. This impedance function can be obtained from Equation 3.46 as

$$\begin{aligned} G_{Vi}(s) &= C(sI_3 - A)^{-1}B_4 + D_4 \\ &= \frac{c_{11}\,b_{14}}{s - a_{11}} + \frac{(c_{13}\,a_{32} + c_{12}(s - a_{33}))b_{24} + (c_{12}\,a_{23} + c_{13}(s - a_{22}))b_{34}}{(s - a_{22})(s - a_{33}) - a_{23}\,a_{32}} + d_4 \end{aligned} \tag{3.47}$$

where s represents complex frequency, I_3 the 3×3 identity matrix, B_4 and D_4 represent the fourth columns of B and D, respectively, and m_{ij} refers to the element in the ith row and jth column of the corresponding matrix M.

3.6.1.1 Sensitivity Function

A sensitivity analysis can be performed on the PEMFC output impedance. This analysis quantitatively characterizes the effect that each particular parameter and operating point variable of the fuel cell has on that impedance. Such information can then be used for prescribing design changes to a fuel cell system in order to improve its dynamic behavior.

The sensitivity function is defined here as the ratio of the partial derivative of the transfer function (3.47) with respect to a particular parameter to the transfer function divided by that parameter, i.e.,

$$S_\theta = \frac{\partial G_{Vi}/\partial\theta}{G_{Vi}/\theta} \tag{3.48}$$

where θ represents any parameter or operating point variable of the fuel cell. It shows how sensitive the system is to different parameters and variables as a function of frequency.

As examples, we present the sensitivity functions for $G_{Vi}(s)$ with respect to H_{2in} and for $G_{Vi}(s)$ with respect to V_a, which were among the simplest of the derived functions, as

$$S_{H_{2in}} = \frac{Num_{H_{2in}}(s)}{Den_{H_{2in}}(s)} \tag{3.49}$$

$$Num_{H_{2in}}(s) = 2^*A_c^*H_{2in}^*K_1^2{}^*K_2^*P_{anode}^*P_{H_2Oa}^2{}^*R^*T^*Y_{H_2} \tag{3.50}$$

$$\begin{aligned} Den_{H_{2in}}(s) = {} & F^*P_{H_2}^*(H_{2in}^*K_1^*P_{H_2Oa}^*I^*K_1^*K_2^*P_{H_2Oa}^*Y_{H_2} \\ & + P_{anode}^2{}^*s^*Y_{H_2})^2{}^*[-4^*L(2^*A_c^*K_1^*K_2^*P_{anode}^*P_{H_2Oa}^*R^*T^*Y_{H_2}) \\ & /(F^*P_{H_2}^*(H_{2in}^*K_1^*P_{H_2Oa} - I^*K_1^*K_2^*P_{H_2Oa}^*Y_{H_2} + P_{anode}^2{}^*s^*Y_{H_2})) \\ & + (A_c^*K_2^*K_3^*P_{cath}^*(P_{H_2Oc} + P_{N_2})^*R^*T^*Y_{O_2}^*(-(P_{cath}^2{}^*s^*Y_{O_2}) \\ & - K_3^*(P_{N_2} - P_{O_2})^*(O_{2in} + I^*K_2^*Y_{O_2}))) \\ & /(F^*P_{O_2}^*(2^*P_{cath}^4{}^*s^2{}^*Y_{O_2}^2 + K_3^*P_{cath}^2{}^*s^*Y_{O_2}^*(2^*O_{2in}^*(P_{H_2Oc} \\ & + 2^*P_{N_2} + P_{O_2}) - I^*K_2^*(P_{H_2Oc} - P_{N_2} - 2^*P_{O_2})^*Y_{O_2}) \\ & + K_3^2{}^*P_{N_2}^*(P_{cath})^*(2^*O_{2in}^2 + I^*K_2^*O_{2in}^*Y_{O_2} + K_2^2{}^*Y_{O_2}^2))) \\ & + (I^*A_c^*K_2^*K_3^*P_{cath}^*(P_{N_2} + P_{O_2})^*R^*T^*Y_{O_2}^*(4^*I^*P_{cath}^2{}^*s^*Y_{O_2} \\ & + K_3^*(P_{H_2Oc} + 2^*P_{N_2})^*(2^*I^*O_{2in} + K_2^*Y_{O_2}))) \\ & /(F^*P_{H_2Oc}^*(2^*P_{cath}^4{}^*s^2{}^*Y_{O_2}^2 + K_3^*P_{cath}^2{}^*s^*Y_{O_2}^*(2^*O_{2in}^*(P_{H_2Oc} \\ & + 2^*P_{N_2} + P_{O_2}) - I^*K_2^*(P_{H_2Oc} - P_{N_2} - 2^*P_{O_2})^*Y_{O_2}) \\ & + K_3^2{}^*P_{N_2}^*(P_{cath})^*(2^*O_{2in}^2 + I^*K_2^*O_{2in}^*Y_{O_2} + K_2^2{}^*Y_{O_2}^2))))] \end{aligned} \tag{3.51}$$

and

$$S_{Va} = \frac{Num_{Va}(s)}{Den_{Va}(s)} \tag{3.52}$$

$$Num_{Va}(s) = -2^*Ac^*K^2{}^*P_{anode}^3{}^*P_{H_2Oa}^*R^2{}^*s^*T^2{}^*V_a \tag{3.53}$$

$$
\begin{aligned}
Den_{VA}(s) = {} & F^*P_{H_2}{}^*(F_{in}{}^*P_{H_2Oa}{}^*R^*T - I^*K_2{}^*P_{H_2Oa}{}^*R^*T + P_{anode}^2{}^*s^*V_a)^{2*}[4^*L \\
& + (A_c{}^*K_2{}^*K_3{}^*P_{cath}{}^*(P_{H_2Oc} + P_{N_2})^*R^*(A_{in}{}^*K_3{}^*(P_{N_2} - P_{O_2}) \\
& + I^*K_2{}^*K_3{}^*(P_{N_2} - P_{O_2}) + P_{cathode}^2{}^*s)^*T) \\
& /(F^*P_{O_2}{}^*(2^*A_{in}^2{}^*K_3^2{}^*P_{N_2}{}^*(P_{cath}) + K_2^2{}^*K_3^2{}^*P_{N_2}{}^*(P_{cath}) \\
& - I^*K_2{}^*K_3{}^*P_{cath}^2{}^*(P_{H_2Oc} - P_{N_2} - 2^*P_{O_2})^*s + 2^*P_{cath}^4{}^*s^2 \\
& + A_{in}{}^*K_3{}^*(I^*K_2{}^*K_3{}^*P_{N_2}{}^*(P_{cath}) + 2^*P_{cath}^2{}^*(P_{H_2Oc} + 2^*P_{N_2} + P_{O_2})^*s))) \\
& + (A_c{}^*K_2{}^*K_3{}^*P_{cath}{}^*(P_{N_2} + P_{O_2})^*R^*(2^*A_{in}{}^*K_3{}^*(P_{H_2Oc} + 2^*P_{N_2}) \\
& - I^*K_2{}^*K_3{}^*\left(P_{H_2Oc} + 2^*P_{N_2}\right) + 4^*P_{cath}^2{}^*s)^*T) \\
& /(F^*P_{H_2Oc}{}^*(2^*A_{in}^2{}^*K_3^2{}^*P_{N_2}{}^*(P_{cath}) + K_2^2{}^*K_3^2{}^*P_{N_2}{}^*(P_{cath}) \\
& - I^*K_2{}^*K_3{}^*P_{cath}^2{}^*(P_{H_2Oc} - P_{N_2} - 2^*P_{O2})^*s + 2^*P_{cath}^4{}^*s^2 \\
& + A_{in}{}^*K_3{}^*(I^*K_2{}^*K_3{}^*P_{N_2}{}^*(P_{cath}) + 2^*P_{cath}^2{}^*(P_{H_2Oc} + 2^*P_{N_2} + P_{O_2})^*s))) \\
& + (2^*A_c{}^*K_2{}^*P_{anode}{}^*P_{H_2Oa}{}^*R^{2*}T^2)/(F^*P_{H_2}{}^*(F_{in}{}^*P_{H_2Oa}{}^*R^*T \\
& - I^*K_2{}^*P_{H_2Oa}{}^*R^*T + P_{anode}^2{}^*s^*V_a))]
\end{aligned}
\tag{3.54}
$$

where

$$
K_1 = \frac{(R^*T)}{V_a},\quad K_2 = 2^*Kr,\quad K_3 = \frac{(R^*T)}{V_c},\quad P_{anode} = P_{H_2} + P_{H_2Oa},
$$
$$
P_{cath} = P_{O_2} + P_{N_2} + P_{H_2Oc},\quad F_{in} = \frac{H_{2in}}{Y_{H_2}},\quad A_{in} = \frac{O_{2in}}{Y_{O_2}} \tag{3.55}
$$

3.6.1.2 Sensitivity Function Plots

The various sensitivity functions for the PEMFC's output impedance were evaluated using MATLAB and then plotted. Since the calculated sensitivity functions are complicated expressions and the parameters and input variables are many, only a few representative plots are presented here.

The baseline parameter and input values that were used are given in Table 3.5. These correspond to a PEMFC stack at the DoE's National Energy Technology Laboratory, Morgantown, WV that underwent testing in 2003. This stack had four cells and ran on air (rather than pure oxygen).

These sensitivity function plots for the DoE National Energy Technology Laboratory (NETL) model were derived for the anode and cathode volume,

TABLE 3.5
Fuel Cell Parameter and Input Values for Sensitivity Analysis

Parameter	Description	Value or Input
N	Number of cells in the stack	4
T	Stack operating temperature	338.6 K
H_{2in}	Anode inlet flow rate of hydrogen	3000 mL/min
Air_{in}	Cathode inlet flow rate of air	10,000 mL/min
I	Cell output current	40.25 A
A_c	Cell active area	136.7 cm^2
i	Cell output current density	2926 A/m^2
V_a	Volume of the cell anode channels	6.495 cm^3
V_c	Volume of the cell cathode channels	12.96 cm^3
L	Sum of cell voltage losses	–36 μV/cell (for a 2.2 A change)
a	Constant associated with cell activation losses (the slope of Tafel line)	0.06 V
b	Constant associated with cell concentration losses	0.05 V
i_n	Cell internal current density corresponding to internal current losses	20 A/m^2
i_o	Cell exchange current density corresponding to activation losses	
i_l	Cell-limiting current density corresponding to concentration losses	9000 A/m^2
r	Cell area-specific resistance corresponding to resistive losses	3 μΩ m^2
Y_{H_2}	Molar fraction of H_2 at the anode inlet	0.9
Y_{O_2}	Molar fraction of O_2 at the cathode inlet	0.189 (= 0.9*0.21)

cell's active area, output current density, cell's temperature, inlet flow of hydrogen, inlet flow of air, and voltage loss constants, based on the NETL fuel cell's parameter values and one set of operating conditions.

We first present the output impedance sensitivity plot with respect to the fuel cell's anode volume (Figure 3.14). From the plot, we can notice that the overall impedance function is more sensitive to changes in the anode volume at frequencies around 1 rad/s.

We next present the output impedance sensitivity plot with respect to the fuel cell's cathode volume (Figure 3.15). From the plot, we can notice that the overall impedance function is more sensitive to changes in the cathode volume at frequencies around 1 rad/s.

We next present the output impedance sensitivity plot with respect to the fuel cell's active area (Figure 3.16). From the plot, we can notice that the overall impedance function is more sensitive to changes in the cell's active area at low frequencies.

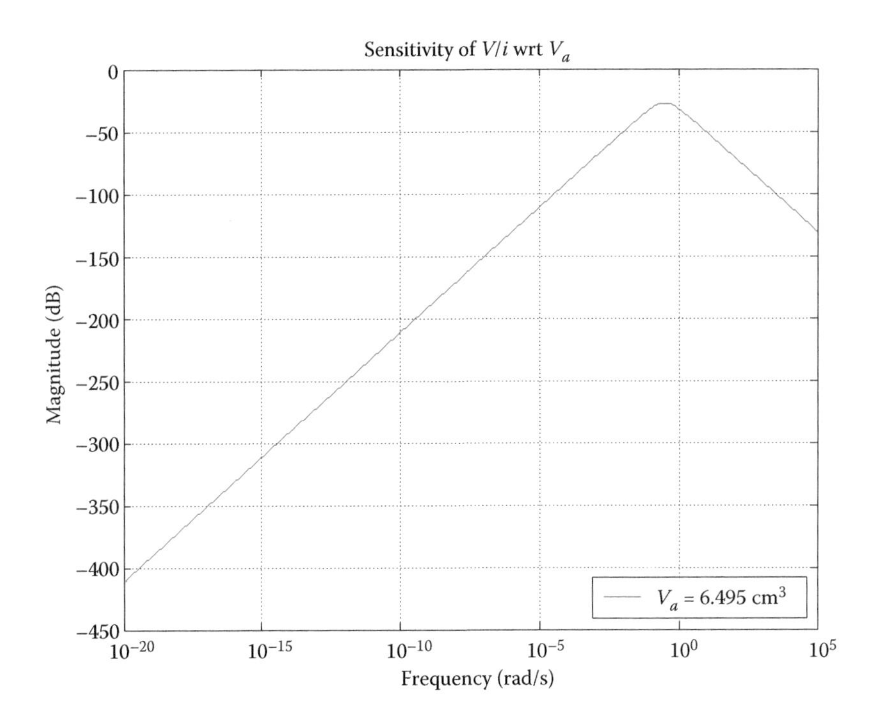

FIGURE 3.14
Output impedance sensitivity plot with respect to cell anode volume.

We next present the output impedance sensitivity plot with respect to the fuel cell's output current density (Figure 3.17). From the plot, we can notice that the overall impedance function is more sensitive to changes in the cell's output current density at frequencies above 10 rad/s.

We next present the output impedance sensitivity plot with respect to the inlet flow rate of hydrogen (Figure 3.18). From the plot, we can notice that the overall impedance function is more sensitive to changes in the flow rate of hydrogen at a frequency range below 10 rad/s.

We next present the output impedance sensitivity plot with respect to the inlet flow rate of air (Figure 3.19). From the plot, we can notice that the overall impedance function is more sensitive to changes in the flow rate of oxygen at frequencies below about 1 rad/s.

We next present the output impedance sensitivity plot with respect to the fuel cell's temperature (Figure 3.20). From the plot, we can notice that the overall impedance function is more sensitive to changes in temperature at frequencies around 0.1 rad/s.

We next present the output impedance sensitivity plot with respect to the voltage losses constant a (Figure 3.21). From the plot, we can notice that the overall impedance function is more sensitive to changes in the voltage losses constant a at frequencies above 10 rad/s.

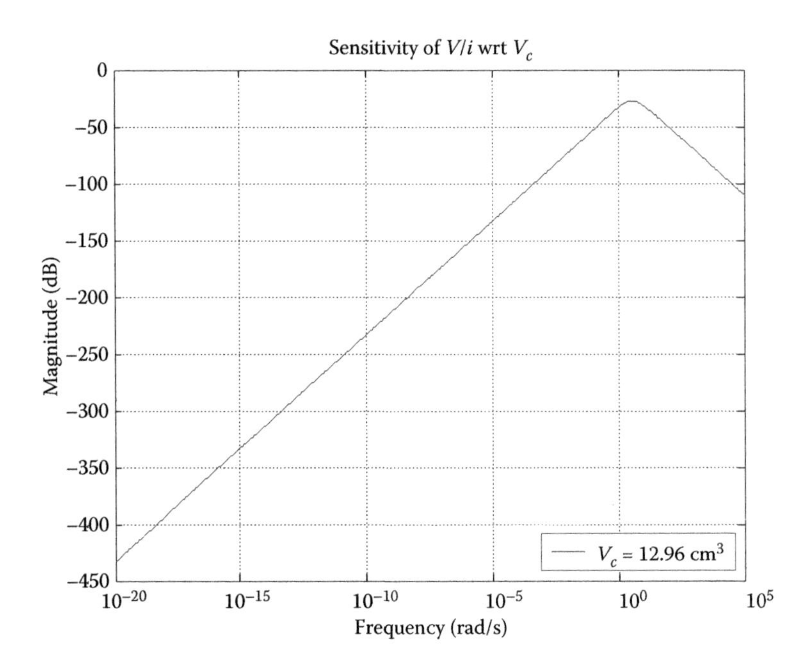

FIGURE 3.15
Output impedance sensitivity plot with respect to cell cathode volume.

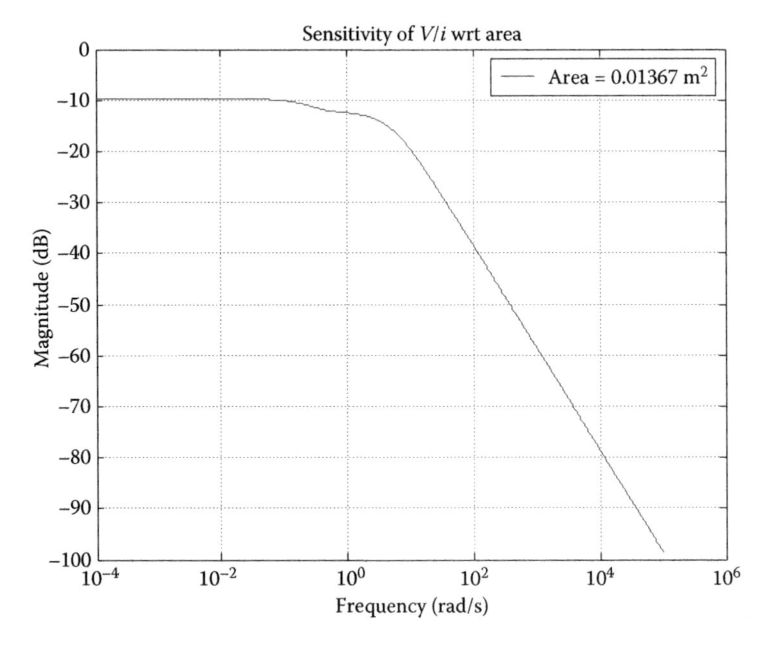

FIGURE 3.16
Output impedance sensitivity plot with respect to cell's active area.

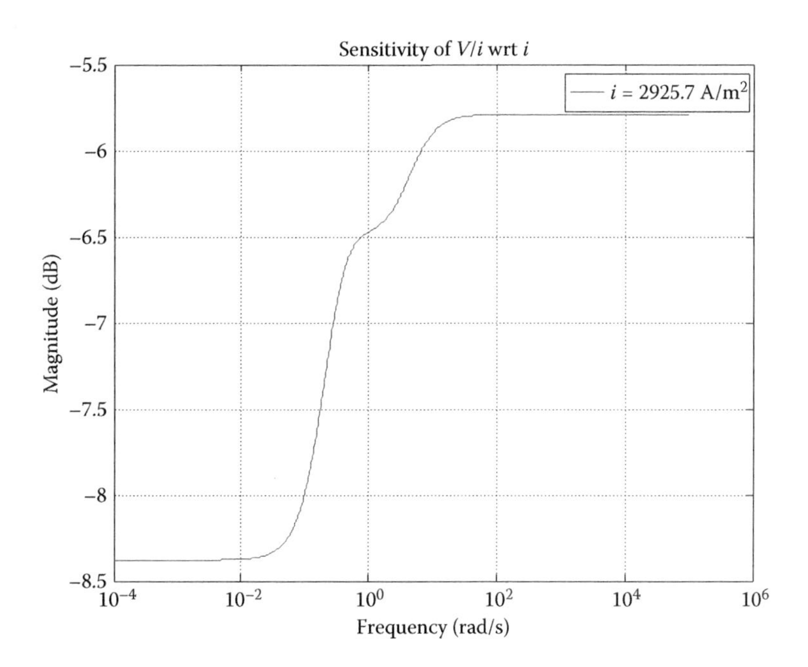

FIGURE 3.17
Output impedance sensitivity plot with respect to output current density.

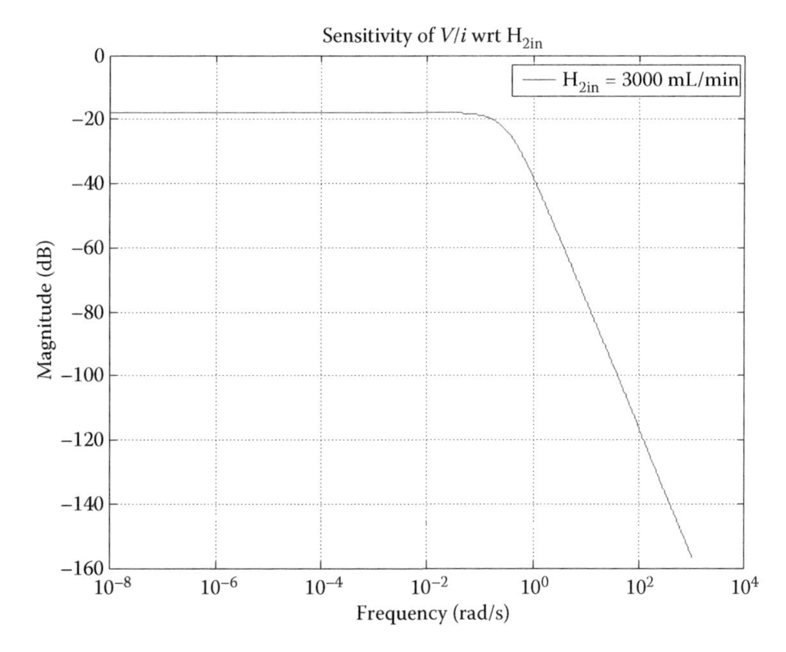

FIGURE 3.18
Output impedance sensitivity plot with respect to the hydrogen inlet flow.

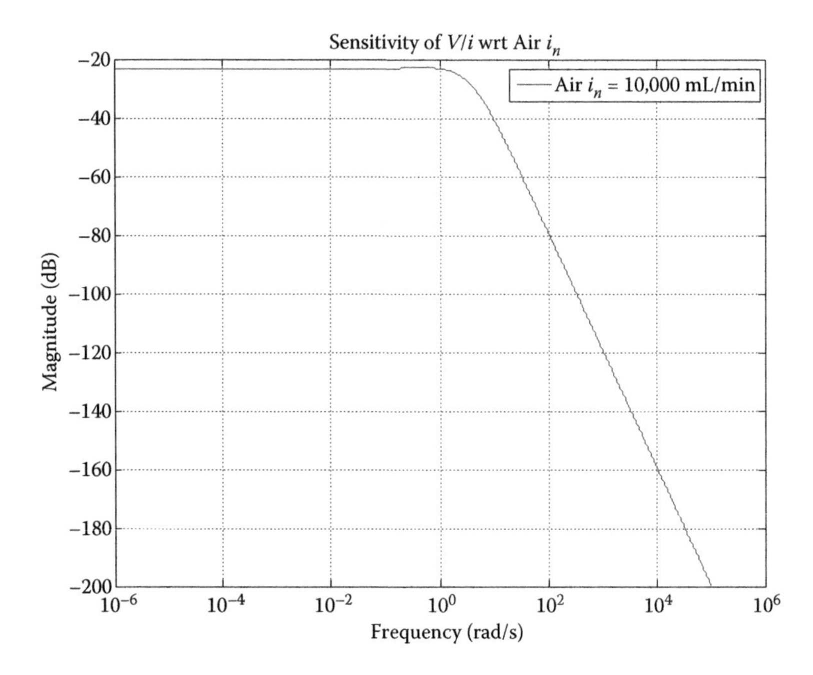

FIGURE 3.19
Output impedance sensitivity plot with respect to the air inlet flow.

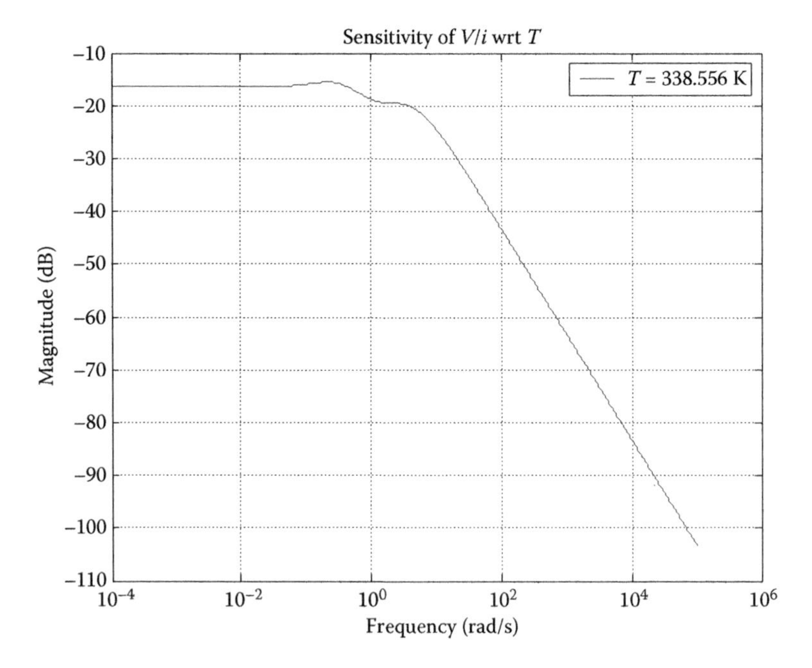

FIGURE 3.20
Output impedance sensitivity plot with respect to the cell temperature.

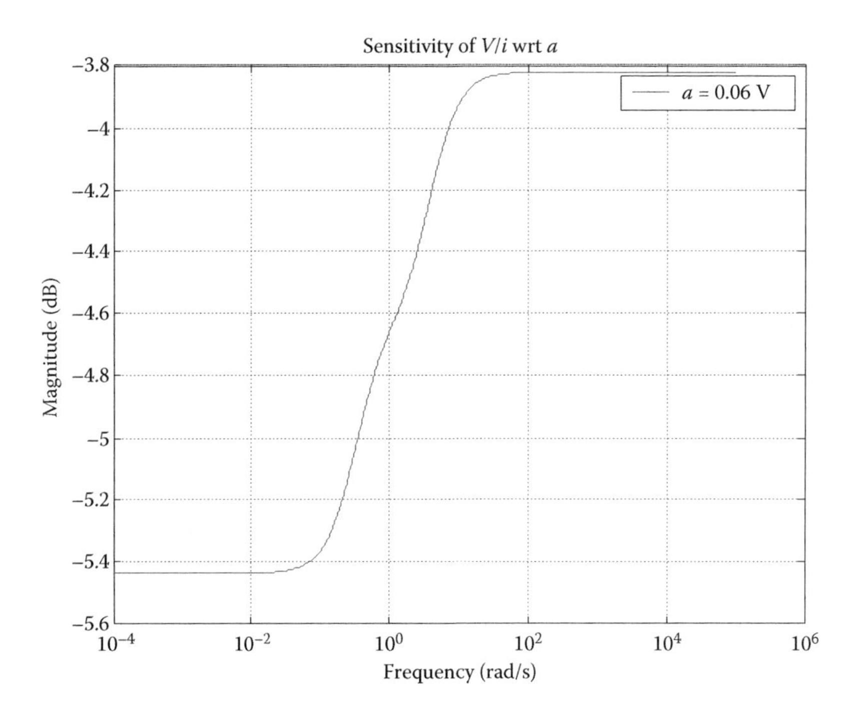

FIGURE 3.21
Output impedance sensitivity plot with respect to the voltage losses constant a.

We next present the output impedance sensitivity plot with respect to the voltage losses constant b (Figure 3.22). From the plot, we can notice that the overall impedance function is more sensitive to changes in the voltage losses constant b at frequencies above 10 rad/s.

We next present the output impedance sensitivity plot with respect to the internal and fuel crossover current density constant i_n (Figure 3.23). From the plot, we can notice that the overall impedance function is more sensitive to changes in the internal current density at frequencies above 10 rad/s.

We next present the output impedance sensitivity plot with respect to the limiting current density constant i_l (Figure 3.24). From the plot, we can notice that the overall impedance function is more sensitive to changes in the limiting current density at frequencies above 10 rad/s.

We next present the output impedance sensitivity plot with respect to the specific area resistance r (Figure 3.25). From the plot, we can notice that the overall impedance function is more sensitive to changes in the specific area resistance r at frequencies above 10 rad/s.

Comparing the plots of the output impedance's sensitivities with respect to the cell's anode volume and to its cathode volume in Figures 3.14 and 3.15, respectively, note that the plots are somewhat similar to each other. However, the magnitude of the function for cathode volume is greater at the higher

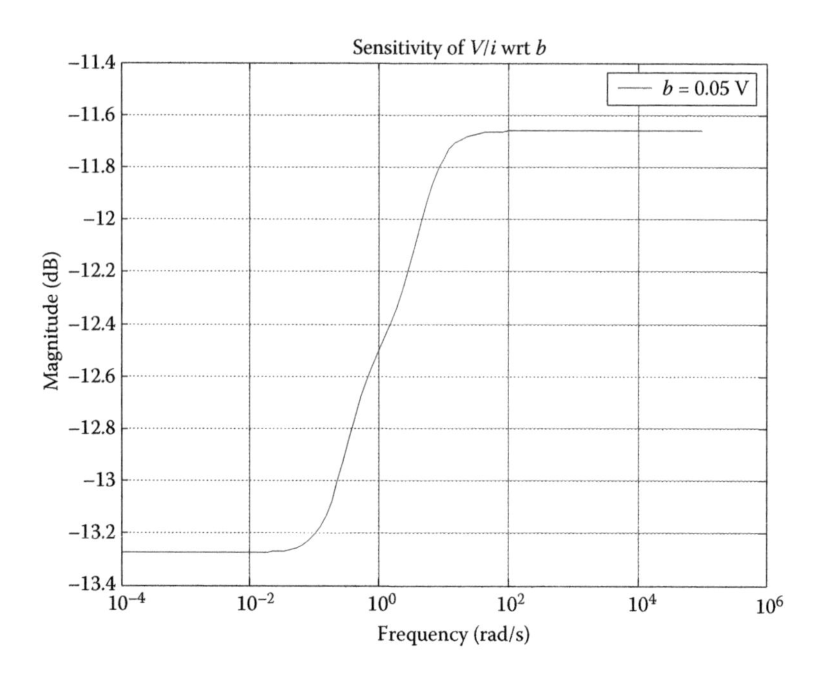

FIGURE 3.22
Output impedance sensitivity plot with respect to the voltage losses constant b.

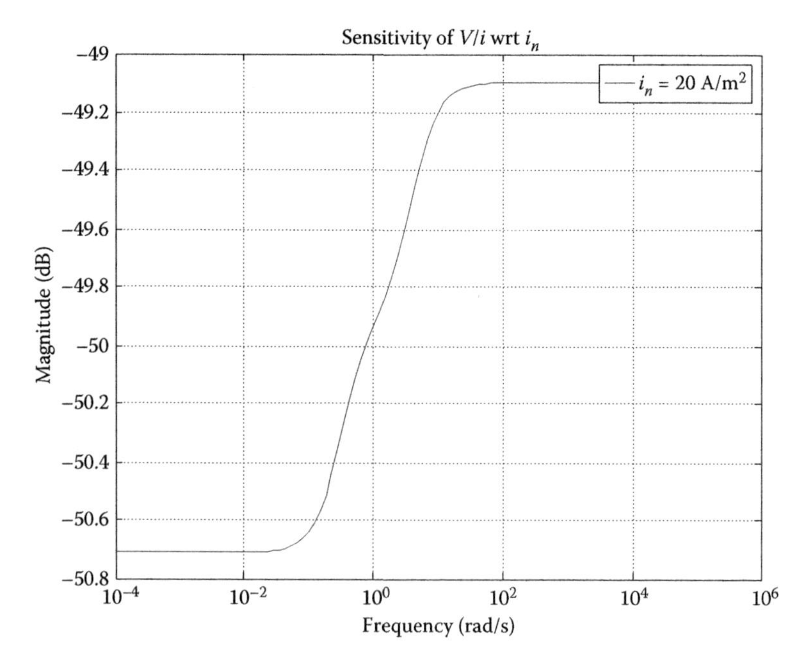

FIGURE 3.23
Output impedance sensitivity plot with respect to the internal current density constant i_n.

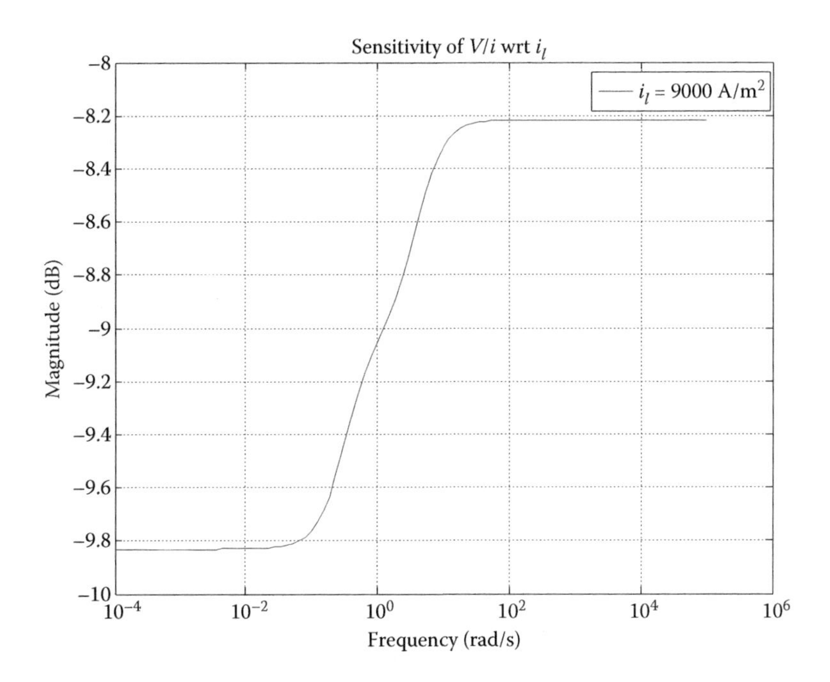

FIGURE 3.24
Output impedance sensitivity plot with respect to the limiting current density constant i_l.

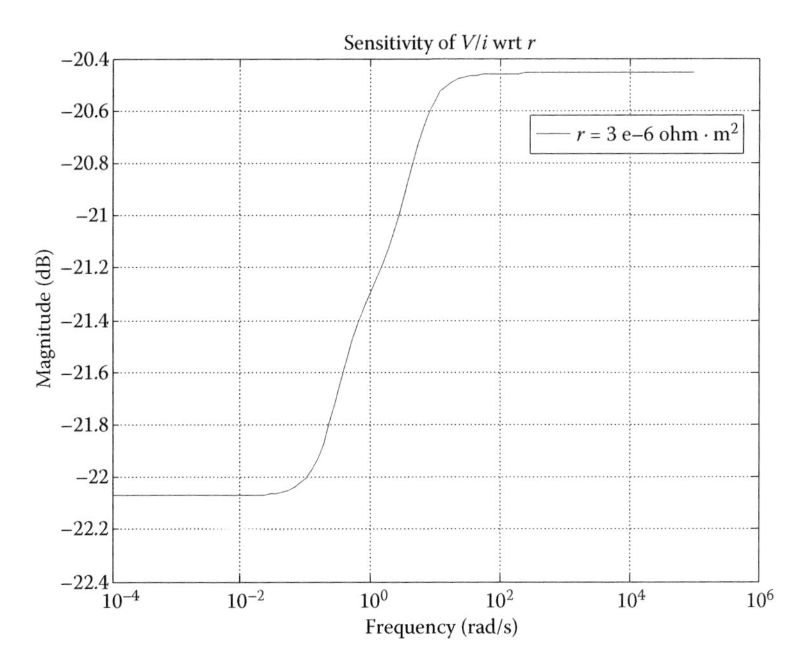

FIGURE 3.25
Output impedance sensitivity plot with respect to the specific area resistance r.

frequencies above 1 rad/s than the corresponding function for anode volume, and vice versa for frequencies below 1 rad/s. Hence, we infer that the high-frequency component of the cell's dynamic response is more sensitive to changes in its cathode volume than to its anode volume under typically expected operating conditions. Considering the impedance sensitivity function with respect to the fuel cell's active area as graphed by Figure 3.16, we can see that the overall impedance function is also more sensitive to changes in the fuel cell's active area at lower frequencies than at frequencies higher than about 1 rad/s. The impedance sensitivity functions with respect to the inlet flow rates of hydrogen and air are shown in Figures 3.18 and 3.19, respectively. From these plots, we can see that the impedance is more sensitive to changes in the flow rate of hydrogen at a frequency range lower than 0.1 rad/s. On the other hand, for the flow rate of air, the impedance has a higher sensitivity at frequencies below about 1 rad/s.

3.6.2 Summary

In this section, we have described the results obtained from a sensitivity analysis performed on an input–output transfer function that is derived from the linear small-signal model of PEMFC dynamics proposed by Chiu et al. [18]. These results provide greater insight into the important design issue of which fuel cell parameters have greater impacts on the cell's dynamic response. As one example, it was determined that the fuel cell's dynamic response at higher frequencies is typically more sensitive to changes in its cathode volume than to its anode volume, which implies that the former is a more significant parameter for influencing the higher-frequency component of transient response than the latter.

3.7 Temperature and Fuel Dependence of an Equivalent Circuit Model of Direct Methanol Fuel Cells' Dynamic Response

This section will focus on DMFCs. As has been mentioned in Chapter 1, methanol is easier to transport than the hydrogen used by PEMFCs and easier to work with; so, DMFCs are an attractive option for various portable and mobile applications, such as powering military equipment, forklifts, and scooters [41–43]. Moreover, methanol can be produced quite easily from biomass [44].

3.7.1 Importance of Equivalent Circuit Model

As is the case from PEMFCs, it is important to properly characterize an equivalent circuit model of the DMFC's dynamic behavior, for performing analysis and optimal design of the power systems based on such sources. Such characterization includes identifying the circuit's components, their connection, and also the dependence of the component values on variable operating conditions such as temperature and fuel flow rate. The characterized model can then be used in several ways, such as for accurately evaluating and then improving how the DMFC interacts with other components of the complete power system (such as ultracapacitors (UCs) and switch-mode power converters), e.g., the effects of DC–AC converter ripple current on the lifespan of the fuel cell. Another way is for adjusting, either statically or dynamically, and in either an open-loop or closed-loop fashion, the temperature and/or fuel flow to obtain a more optimal dynamic performance of DMFCs in applications with fast, frequent, and/or significant load changes.

In the following sections, we will describe two such equivalent circuits, as well as the experimental dynamic response data we obtained that determined the particular circuit we would use for our study. Thereafter, we present the procedure used to estimate the component values of that equivalent circuit model and discuss the ensuing results, followed by several conclusions.

3.7.2 Equivalent Circuit Models of DMFC Dynamic Response

Several equivalent circuit models have been proposed to model the dynamic response of H_2-fueled PEMFCs as has been described in the previous sections of this chapter. Moreover, some of these models have been fit to, and its component values estimated from, measured frequency-domain (AC impedance) data with varying degrees of agreement [45,46]. On the other hand, only a couple of equivalent circuits have been proposed specifically to model the dynamic response of DMFCs. Other kinds of models, such as those presented in References 47–49, while important in their own right, are usually not as useful as circuit-type models to electrical engineers and control system engineers when they perform diagnostics, design, or analysis, e.g., to determine the amount of AC (ripple) current being drawn from the fuel cell by a given switch-mode DC–AC converter, which affects the cell's lifespan.

The circuit model described in Reference 50 for DMFCs is a first-order model (see Figure 3.26) that has the following components: a source E for the open-circuit voltage, which depends on the methanol and oxygen feed

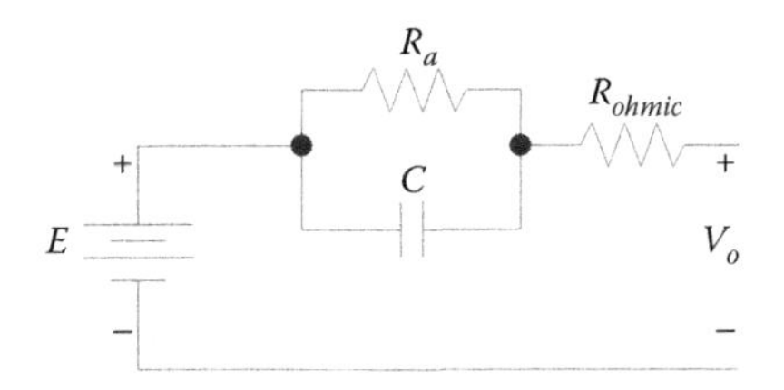

FIGURE 3.26
First-order equivalent circuit model of DMFC dynamic response, described in Reference 53.

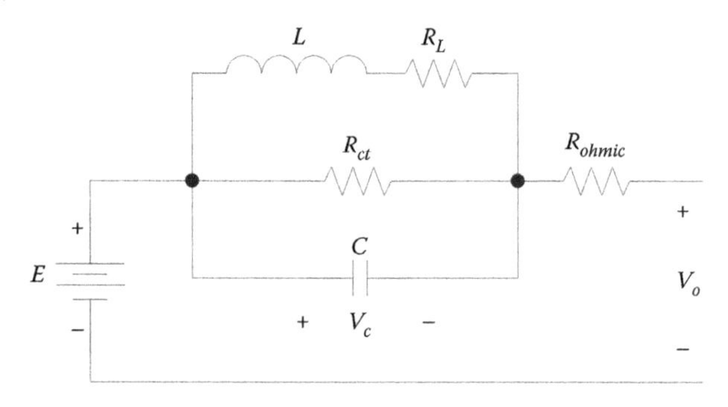

FIGURE 3.27
Second-order equivalent circuit model of DMFC dynamic response, described in Reference 54.

concentrations; a resistance R_a representing the combined activation and mass-transport losses; a capacitance C for the double-layer capacitance of the electrodes; and a resistance R_{ohmic} representing the ohmic loss. In addition, second-order equivalent circuit models were proposed in References 51 and 52 (see Figures 3.27 and 3.28, respectively). Note that the model described in Reference 51 is identical to that in Reference 52, except for the additional resistor representing the cell's ohmic resistance. The other components described in Reference 52 to model the DMFC anode impedance are an inductance L to represent inductive behavior (phase delay) that can be explained using kinetic theory [53] for the reaction mechanism for methanol electro-oxidation involving intermediate adsorbates, a behavior confirmed by others such as in Reference 49; a resistance R_o that serves to modify the phase delay according to the reaction scheme; a resistance R_∞ that is associated with the part of the current response occurring without change in adsorbate coverage; and a capacitance C_d that is believed to be associated with the redistribution of charge at the anode (it depended on parameters such as current density) instead of double-layer capacitances of the DMFC anode and cathode, since double-layer capacitance values for similar electrodes have been reported to be an order of magnitude lower than values calculated for this C_d. This circuit models well the frequency-domain impedance spectra of a DMFC

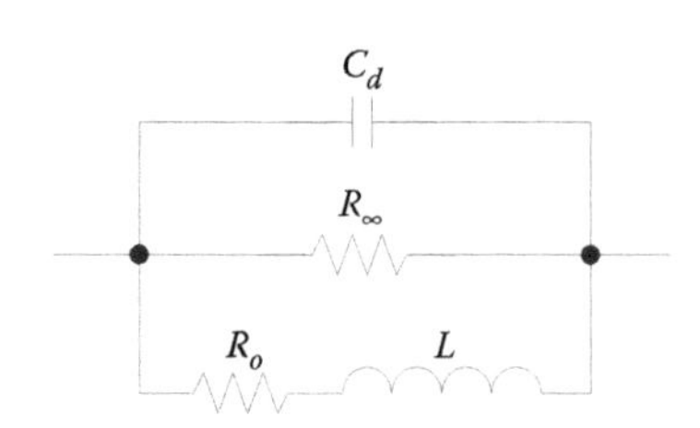

FIGURE 3.28
Second-order equivalent circuit model of DMFC dynamic response, described in Reference 55.

operating with fuel flow at several times the stoichiometric rate, so excluding mass-transport limitations [52].

The model described in Reference 51, based on the model in Reference 52, includes an ohmic resistance R_{ohmic}, and a capacitance C similar to C_d in Reference 52; both of these values are supposedly constant. On the basis of impedance spectra measured with the cell output voltage at various levels from 0.5 V to 0.1 V, Wang et al. [51] also proposed nonlinear functions of current for the values of the parameters R_{ct}, R_L, and L, which are similar to R_∞, R_o, and L, respectively, in Reference 52.

3.7.3 Testing of DMFC to Ascertain Equivalent Circuit Model Parameters

As described in Reference 54, testing of DMFCs can be performed to ascertain the equivalent circuit model parameters describing their dynamic response. Only a summary of the procedures followed therein for DMFC testing and for parameter identification is now provided, as is its conclusions.

A series of experiments were conducted at different combinations of temperature and fuel flow rate using the experimentation setup diagrammed in Figure 3.29. Tests were performed with temperature of the DMFC controlled to 30°C, 50°C, and 70°C on separate occasions with the aid of a thermocouple and electric heating elements embedded in the DMFC end plates, allowing for closed-loop control by the fuel cell test station (FCTS) software. The fuel flow rate was also varied, at each temperature setting, using the same software interface to be 0.2, 0.4, and then 0.6 mL/min on separate tests; however, the flow rate of air was fixed at 1 L/min. Tests were conducted first with fuel having a concentration of 1 mol/L, and then with fuel having a concentration of 0.5 mol/L.

The tests consisted of applying a fixed load resistance (16.8 Ω) to the terminals of the fuel cell, thus drawing a current of about 40 mA while the FCTS software recorded the voltage values at these terminals at a 1 Hz rate. The test data were first of the voltage values across the DMFC terminals after the load resistance was connected to the fuel cell; this part of the fuel cell response will be referred to as the "load-on" response. After an interval of 5 min, the load resistance was disconnected to produce an open circuit across the terminals, but the values for DMFC terminal voltage continued being recorded for another 5 min; this portion of the test will be referred to as the "load-off" response. This load-on, load-off test was repeated several times in order to ensure repeatability and also with the intent to average out the effect of random phenomena, such as electrical noise.

Note that, as compared to circuit modeling using electrochemical impedance spectra data, which is a frequency-domain approach, this test procedure does not require an expensive AC impedance (frequency response) analyzer. Instead, only time-domain data are collected. Moreover, that the choice of test conditions (temperature, fuel flow) was aimed at obtaining responses over a fairly wide range of possible operating conditions. Finally, the choice

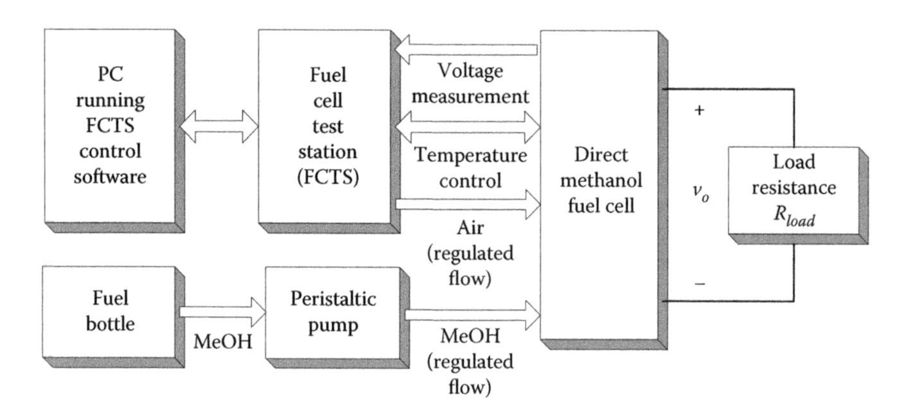

FIGURE 3.29
Diagram of DMFC experimentation setup in Reference 54.

of concentrations of 1 and 0.5 mol/L, was mainly determined for the ease of obtaining those values, given the limited precision of the volume measurements that could be made.

Figure 3.30 is illustrative of the measured load-on, load-off responses at the various operating conditions, although it was noted that the test data obtained at the 1 mol/L, 70°C, and 0.4 mL/min condition had somewhat greater variability (reduced repeatability) than the data at other test conditions (the R^2 values comparing the data sets at a given condition to each other, all being greater than 0.95, where R^2 has its usual definition as the ratio of the sum of squares of the regression to the total sum of squares), which

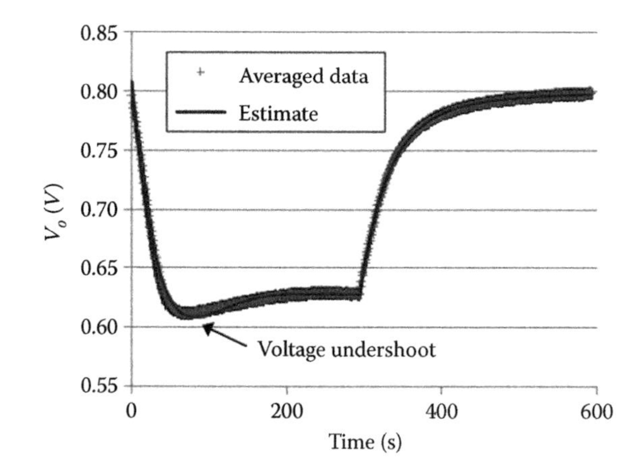

FIGURE 3.30
Comparison of averaged data (with error bars) of load-on/load-off response of DMFC output voltage for an input of 1 mol/L concentration methanol at 0.4 mL/min, 50°C, to its least-squares estimate.

may explain certain results described in the next section; this was very likely due to random effects, and was not especially grievous (the R^2 values comparing the three data sets at this one-test condition to each other being 0.89, 0.98, and 0.84). It was observed from these responses of the fuel cell output voltage that the load-on response typically had an undershoot—meaning the response's minimum value is less than its steady-state value. This type of response requires a mathematical function with at least two exponential terms to describe, which a first-order circuit cannot produce—its response, by definition, is a single exponential function of time [55]. Furthermore, small overshoots of a few millivolts—overshoot meaning the response's maximum value is greater than its steady-state value—were observed from the 0.5 mol/L tests at 70°C. It is uncertain, at this time, which of the several difference(s) between these tests and those in Reference 50 result(s) in load-on responses with undershoots documented in Reference 54 but (first-order) responses without undershoots in Reference 50. But these observed undershoots meant that Singh et al. [54] focused on the second-order equivalent circuit model of Figure 3.27 for their study.

It is well known that second-order circuits can yield a transient response that is either overdamped (sum of two real exponential responses) or underdamped (sum of two complex exponential responses, resulting in an exponentially damped sinusoidal response). From the data collected by Singh et al. [54], it was clear that the load-off responses belonged to the overdamped category while the load-on responses belonged to the underdamped category; so, the corresponding capacitor voltage can be expressed as shown below

$$v_c(t) = a\,e^{bt} + c\,e^{dt} \tag{3.56}$$

$$v_c(t) = \chi e^{-\alpha t}\cos(\omega t + \delta) + \Delta v \tag{3.57}$$

where $v_c(t)$ is the capacitor voltage as indicated in Figure 3.27, t time, Δv the voltage to which C is charged to during the load-on response, and the rest are constants that need to be determined from each test response. For both cases, the output voltage can be readily related to the capacitor voltage $v_c(t)$ by Kirchoff's voltage law, and so the load-on response is given by

$$v_o(t) = E - (a\,e^{bt} + c\,e^{dt}) - (v_o(t)R_L)R_{ohmic} \tag{3.58}$$

and the load-off response is given by

$$v_o(t) = E - (\chi e^{-\alpha t}\cos(\omega t + \delta) + \Delta v) \tag{3.59}$$

However, the crux of the problem is centered on estimating the parameters a, b, c, d or $\alpha, \omega, \chi, \delta$ embedded in the capacitor voltage expression, since the voltage E and resistance R_{ohmic} can be readily obtained. Therefore, this guided their analysis, as described next, to relate the model component values to the parameters a, b, c, d of Equation 3.56 or $\alpha, \omega, \chi, \delta$ of Equation 3.57.

3.7.4 Model Component Value Estimation

First, the (either load-on or load-off) responses were averaged and then approximated by nonlinear least-squares curve fitting using the MATLAB Curve Fitting Toolbox to determine the four response parameters, either a, b, c, d of (3.56) or $\alpha, \omega, \chi, \delta$ of (3.57). Twenty-five iterations of each fitting were performed with different initial guesses of the parameter values to greatly increase the likelihood of obtaining the globally optimal solution. After selecting the best outcome of these iterations, it was found that all of the 1 mol/L load-on curves were fitted with $R^2 > 0.990$ with the exception of the 70°C, 0.4 mL/min case where the best fit yielded $R^2 = 0.961$. In addition, all of the 0.5 mol/L load-on curves were fitted with $R^2 > 0.987$. Moreover, all of the 1 mol/L load-off curves were fitted with $R^2 > 0.997$ with the exception of the 70°C, 0.4 mL/min case where the best fit yielded $R^2 = 0.990$. All of the 0.5 mol/L load-off curves were fitted with $R^2 > 0.997$.

Next, the values for the passive elements of the model were derived. Before analyzing this model, it was checked and verified that the model would work as expected. In particular, at steady-state, the capacitor acts as an open circuit and the inductor L acts as a short circuit, allowing all the current to flow through it. Then, when the R_{load} resistor is either connected or disconnected from the terminals of the fuel cell, the current through the inductor, and also the voltage drop across the capacitor, cannot change instantaneously.

To find the values for the capacitance, inductance, and resistances in the circuit of Figure 3.27, Laplace-domain circuit analysis was performed with the appropriate initial inductor current and initial capacitor voltage. For the load-off response, this analysis yielded an expression for the capacitor voltage that was of the form

$$V_c(s) = \frac{A_1 s + A_0}{s^2 + J_1 s + J_0} \tag{3.60}$$

where A_1, A_0, J_1, and J_0, are functions of R_{ct}, R_L, L, C, and/or E. Whereas for the load-on response, the circuit analysis yielded an expression for the capacitor voltage that was of the form

$$V_c(s) = \frac{-B_1 s + B_0}{s^2 + K_1 s + K_0} + \frac{B_1}{s} \tag{3.61}$$

where B_1, B_0, K_1, and K_0, are functions of R_{ct}, R_L, L, C, and/or E. Also, note that the Laplace transform of Equation 3.56 yields

$$V_c(s) = \frac{(a+c)s-(ad+bc)}{s^2-(b+d)s+bd} \tag{3.62}$$

which is comparable to Equation 3.60, whereas the Laplace transform of Equation 3.57 yields

$$V_c(s) = \frac{\chi\cos(\delta)s+\chi[\alpha\cos(\delta)-\omega\sin(\delta)]}{s^2+2\alpha s+(\alpha^2+\omega^2)}+\frac{\Delta v}{s} \tag{3.63}$$

which is comparable to Equation 3.61. For the load-on responses, R_{ohmic} was easily calculated from the initial instantaneous drop in $v_o(t)$, since $v_c(t)$ cannot change instantaneously from its initial zero value, while matching expressions (3.61) through (3.63) resulted in four equations that nonlinearly depend on the to-be-determined R_{ct}, R_L, L, and C. They also depend on the known value of E, which is the open-circuit voltage and also the initial value of the load-on response, and on the known value of Δv, which is the difference between the initial and final (steady-state) values of the load-on, and ideally also the load-off, responses. These component values were then estimated by the nonlinear least-squares approach, making use of the MATLAB® Optimization Toolbox, and the results obtained are shown in Tables 3.6 and 3.7. Note that the estimations resulted in the sum of squared errors (SSE) being less than 2.63×10^{-4} and 2.86×10^{-5} for each of the nine 1 mol/L and nine 0.5 mol/L operating conditions, respectively.

For the load-off response, R_{ohmic} could not be calculated, whereas matching expressions (3.60) and (3.62) resulted in one equation giving an exact solution for R_{ct} and three equations that nonlinearly depended on the obtained R_{ct}, and the to-be-determined R_L, L, and C. The values of R_L, L, and C were then estimated by the nonlinear least-squares approach, making use of the MATLAB Optimization Toolbox again, and the results obtained are shown in Tables 3.8 and 3.9. Note that the estimations resulted in the SSE being less than 3.92×10^{-7} and 3.55×10^{-7} for each of the nine 1 mol/L and nine 0.5 mol/L operating conditions, respectively.

3.7.5 Estimated Model Component Value Results

3.7.5.1 For 1 mol/L Fuel Concentration

For the load-on response, the estimated resistance R_{ct} was observed to be essentially independent of the fuel flow rate between 0.2 and 0.6 mL/min, but increasing significantly with temperature between 30°C and 70°C. A mesh plot of the variation of R_{ct} is shown in Figure 3.31 for visualization purposes. The dependence of R_L on temperature and fuel flow rate is quite similar to that of R_{ct} except it showed a large abrupt increase at the 70°C, 0.4 mL/min operating condition; recall that the test data at this condition had somewhat greater variability (reduced repeatability) than the data at other

TABLE 3.6

Second-Order Equivalent Circuit Model Values

	R_{ct} (Ω)			R_L (Ω)			L (H)			C (F)		
	0.2 mL/min	**0.4 mL/min**	**0.6 mL/min**	**0.2 mL/min**	**0.4 mL/min**	**0.6 mL/min**	**0.2 mL/min**	**0.4 mL/min**	**0.6 mL/min**	**0.2 mL/min**	**0.4 mL/min**	**0.6 mL/min**
30°C	1.495	1.555	2.502	45.22	52.14	88.73	3.611	3.336	2.456	43,945	18,041	36,807
50°C	4.676	5.847	7.959	141.2	173.1	308.0	2.534	2.689	4.122	2015	766.3	22.85
70°C	18.39	21.64	15.23	578.2	1006	341.1	1.651	2.717	1.209	34.80	29.80	364.8

	R_{ohmic} (Ω)		
	0.2 mL/min	**0.4 mL/min**	**0.6 mL/min**
30°C	0.1512	0.1277	0.1073
50°C	0.1300	0.1558	0.1699
70°C	0.1220	0.1241	0.1613

Note: Component values estimated from DMFC output voltage load-on responses at various temperatures and (1 mol/L concentration) fuel flow rates.

TABLE 3.7

Second-Order Equivalent Circuit Model Values

	R_{ct} (Ω)			R_L (Ω)			L (H)			C (F)		
	0.2 mL/min	0.4 mL/min	0.6 mL/min	0.2 mL/min	0.4 mL/min	0.6 mL/min	0.2 mL/min	0.4 mL/min	0.6 mL/min	0.2 mL/min	0.4 mL/min	0.6 mL/min
30°C	1.020	1.023	1.024	21.14	26.51	18.16	1895	2569	1653	67.88	69.74	68.93
50°C	1.006	1.003	1.007	84.44	79.55	84.68	3924	3984	4738	36.55	37.72	36.25
70°C	0.9832	0.9664	0.9835	85.80	97.24	89.78	2304	10,124	2813	22.50	22.48	20.92

Note: Component values estimated from DMFC output voltage load-off responses at various temperatures and (1 mol/L concentration) fuel flow rates.

TABLE 3.8

Second-Order Equivalent Circuit Model Values

	R_{ct} (Ω)			R_L (Ω)			L (H)			C (F)		
	0.2 mL/min	**0.4 mL/min**	**0.6 mL/min**	**0.2 mL/min**	**0.4 mL/min**	**0.6 mL/min**	**0.2 mL/min**	**0.4 mL/min**	**0.6 mL/min**	**0.2 mL/min**	**0.4 mL/min**	**0.6 mL/min**
30°C	6.672	23.68	11.08	0.2688	0.8142	0.5192	26.27	36.36	42.34	6.653	5.130	4.524
50°C	5.276	32.09	33.72	0.3887	0.6756	0.8233	21.27	18.66	23.17	9.329	9.320	8.463
70°C	1.705	2.921	6.951	4.117	0.9121	3.276	102.8	22.86	63.15	19.30	9.141	5.885

	R_{ohmic} (Ω)		
	0.2 mL/min	**0.4 mL/min**	**0.6 mL/min**
30°C	0.1226	0.1456	0.1575
50°C	0.1553	0.1397	0.1272
70°C	0.1583	0.1827	0.1660

Note: Component values estimated from DMFC output voltage load-on responses at various temperatures and (0.5 mol/L Concentration) fuel flow rates.

TABLE 3.9

Second-Order Equivalent Circuit Model Values

	R_{ct} (Ω)			R_L (Ω)			L (H)			C (F)		
	0.2 mL/min	**0.4 mL/min**	**0.6 mL/min**	**0.2 mL/min**	**0.4 mL/min**	**0.6 mL/min**	**0.2 mL/min**	**0.4 mL/min**	**0.6 mL/min**	**0.2 mL/min**	**0.4 mL/min**	**0.6 mL/min**
30°C	0.9997	0.9566	0.9606	14.33	0.7226	4.956	2093	12.54	65.09	84.17	228.0	118.6
50°C	0.9976	1.001	0.9999	98.99	99.11	99.44	5014	4899	5451	39.74	38.49	38.06
70°C	0.9831	0.9770	0.9396	99.09	99.50	97.64	3873	2081	13,024	22.73	20.23	24.74

Note: Component values estimated from DMFC output voltage load-off responses at various temperatures and (0.5 mol/L Concentration) fuel flow rates.

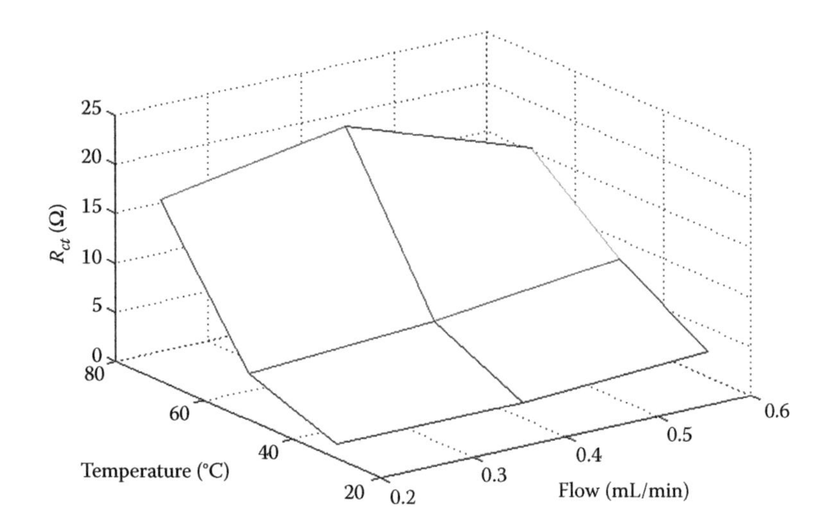

FIGURE 3.31
Dependence of resistance R_{ct} values, estimated from load-on step responses, on temperature and (1 mol/L concentration) fuel flow rate.

test conditions, so that this increase may be slightly suspect. Interestingly, L did not show much variation (in absolute terms) over the studied ranges of temperature and fuel flow rate. On the other hand, the estimates of C varied widely in value and without a discernible trend with respect to temperature or fuel flow rate.

For the load-off response, the estimated resistance R_{ct} was observed to be essentially unchanged over the tested ranges of temperature and fuel flow rate. Whereas R_L increased rather abruptly with temperature between 30°C and 50°C and then leveled off up to 70°C, but was relatively independent of fuel flow rate. On the other hand, C was observed to be essentially independent of fuel flow conditions but had an inverse dependency on temperature. L also increased with temperature between 30°C and 50°C, but was relatively independent of fuel flow rate at these two temperatures; while at 70°C, its values at both 0.2 and 0.6 mL/min decreased from their values at 50°C, but its value at 0.4 mL/min instead increased from the value at 50°C. Since it was noted that the test data at this 70°C, 0.4 mL/min condition had somewhat greater variability (reduced repeatability) than the other test data, this inconsistent behavior (dissimilar from the 0.2 and 0.6 mL/min values) is again slightly suspect, and needs to be confirmed in future studies.

3.7.5.2 For 0.5 mol/L Fuel Concentration

For the load-on response when using fuel of lower concentration, the resistances R_{ct} and R_{ohmic} were essentially independent of the fuel flow rate between 0.2 and 0.6 mL/min, and of the temperature between 30°C and

70°C, although R_{ct} varied somewhat more than R_{ohmic}. Furthermore, the R_{ohmic} values were very much consistent with those obtained from the 1 mol/L tests, as expected. The L values varied slightly more than the C values (in absolute and percentage terms) over the studied ranges of temperature and fuel flow rate, although both were without a discernible trend with respect to temperature or fuel flow rate. Compared to the 1 mol/L results, the L values had increased by an order of magnitude, while the C values had decreased by at least one order of magnitude; so, these were significantly affected by the fuel concentration.

For the load-off response, the resistance R_{ct} was essentially independent over the tested ranges of temperature and fuel flow rate, and also very much consistent with those obtained from the 1 mol/L tests. Whereas R_L increased rather abruptly with temperature between 30°C and 50°C and then leveled off up to 70°C; however, it was relatively independent of fuel flow rate at 50°C and 70°C, just like the 1 mol/L case. On the other hand, C was essentially independent of fuel flow conditions at 50°C and 70°C, but had an inverse dependency on temperature (see Figure 3.32), just like the 1 mol/L cases. The L values also increased with temperature between 30°C and 50°C to become somewhat independent of fuel flow rate at 50°C, although curiously its values for the 30°C 0.4 mL/min and 30°C 0.6 mL/min cases were 2 orders of magnitude lower than for the other cases, of which values were in line with the L values for the 1 mol/L condition.

Finally, in broadly comparing the component values for the 1 mol/L cases to the 0.5 mol/L cases, it was noted that the higher fuel concentration resulted in greater variability of those values. This comparison was between

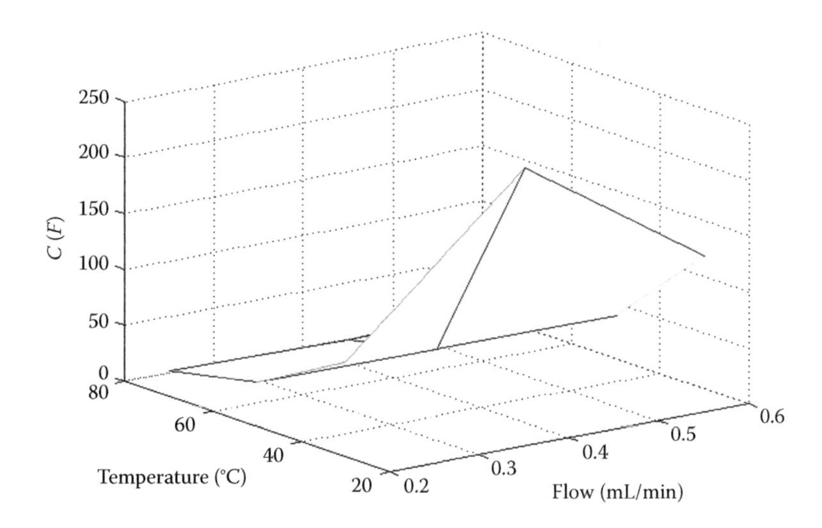

FIGURE 3.32
Dependence of capacitance C values, estimated from load-off step responses, on temperature and (0.5 mol/L concentration) fuel flow rate.

the values for the various temperature and fuel flow rate conditions, as well as between the values for the load-on and load-off responses.

3.7.6 Further Discussion

Muller et al. [52], presented values of $C = 0.135$ F/cm^2, $R_{ct} = 1.705\ \Omega$ cm^2, $R_L = 0.863\ \Omega$ cm^2, and $L = 0.716$ H/cm^2 for their proposed second-order model (Figure 3.28), obtained under load, and estimated using frequency-domain (AC impedance) data and technique. For a 25 cm^2 active area membrane as in Reference 54, these would correspond to values of $C = 3.375$ F, $R_{ct} = 68.2$ mΩ, $R_L = 34.5$ mΩ, and $L = 17.9$ H. Comparison of these values to the corresponding estimates in Tables 3.6 and 3.8 (under load) yielded average percentage errors of {12,819%, 880,315%, –85%, –335,790%} and {20,118%, 3,699%, 122%, 156%}, respectively, for R_{ct}, R_L, L, and C. The likely explanation for the large discrepancies is that the operating conditions in Reference 52 differed significantly in a few key respects from the operating conditions used in Reference 54; in particular, the lowest current density in the tests by Muller et al. was 100 mA/cm^2, although it is not completely clear from Reference 52 which set of data (and operating conditions) corresponded to their estimates as given above, while the tests described by Singh et al. [54] were at a current density of about 2 mA/cm^2. These parameters' dependence on current was described in Reference 51.

On the other hand, let us now consider the output impedances corresponding to the obtained parameter estimates. The output impedance of the studied second-order equivalent circuit model (Figure 3.27) is expressed as

$$Z = R_{ohmic} + \frac{R_{ct} L s + R_{ct} R_L}{R_{ct} L C s^2 + (L + R_{ct} R_L C)s + (R_{ct} + R_L)} \tag{3.64}$$

The Nyquist plots using the estimates presented in Tables 3.6 and 3.7 are shown in Figure 3.33, while the Nyquist plots using the estimates presented in Tables 3.8 and 3.9 are shown in Figure 3.34. These plots compare well to the DMFC impedance plots shown in Reference 49, which have similar shapes, and Re(Z) values ranging from 0 to 5 Ω and Im(Z) values ranging from –2.5 to 1 Ω. These similarities provide a measure of support for the validity of the method proposed by Singh et al. [54] and the resulting model component estimates.

3.7.7 Conclusions

This section has described the equivalent circuit modeling of DMFC dynamics from measured temporal (time-domain) responses during step changes in load current at various temperature and fuel flow-rate operating conditions as proposed by Singh et al. [54]. The simulated responses of the assumed

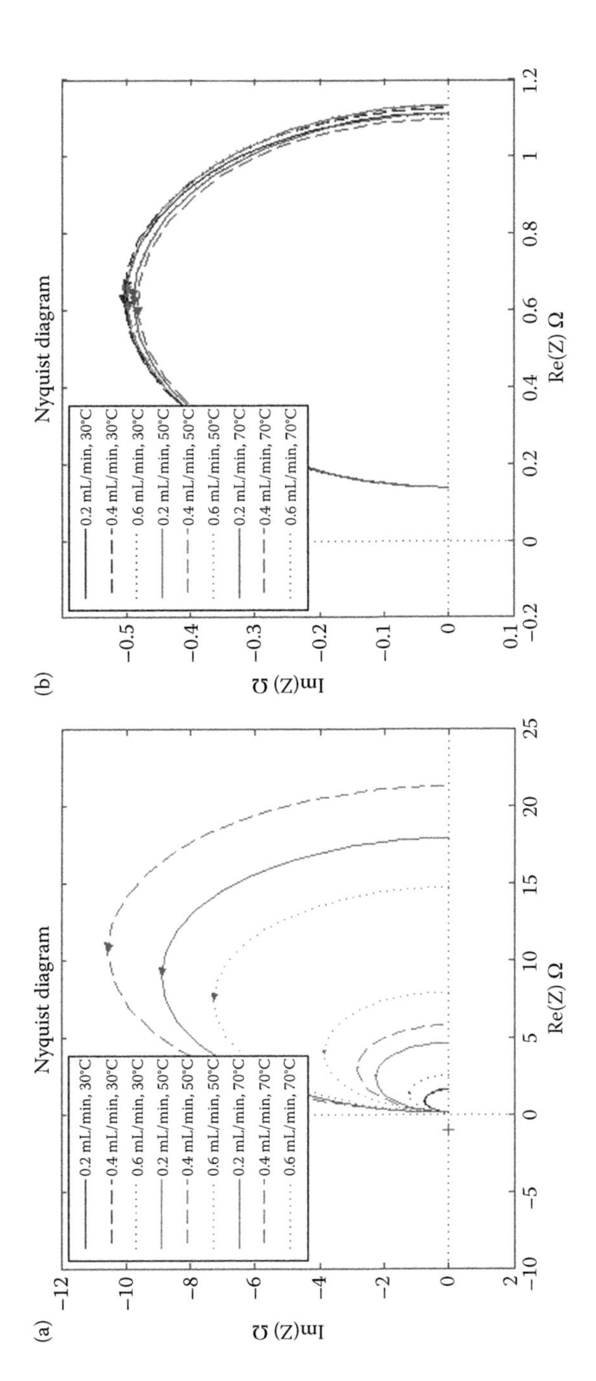

FIGURE 3.33
Nyquist plots of second-order equivalent circuit model's output impedance using component values estimated from (a) load-one step responses and (b) load-off step responses; 1 mol/L concentration fuel.

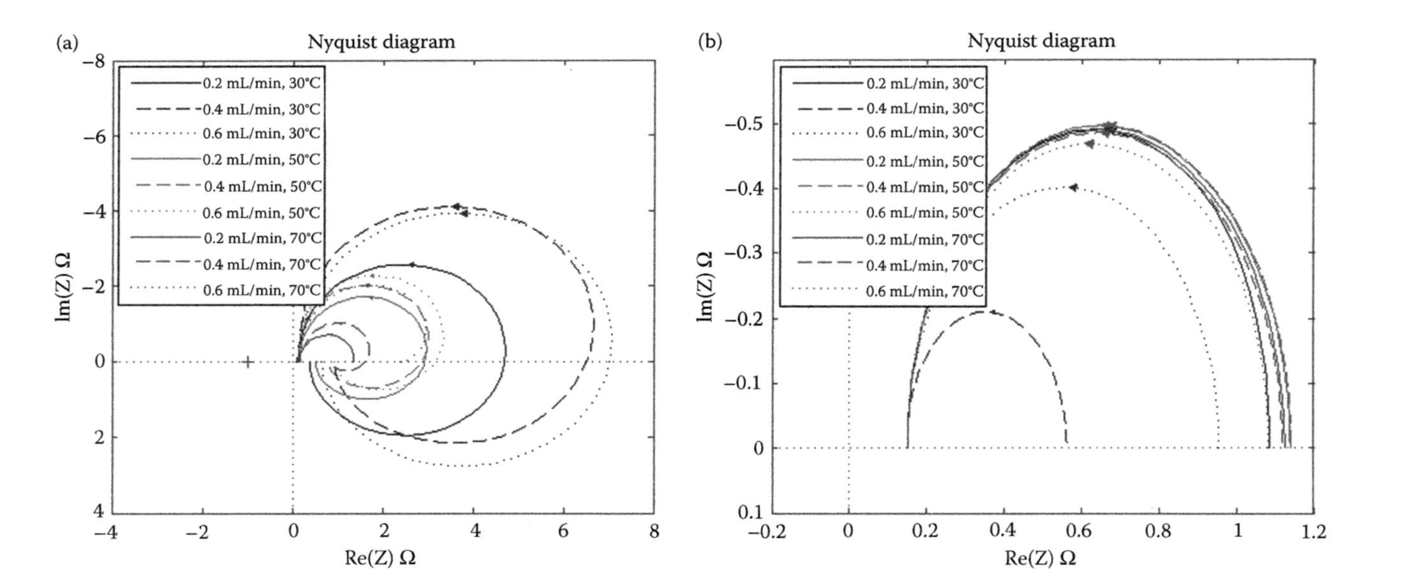

FIGURE 3.34
Nyquist plots of second-order equivalent circuit model's output impedance using component values estimated from (a) load-on step responses and (b) load-off step responses; 0.5 mol/L concentration fuel.

second-order circuit model, a first-order model being determined to be less accurate for reproducing the observed response undershoots and overshoots, with its estimated component values yielded excellent fits with the experimental data for the 30–70°C temperature range and the 0.2–0.6 mL/min fuel flow range under study. The results provide all the information that is obtained by an AC impedance measurement, without requiring an impedance (frequency response) analyzer, and also determines the contribution of each component to the overall impedance. One can use these modeling results in ways such as: (1) diagnostically—to determine which parts of a fuel cell are degrading or have degraded and (2) prospectively—to guide changes to the fuel cell's design and fabrication (such as the MEA's various components) that would improve its output impedance, dynamic behavior, etc. Perhaps, even more importantly, this equivalent circuit model can be used to evaluate and then improve upon how the DMFC interacts with other components of the complete power system (such as UCs and switch-mode power converters), e.g., the effects of DC–AC converter ripple current on the lifespan of the fuel cell, similar to Reference 45 for H_2-fueled PEMFCs.

It was further determined in Reference 54 that the model's capacitor, inductor, and resistor values are quite sensitive to temperature but relatively insensitive to fuel flow rates—more so for load-off than for load-on responses—for the same fuel concentration. Moreover, each component's dependencies on temperature and fuel flow rate differ from those of the other components. Such dependencies could mean that in practice, they should be adjusted either statically or dynamically, and in either an open-loop or closed-loop fashion, to obtain a more optimal performance of DMFCs in applications with fast, frequent, and/or significant load changes. In addition, different fuel concentrations resulted in greater variability of the parameter values during the load-on responses as compared to the load-off responses, which is expected since the former correspond to when current is flowing and thus fuel being consumed to produce that current.

Nomenclature

E_0	Cell open-circuit potential at standard pressure
N	Number of cells in the stack
E	Fuel cell (stack) output voltage
L	Sum of cell voltage losses
i	Cell output current density
i_n	Cell internal current density corresponding to internal current losses
i_o	Cell exchange current density corresponding to activation losses

i_l	Cell-limiting current density corresponding to concentration losses
r	Cell area-specific resistance corresponding to resistive losses
a	Constant associated with cell activation losses (slope of Tafel line)
b	Constant associated with cell concentration losses
T	Stack operating temperature
P_{H_2}	Partial pressure of hydrogen inside the cell anode
P_{O_2}	Partial pressure of oxygen inside the cell cathode
$P_{H_2O_a}$	Partial pressure of water vapor inside the cell anode
$P_{H_2O_c}$	Partial pressure of water vapor inside the cell cathode
P_{N_2}	Partial pressure of nitrogen inside the cell cathode
$P_{cathode}$	Summation of partial pressures inside the cell cathode
R	Universal gas constant (8.3144 J/mol K)
F	Faraday's constant (96,439 C/mol)
P_{std}	Standard pressure (101,325 Pa)
P_{op}	An assumed constant cell operating pressure
H_{2in}, H_2O_{ain}	Anode inlet molar flow rates of hydrogen and water vapor
$Anode_{in}$	Summation of anode inlet molar flow rates
O_{2in}, N_{2in}, H_2O_{cin}	Cathode inlet molar flow rates of oxygen, nitrogen, and water vapor
$Cath_{in}$	Summation of cathode inlet molar flow rates
H_{2out}, H_2O_{aout}	Anode outlet molar flow rates of hydrogen and water vapor
O_{2out}, N_{2out}, H_2O_{cout}	Cathode outlet molar flow rates of oxygen, nitrogen, and water vapor
H_{2used}, O_{2used}, $H_2O_{cproduced}$	Molar usage and production rates of hydrogen, oxygen, and water vapor (in the cathode)
I	Cell output current
A_c	Cell active area
V_a	Volume of the cell anode channels
V_c	Volume of the cell cathode channels
F_{H_2}, F_{O_2}, $F_{H_2O_c}$	Pressure fractions of hydrogen, oxygen, and water vapor (in the cathode) inside the fuel cell
Y_{H_2}	Molar fraction of H_2 at the anode inlet
Y_{O_2}	Molar fraction of O_2 at the cathode inlet
$Y_{H_2O_c}$	Molar fraction of water vapor at the cathode inlet
ΔP_{H_2}, ΔP_{O_2}, $\Delta P_{H_2O_c}$	Perturbations of the partial pressures of hydrogen, oxygen, and water vapor (in the cathode) inside the fuel cell

ΔH_{2in}, ΔO_{2in}, ΔH_2O_{cin}	Perturbations of the inlet flow rates of hydrogen, oxygen, and water vapor (to the cathode)
Δi	Perturbation of the output current density
ΔE	Perturbation of the fuel cell stack output voltage

References

1. A.M. Borbely and Jan G. Kreider, *Distributed Generation: The Power Paradigm for the New Millennium*, CRC Press, New York, 2001.
2. F. Barbir, *PEM Fuel Cells: Theory and Practice*, Elsevier Academic Press, Burlington, MA, 2005.
3. J. Purkrushpan, A.G. Stefanopoulou, and H. Peng, Control of fuel cell breathing, *IEEE Control Systems Magazine*, April, 2004.
4. J. Larminie and A. Dicks, *Fuel Cell Systems Explained*, Wiley, New York, 2002.
5. Rajashekrara, K., Propulsion system strategies for fuel cell vehicles, SAE paper 2000-01-0369.
6. F. Barbir and T. Gomez, Efficiency and economics of PEM fuel cells, *International Journal of Hydrogen Energy*, 22(10/11), 1027–1037, 1997.
7. J. Purkrushpan and H. Peng, *Control of Fuel Cell Power Systems: Principles, Modeling, Analysis and Feedback Design*, Springer, Germany, 2004.
8. J. Purkrushpan, A.G. Stefanopoulou, and H. Peng, Modeling and control for PEM fuel cell stack systems, *Proceedings of the American Control Conference*, Anchorage, Alaska, pp. 3117–3122, 2002.
9. J.C. Amphlett, R.M. Baumert, R.F. Mann, B.A. Peppy, P.R. Roberge, and A. Rodrigues, Parametric modeling of the performance of a 5-kW proton exchange membrane fuel cell stack, *Journal of Power Sources*, 49, 349–356, 1994.
10. X.D. Xue, K.W.E. Cheng, and D. Sutanto, Unified mathematical modeling of steady-state and dynamic voltage–current characteristics for PEM fuel cells, *Electrochimica Acta*, 52(3), 1135–1144, 2006.
11. M.J. Khan and M.T. Labal, Modeling and analysis of electro chemical, thermal, and reactant flow dynamics for a PEM fuel cell system, *Fuel cells*, 4(1), 463–475, 2005.
12. Ballard Power System, Inc., Canada, at http://www.ballard.com.
13. M.Y. El-Sharkh, A. Rahman, M.S. Alamm, A.A. Sakla, P.C. Byrne, and T. Thomas, Analysis of active and reactive power control of a stand-alone PEM fuel cell power plant, *IEEE Transactions on Power Delivery*, 19(4), 2022–2028, 2004.
14. J.J.E. Slotine and W. Li, *Applied Nonlinear Control*. Prentice-Hall, Englewood Cliffs, New Jersey, 1991.
15. Woonki Na and Bei Gou, Feedback linearization based nonlinear control for PEM fuel cells, *IEEE Transactions on Energy Conversion*, 23(1), 179–190, 2008.
16. J. Sun and V. Kolmannovsky, Load governor for fuel cell oxygen starvation protection: A robust nonlinear reference governor approach, *IEEE Transactions on Control Systems Technology*, 3(6), 911–913, 2005.
17. S. Basu, M.W. Renfro, and B.M. Cetegen, Spatially resolved optical measurements of water partial pressure and temperature in a PEM fuel cell under dynamic operating conditions, *Journal of Power Sources*, 162(1), 286–293, 2006.

18. L.Y. Chiu, B. Diong, and R.S. Gemmen, An improve small-signal model of the dynamic behavior of PEM fuel cells, *IEEE Transactions on Industry Applications*, 40(4), 970–977, 2004.
19. R.E. Sonntag, C. Borgnakke, and G.J.V. Wylen, *Fundamentals of Thermodynamics*. Wiley, New York, 1998.
20. M.P. Nielsen, P. Pedersen, C.A. Andesen, M.O. Christen, and A.R. Korgaard, Design and control of fuel cell system for transport application, Aalborg University, Project Report, 2002.
21. C. Wang, M.H. Nehrir, and S.R. Shaw, Dynamic models and model validation for PEM fuel cells using electrical circuits. *IEEE Transactions on Energy Conversion*, 20(2), 442–451, 2005.
22. J.M. Correa, F.A. Farret, and L.N. Canha, An analysis of the dynamic performance of proton exchange membrane fuel cells using an electrochemical model, *IECON '01, The 27th Annual Conference of the IEEE Industrial Electronics Society*, Denver, CO, Vol. 1, pp. 141–146, 2001.
23. C.J. Hatiziadoniu, A.A. Lobo, F, Pourboghrat, and M. Daneshdoot, A simplified dynamic model of grid connected fuel-cell generators, *IEEE Transactions on Power Delivery*, 17(2), 467–473, 2002.
24. A. Sakhare, A. Davari, and A. Feliachi, Fuzzy logic control of fuel cell for stand-alone and grid connection, *Journal of Power Sources*, 135(1–2), 165–176, 2004.
25. J.C. Amphlett, R.M. Baumert, R.F. Mann, B.A. Peppley, and P.R. Roberge, Performance modeling of the Ballard Mark IV solid polymer electrolyte fuel cell, 1. Mechanistic model development, *Journal of the Electrochemical Society*, 142(1), 1–8, 1995.
26. J.C. Amphlett, E.K. De Oliveira, R.F. Mann, P.R. Roberge, and A. Rodrigues, Dynamic interaction of a proton exchange membrane fuel cell and a lead-acid battery, *Journal of Power Sources*, 65, 173–178, 1997.
27. F. Grasser and A. Rufer, A fully analytical PEM fuel cell system model for control applications, *IEEE Transactions on Industry Applications*, 43(6), 1499–1506, 2007.
28. S. Pasricha, M. Keppler, S.R. Shaw, and M.H. Nehrir, Comparison and identification of static electrical terminal fuel cell models, *IEEE Transaction on Energy Conversion*, 22(3), 746–754, 2007.
29. R.S. Gemmen, Analysis for the effect of inverter ripple current on fuel cell operating condition, *Journal of Fluids Engineering*, 125(3), 576–585, 2003.
30. S.C. Page, A.H. Anbuky, S.P. Krumdieck, and J. Brouwer, Test method and equivalent circuit modeling of a PEM fuel cell in a passive state, *IEEE Transaction on Energy Conversion*, 22(3), 764–773, 2007.
31. M. Uzunoglu and M.S. Alam, Dynamic modeling, design, and simulation of a combined PEM fuel cell and ultracapacitor system for stand-alone residential applications, *IEEE Transaction on Energy Conversion*, 21(3), 767–775, 2006.
32. S. Pasricha and S.R. Shaw, A dynamic PEM fuel cell model, *IEEE Transaction on Energy Conversion*, 21(2), 484–490, 2006.
33. J.M. Correa, F.A. Farret, L.N. Canha, and M.G. Simoes, An electrochemical-based fuel-cell model suitable for electrical engineering automation approach, *IEEE Transactions on Industrial Electronics*, 51(5), 1103–1112, 2004.
34. J.M. Correa, F.A. Farret, V.A. Popov, and M.G. Simoes, Sensitivity analysis of the modeling parameters used in simulation of proton exchange membrane fuel cells, *IEEE Transaction on Energy Conversion*, 20(1), 211–218, 2005.

35. BCS Technology Co., *Data Sheet for a 500 W FC Stack*, 2001.
36. L. Lorandi, E. Hernandez, and B. Diong, Parametric sensitivity analysis of fuel cell dynamic response, *SAE Transactions Journal of Aerospace*, 113-1, 2024–2032, 2004.
37. E. Hernandez, Parametric sensitivity analysis and circuit modeling of PEM fuel cell dynamics, M.S. thesis, University of Texas at El Paso, December 2004.
38. R.S. Gemmen, E. Liese, J.G. Rivera, and J. Brouwer, Development of dynamic modeling tools for solid oxide and molten carbonate hybrid fuel cell gas turbine systems, *Proceedings of the ASME/IGTI Turbo Expo*, Munich, Germany, 2000.
39. K. Sedghisigarchi and A. Feliachi, Control of grid-connected fuel cell power plant for transient stability enhancement, *Proceedings of the IEEE Power Engineering Society Winter Meeting*, 2002, 383–388.
40. E. Liese and R.S. Gemmen, Dynamic modeling results of a 1 MW molten carbonate fuel cell/gas turbine power system, *Proceedings of the ASME/IGTI Turbo Expo*, 2002, 341–349.
41. http://www.sfc.com/en/defense.html *Defense* (last accessed on July 22, 2011).
42. http://www.oorjaprotonics.com/. *Oorja Protonics—Enabling Power* (last accessed on July 22, 2011).
43. http://www.wired.com/gadgetlab/2007/10/yamaha-fc-dii-f/. *Yamaha FC-Dii Fuel Cell Bike Prototype to Premiere at Tokyo Motor Show* (last accessed on July 22, 2011).
44. C.N. Hamelinck and A.P.C. Faaij, Future prospects for production of methanol and hydrogen from biomass, *Journal of Power Sources*, 111(1), 1–22, 2002.
45. W. Choi, P.N. Enjeti, and J.W. Howze, Development of an equivalent circuit model of a fuel cell to reduce current ripple, *Proceedings of the IEEE Applied Power Electronics Conference*, Anaheim, California, pp. 355–361, February 2004.
46. X. Yan, M. Hou, L. Sun, D. Liang, Q. Shen, H. Xu, P. Ming, and B. Yi, AC impedance characteristics of a 2 kW PEM fuel cell stack under different operating conditions and load changes, *International Journal of Hydrogen Energy*, 32, 4358–4364, 2007.
47. A.A. Kulikovsky, *Analytical Modelling of Fuel Cells*, Elsevier Ltd., Amsterdam, 2010.
48. T.S. Zhao, K.D. Kreuer, and T.V. Nguyen, *Advances in Fuel Cells*, Elsevier Ltd., New York, 2007.
49. U. Krewer, System-oriented analysis of the dynamic behaviour of direct methanol fuel cells, Dr.-Ing. Dissertation, Otto von Guericke University, Magdeburg, Germany, 2005.
50. M. Ordonez, P. Pickup, J.E. Quaicoe, and M.T. Iqbal, Electrical dynamic behavior of a direct methanol fuel cell, *IEEE Power Electronics Society Newsletter*, 19(1), 10–15, 2007.
51. Y. Wang, J.P. Zheng, G. Au, and E.J. Plichta, A novel method for electrically modeling of fuel cell using frequency response technology: Executed on a direct methanol fuel cell, *Proceedings of the 211th ECS Meeting*, Chicago, Illinois, May 6–11, 2007.
52. J.T. Muller, P.M. Urban, and W.F. Holderich, Impedance studies on direct methanol fuel cell anodes, *Journal of Power Sources*, 84(2), 157–160, 1999.
53. D.A. Harrington and B.E. Conway, AC impedance of Faradaic reactions involving electrosorbed intermediates: Kinetic theory, *Electrochimica Acta*, 32(12), 1703–1712, 1987.

54. K. Singh, B. Diong, D. Estrada, and B. Nguyen, Temperature, fuel-flow-rate and fuel-concentration dependence of the parameters of an equivalent circuit model of DMFC dynamic response, *Alternative Energy*, 3(3), 1–9, 2014.
55. W.H. Hayt, Jr., J.E. Kemmerly, and S.M. Durbin, *Engineering Circuit Analysis*, 7th edition, McGraw-Hill, Inc., New York, 2007.

4

*Linear and Nonlinear Control Design for Fuel Cells**

4.1 Introduction

Fuel cell-based power systems present a wide range of challenging problems for control and system integration. Considerable research and development efforts have been devoted to fuel cell operation and systems, in particular, in the area of proton exchange membrane (PEM) fuel cells for mobile and stationary applications. Control plays a critical role in the fuel cell operation and systems. The objective of a control system is to modify the natural response of a fuel cell electrochemical reactor and maintain the desired operation in the presence of uncertainties and disturbances. Moreover, the control system attempts to diagnose abnormal operation, monitors the health of the stack, and adapts the operation to address material and electrode degradation. A wide range of operating conditions that the fuel cell system needs to operate also makes the controls challenging.

Due to the complexity of a fuel cell system, control strategies are applied in various aspects, including compressor motor control, pressure control, humidity control, temperature control, ratio control, control of output voltage and current, etc. Both linear and nonlinear control approaches have been developed for fuel cell systems in the literature. This chapter will introduce the linear and nonlinear control designs for PEM fuel cells, and the modeling of fuel cells will also be covered accordingly.

4.2 Linear Control Design for Fuel Cells

A robust control strategy is critical to satisfy high-power density demand in PEM fuel cells. In this section, we will introduce a linear ratio control strategy

* This chapter was mainly prepared by Dr. Bei Gou, Smart Grid LLC, USA and Dr. Woonki Na, California State University, Fresno, USA.

based on the distributed parameter model of fuel cells, which includes the effect of spatial variations.

4.2.1 Distributed Parameter Model of Fuel Cells

Figure 4.1 shows the distributed parameter model of fuel cells proposed in Reference 1. In the model, a fuel cell is modeled along its channel. The model considers heat transfer between the solid and the two gas channels, and between the solid and cooling water. The water content is also modeled to calculate the condensation and evaporation, water drag through the membrane, and water generation at the cathode. The energy balance on the solid is modeled dynamically, but all the other equations are assumed to be at quasi-steady state for a given solid temperature profile.

Based on the work in References 1 and 2, the energy balance equation is given by [3]

$$\rho_s C_{p,s}\frac{\partial T_s}{\partial t} = k_s\frac{\partial^2 T_s}{\partial x^2} + \frac{U_g A_g}{f}(T_a + T_C - 2T_s) + \frac{U_W A_{cool}}{f}(T_{cool} - T_s) - \frac{e}{f}\left(\frac{\Delta H}{2F} + V_{cell}\right)$$
$$+ \frac{1}{f}\Delta H_{vap}(T_s)\left(\frac{dM^1_{W,a}(x)}{dx} + \frac{dM^1_{W,C}(x)}{dx}\right)$$

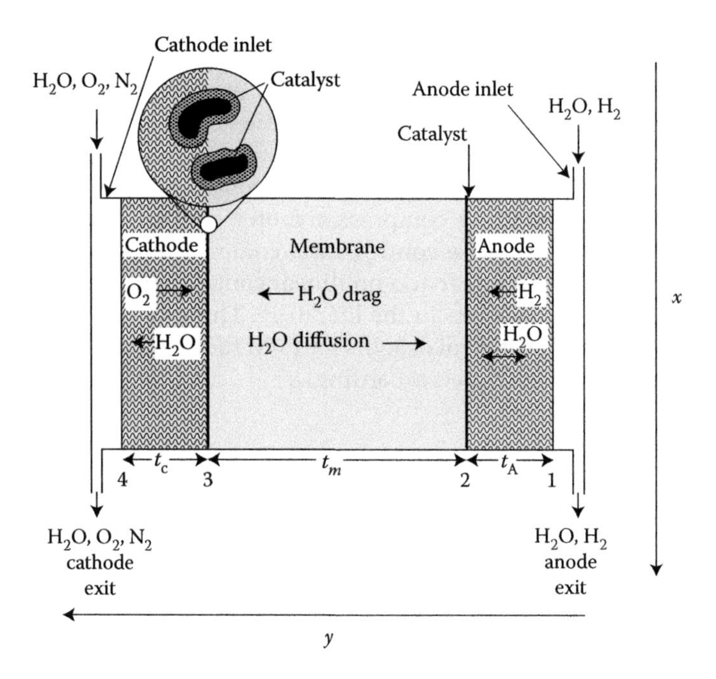

FIGURE 4.1
Schematic diagram of a fuel cell channel.

With boundary conditions

$$k_s \frac{\partial T_s}{\partial x}|_{x=0} = U_C(T_s - T_{\text{inf}})$$

$$k_s \frac{\partial T_s}{\partial x}|_{x=L} = -U_C(T_s - T_{\text{inf}})$$

The V_{cell} is given as

$$V_{cell} = V_{OC}^0 + \frac{R^*T_s}{nF}\ln\left(\frac{P_{H_2}P_{O_2}^{0.5}}{P_{H_2O}}\right) - \frac{R^*T_s}{F}\ln\left(\frac{I(x)}{i_0P_{O_2}(x)}\right) - \frac{I(x)t_m}{\sigma_m(x)}$$

4.2.2 Linear Control Design and Simulations for Fuel Cells

4.2.2.1 Power Control Loop

For the power control loop, the average power density in the PEMFC was controlled by manipulating the inlet molar flow rate of hydrogen and the inlet molar flow rate of oxygen, respectively. Here, we only cover the control using the inlet molar flow rate of hydrogen [3].

The control scheme is given in Figure 4.2. The controller for the hydrogen flow to power density loop is designed by using the internal model control (IMC)-based proportional–integral–derivative (PID) method. The resulting controller is of PI (proportional–integral) form. The performance of the PI controller along with the manipulated variable actions is shown in Figure 4.3. The settling time for this control strategy is approximately 275 s. This large settling time could be attributed to severe nonlinearities present in the PEMFC.

4.2.2.2 Power and Solid Temperature Control Loop

The rate of reaction has to be kept at a high value to achieve maximum power density. However, the increased rate of reaction increases the heat

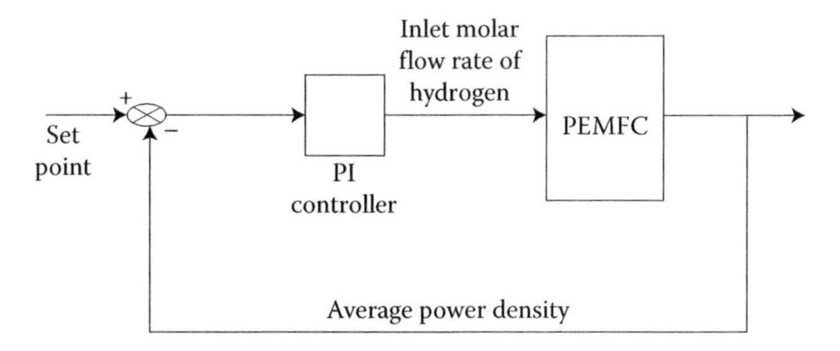

FIGURE 4.2
Schematic of a power control loop using inlet molar flow rate of hydrogen as the manipulated variable.

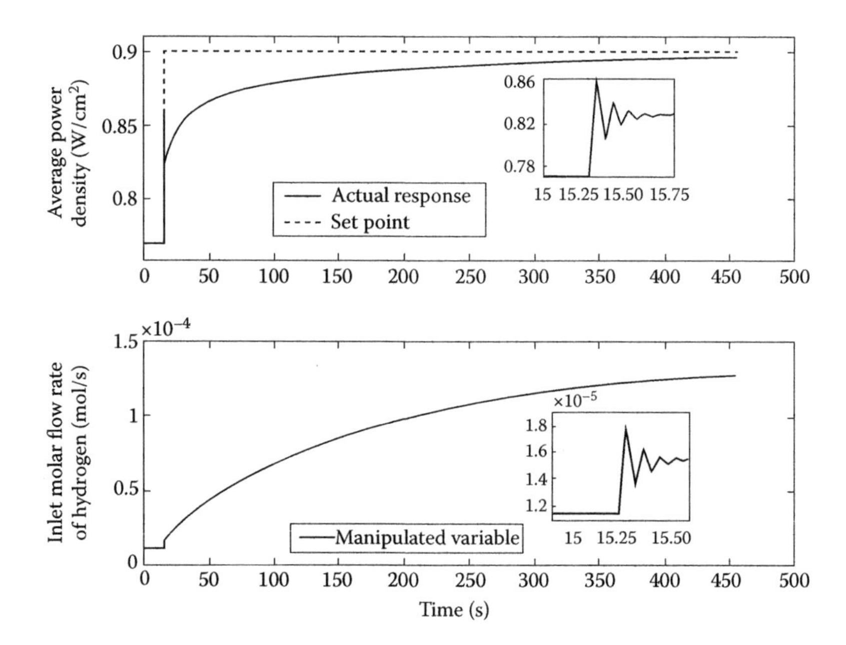

FIGURE 4.3
Performance of the power control loop using the inlet molar flow rate of hydrogen as the manipulated variable.

produced from the reaction, which finally increases the solid temperature of the PEMFC. Furthermore, increasing the solid temperature beyond a specific value will adversely affect the conductivity of the membrane and also the catalyst activity, which in turns affects the rate of reaction. Therefore, it is necessary to control the average solid temperature within specified limits.

The linear control design is given in Figure 4.4. To control the average solid temperature, the steady-state relative gain array (RGA) analysis recommends

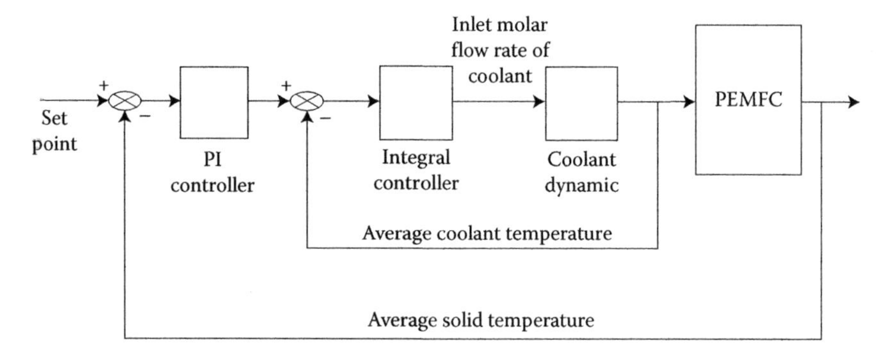

FIGURE 4.4
Schematic of the cascade control loop using inlet molar flow rate of coolant as the manipulated variable.

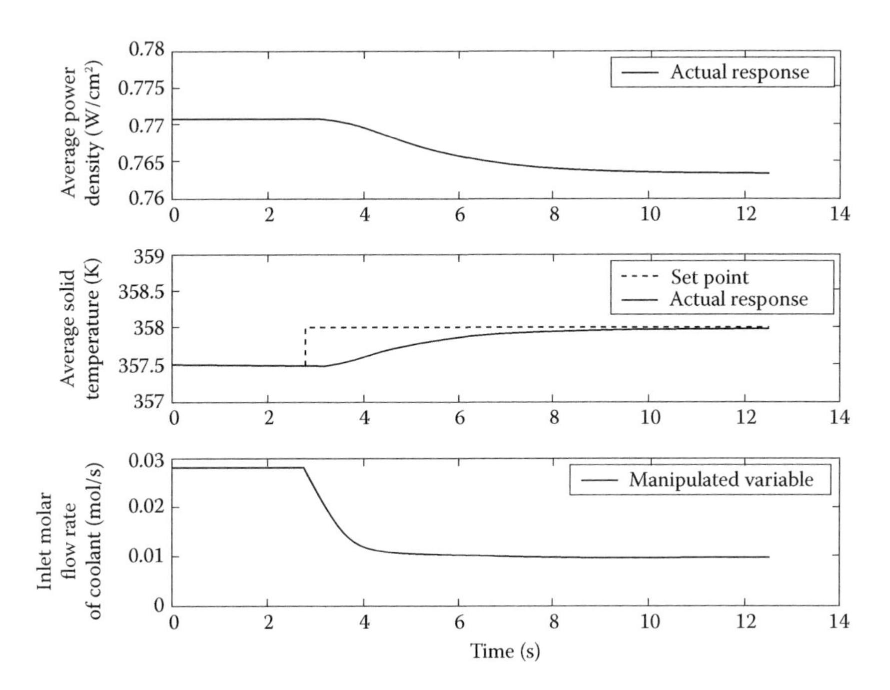

FIGURE 4.5
Performance of cascade control using the inlet molar flow rate of coolant as the manipulated variable.

the inlet coolant flow rate to be the manipulated variable. The average solid temperature is controlled by using a cascade control loop with the inlet molar flow rate of coolant as the manipulated variable.

This strategy in Figure 4.4 provides an additional advantage of disturbance rejection. The performance of the cascade control design is shown in Figure 4.5. Further simulation results can be found in Reference 3.

4.2.2.3 Multi-Input and Multi-Output Control Strategy

The PEM fuel cells can also be considered as a multi-input and multi-output (MIMO) system where average power density and average solid temperature are the two controlled outputs. The MIMO control design is shown in Figure 4.6.

Figure 4.7 shows that the response of the MIMO controller is faster than that of the single-input and single-output (SISO) controller. The settling time for the MIMO controller is approximately 90 s while the settling time for the SISO controller is 275 s. In addition, the MIMO control strategy also avoids the unwanted effect of rise in temperature.

A simulation using air instead of oxygen at the cathode was taken and its performance is shown in Figure 4.8. The results demonstrate that the MIMO

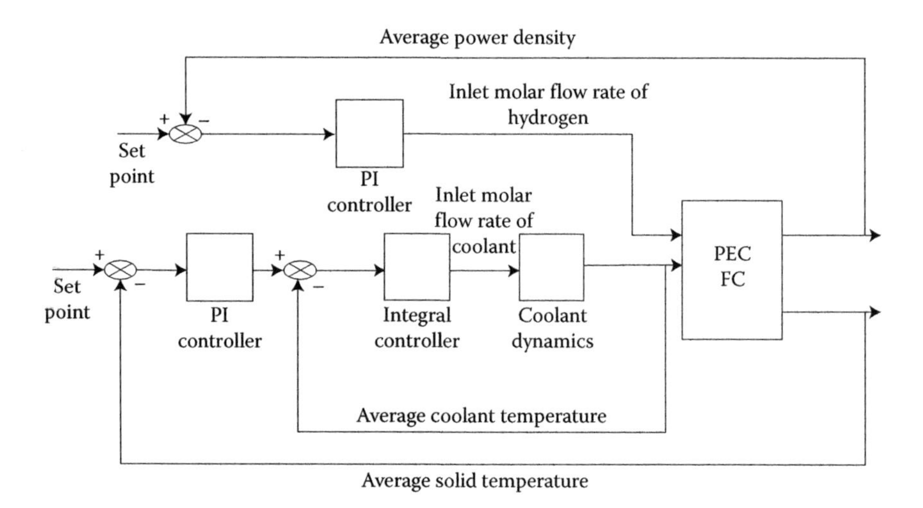

FIGURE 4.6
MIMO control schematic.

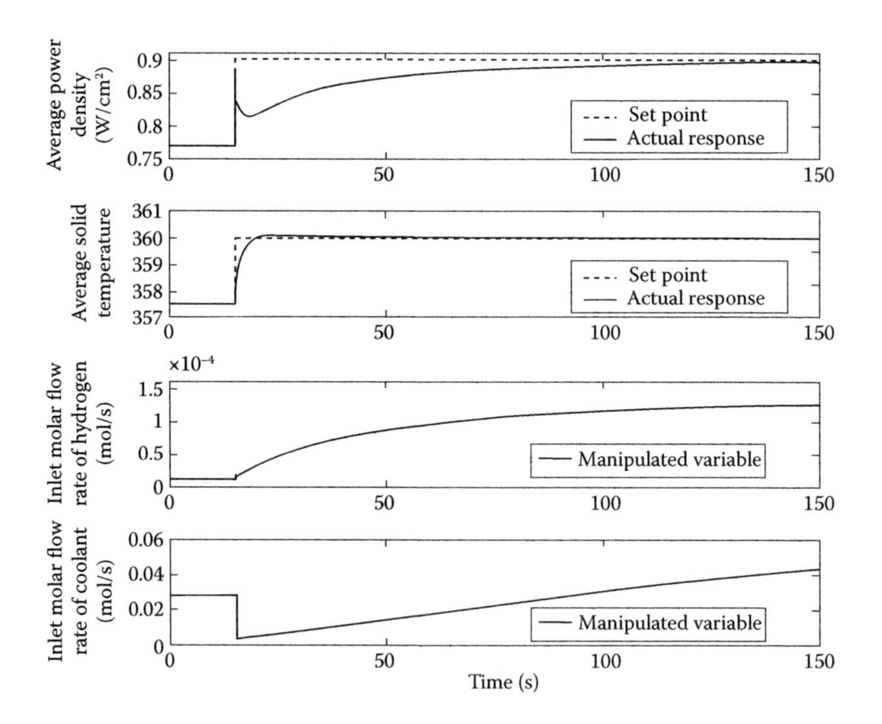

FIGURE 4.7
Performance of the proposed MIMO control strategy in response to changes in the set points for the average power density and the average solid temperature.

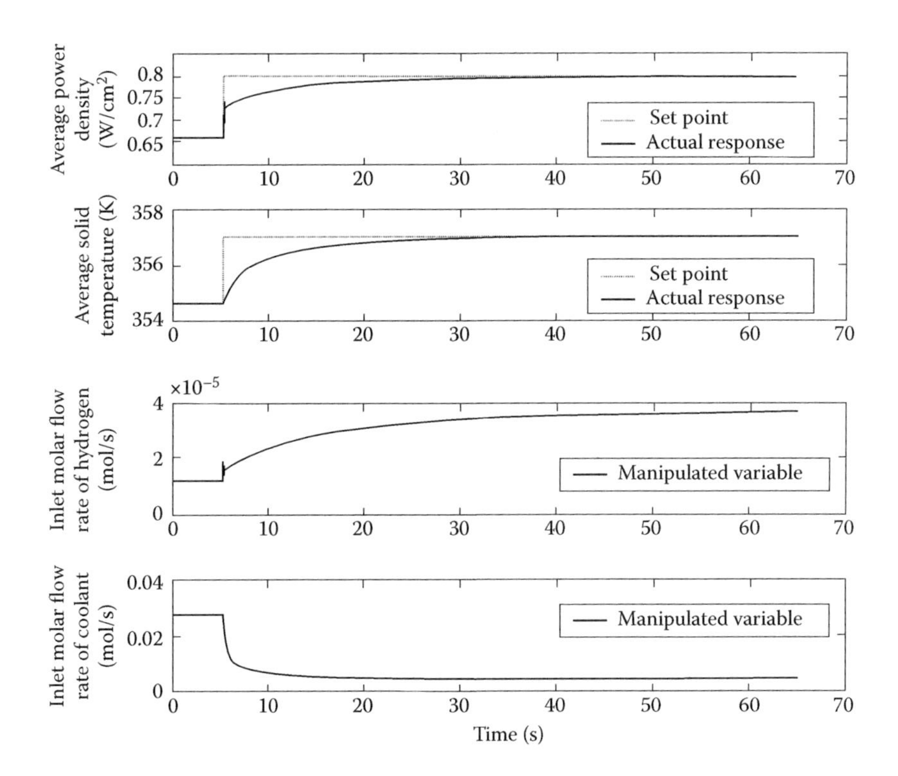

FIGURE 4.8
Performance of the proposed MIMO control strategy in response to changes in the set points for the average power density and the average solid temperature using air at the cathode.

control strategy can provide satisfactory control performance for hydrogen–air fuel cells.

4.2.2.4 Ratio Control

A ratio control strategy was also implemented in Reference 3 to avoid the problem of oxygen starvation and satisfy maximum power density demand. In the ratio control design, the inlet molar flow rate of oxygen is used as a dependent manipulated variable and changed in a constant ratio with respect to the inlet molar flow rate of hydrogen.

The ratio control design is given in Figure 4.9. In this strategy, the measured manipulated variable is the inlet molar flow rate of hydrogen. The inlet flow rate of oxygen is changed in proportion with the inlet flow rate of hydrogen.

Figure 4.10 shows the performance of the ratio control strategy. To improve the performance of ratio control, the average solid temperature is also controlled using the previously described cascade control strategy. Simulation results show that the response of this ratio controller is faster than the MIMO

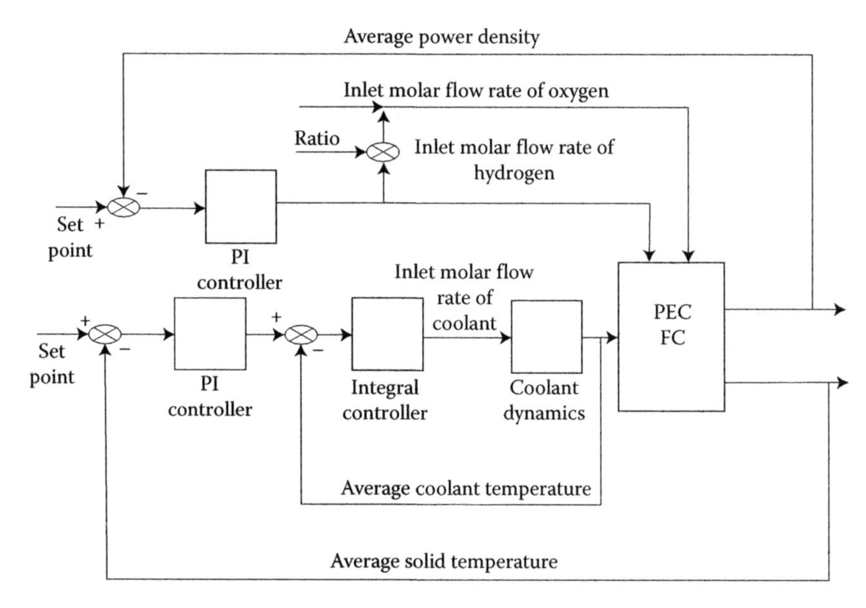

FIGURE 4.9
Schematic for ratio control along with cascade control for temperature.

strategy using only hydrogen as the manipulated variable. The ratio controller is faster because on loop closure, the initial input molar flow rate of oxygen is lower than that of the base case flow rate. Figure 4.11 also shows that in the proposed ratio control strategy, the problem of oxygen starvation has been circumvented due to an increase in the input molar flow rate of oxygen.

In summary, simulation results show that the ratio control strategy provides a faster response than a MIMO control strategy. This ratio control strategy is able to circumvent the problem of oxygen starvation, and the increase in average solid temperature is small as compared to the MIMO control strategy.

Methekar et al. [3] state that a ratio control strategy is able to overcome the problem of oxygen starvation; however, the performance of the linear controllers is slow due to the presence of nonlinearities in the dynamic response of the PEMFC. Hence, a nonlinear controller is essential and necessary for effective control of the PEMFC over a wide range of power densities.

4.3 Nonlinear Control Design for Fuel Cells

A MIMO dynamic nonlinear model of PEMFC is derived in this section, and it is then used to design a nonlinear controller by using feedback

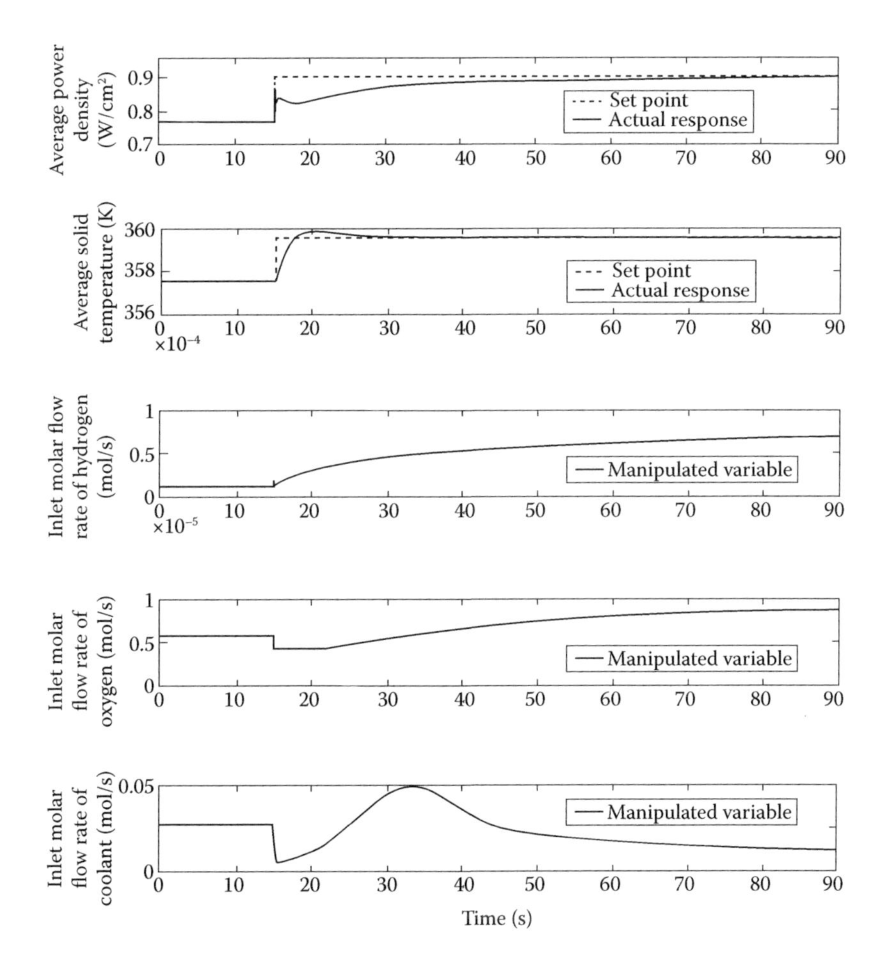

FIGURE 4.10
Performance of the multivariable controller with ratio control for oxygen in response to changes in the set points of the average power density and the average solid temperature.

linearization in order to minimize the difference ΔP between the hydrogen and oxygen partial pressures. The main purpose of keeping ΔP in a certain small range is to protect the membrane from damage and therefore prolong the fuel cell stack life [4–5]. In addition, the pressures have a bigger impact on the performance of fuel cells than other parameters [4–5]. Because the fuel cell voltage is a function of the pressures, each pressure needs to be appropriately controlled to avoid a detrimental degradation of the fuel cell voltage. To achieve this goal, it is necessary to minimize the pressure deviation between the anode and the cathode by using precise actuators such as an accurate valve controller. Normally the optimal pressure controller consists of a pressure sensor and a solenoid flow-control valve. In this book, we mainly focus on developing a pressure-control algorithm for

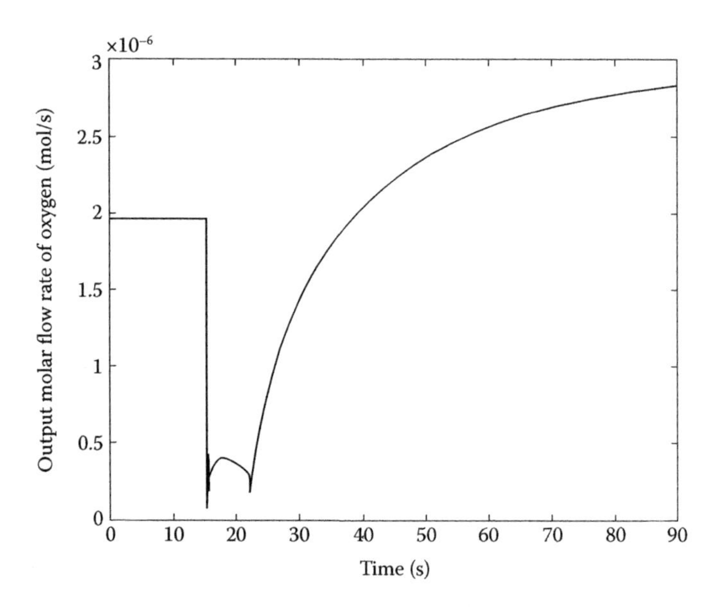

FIGURE 4.11
Output flow rate of oxygen from the PEMFC with ratio control, corresponding to the results shown in Figure 4.27.

the whole fuel cell system instead of designing the pressure sensors and flow controllers inside the fuel cell. Furthermore, the fuel cell voltage is not considered as a control output because of the characteristics of the *V–I* polarization curve, which makes it difficult to perform the voltage control in a short period of time and normally the fuel cell system needs a secondary energy storage for a short period of time control. Therefore, only the partial pressures of hydrogen and oxygen are chosen as the outputs. The stack current is considered as a disturbance to the system instead of an external input [5].

Consider the following MIMO nonlinear system with a disturbance:

$$\dot{X} = f(X) + \sum_{i=1}^{m} g_i(X)u_i + p(X)d, \quad i = 1,\ 2,\ldots,\ m$$

$$\begin{aligned} y_1 &= h_1(X) \\ &\vdots \\ y_m &= h_m(X) \end{aligned} \tag{4.1}$$

where $X \subset R^n$ is the state vector, $U \subset R^m$ the input or control vector, $y \subset R^p$ the output vector, and $f(x)$ and $g(x)$, $i = 1, 2, \ldots, m$, are n-dimensional smooth-vector fields. d represents the disturbance variables and $p(x)$ the dimensional vector field directly related to the disturbance.

Consider the following MIMO nonlinear system with a disturbance, and the nonlinear dynamic system model of PEMFC is rewritten as follows:

$$\dot{X} = f(x) + g_1(x)u_1 + g_2(x)u_2 + p(x)d$$

$$\begin{bmatrix} y_1 \\ y_2 \end{bmatrix} = \begin{bmatrix} x_1 \\ x_3 \end{bmatrix} = \begin{bmatrix} h_1(x) \\ h_2(x) \end{bmatrix} \tag{4.2}$$

where

$$X = \begin{bmatrix} P_{H_2} \\ P_{H_2O_a} \\ P_{O_2} \\ P_{N_2} \\ P_{H_2O_c} \end{bmatrix}; \quad U = \begin{bmatrix} u_a \\ u_c \end{bmatrix}; \quad Y = \begin{bmatrix} P_{H_2} \\ P_{O_2} \end{bmatrix}; \quad d = I_{fct}$$

$$f(x) = 0; \quad g_1(x) = R^*T \cdot \lambda_{H_2} \begin{bmatrix} \dfrac{k_a \cdot Y_{H_2}}{V_a} - \dfrac{k_a}{V_a} \dfrac{x_1}{x_1 + x_2} \\ \dfrac{k_a \cdot \varphi_a P_{vs}}{V_a(x_1 + x_2 - \varphi_a P_{vs})} - \dfrac{k_a}{V_a} \dfrac{x_1}{x_1 + x_2} \\ 0 \\ 0 \\ 0 \end{bmatrix}$$

$$g_2(x) = R^*T \cdot \lambda_{air} \begin{bmatrix} 0 \\ 0 \\ \dfrac{k_c \cdot Y_{O_2}}{V_c} - \dfrac{k_c}{V_c} \dfrac{x_3}{x_3 + x_4 + x_5} \\ \dfrac{k_c \cdot Y_{N_2}}{V_c} - \dfrac{k_c}{V_c} \dfrac{x_4}{x_3 + x_4 + x_5} \\ \dfrac{k_c \cdot \varphi_c P_{vs}}{V_c(x_3 + x_4 + x_5 - \varphi_c P_{vs})} - \dfrac{k_c}{V_c} \dfrac{x_5}{x_3 + x_4 + x_5} \end{bmatrix}$$

$$p(x) = R^*T \begin{bmatrix} -\dfrac{C_1}{V_a} + \dfrac{C_1 \cdot x_1}{V_a(x_1 + x_2)} \\ \dfrac{C_1 \cdot x_2}{V_a(x_1 + x_2)} - \dfrac{C_1}{V_a} \\ -\dfrac{C_1}{2V_c} + \dfrac{C_1 \cdot x_2}{2V_c(x_3 + x_4 + x_5)} \\ 0 \\ -\dfrac{C_1}{V_c} - \dfrac{C_1 \cdot x_5}{V_c(x_3 + x_4 + x_5)} - \dfrac{C_2 \cdot x_5}{V_c(x_3 + x_4 + x_5)} + \dfrac{C_2}{V_c} \end{bmatrix}$$

Equation 4.2 implies that the input–output behavior of the system is nonlinear and coupled. Two steps are needed in order to achieve the control objective

- Obtaining a nonlinear control law that not only can compensate nonlinearities but also can decouple and linearize the input and output behaviors
- Imposing the poles of the closed loop so that the outputs P_{H_2} and P_{O_2} track asymptotically the desired trajectory by adding a proportional integral controller

Equation 4.2 presents a MIMO nonlinear system that makes us ready to develop a nonlinear control law. Normally, the disturbance d in Equation 4.2 cannot be directly used in the control design because an additional necessary condition—that the disturbance is able to be measured and the feed-forward action is allowed—has to be satisfied [6–7]. Otherwise, the linearized map between the new input v and the output y does not exist. The condition renders the following control law by using the measurement of the disturbance:

$$U = -A^{-1}(x)b(x) + A^{-1}(x)v - A^{-1}(x)p(x)d \tag{4.3}$$

As shown in Equation 4.2, $f(x) = 0$ leads to $b(x) = L_f^r h(x) = 0$, and so the control law is written as

$$U = A^{-1}(x)v - A^{-1}(x)p(x)d \tag{4.4}$$

Because each control variable u shows up after the first derivative of each $y_1 = x_1$ and $y_2 = x_3$, the relative degree vector $[r_1\ r_2]$ is [11] and the decoupling matrix $A(x)$ is defined as

$$A(x) = \begin{bmatrix} L_{g_1}h_1(x) & L_{g_2}h_1(x) \\ L_{g_1}h_2(x) & L_{g_2}h_2(x) \end{bmatrix} \tag{4.5}$$

where

$$A(x) = \begin{bmatrix} \frac{k_a \cdot Y_{H2} \cdot \lambda_{H2}}{V_a} - \frac{k_a \cdot \lambda_{H2}}{V_a} \frac{x_1}{x_1 + x_2} & 0 \\ 0 & \frac{k_c \cdot Y_{O_2} \cdot \lambda_{air}}{V_c} - \frac{k_c \cdot \lambda_{air}}{V_c} \frac{x_3}{x_3 + x_4 + x_5} \end{bmatrix}$$

which is nonsingular at $x = x_0$. Additionally, the matrix v and $p(x)$ in Equation 4.4 are given as follows:

$$v = \begin{bmatrix} \dot{y}_1 \\ \dot{y}_2 \end{bmatrix}; \quad p(x) = R^*T \begin{bmatrix} -\frac{C_1}{V_a} + \frac{C_1 \cdot x_1}{V_a(x_1 + x_2)} \\ -\frac{C_1}{2V_c} + \frac{C_1 \cdot x_2}{2V_c(x_3 + x_4 + x_5)} \end{bmatrix} \tag{4.6}$$

The control law given in Equation 4.4 yields a decoupled and linearized input–output behavior (see Figure 4.1)

$$\begin{aligned} \dot{P}_{H_2} &= v_1 \\ \dot{P}_{O_2} &= v_2 \end{aligned} \tag{4.7}$$

The outputs P_{H_2} and P_{O_2} are decoupled in terms of the new inputs v_1 and v_2. Thus, two linear subsystems, which are between the input v_1 and the hydrogen partial pressure $y_1 = P_{H_2}$, and between the input v_2 and the oxygen partial pressure $y_2 = P_{O_2}$, are obtained. Furthermore, note that $\dot{y}_1 = \dot{x}_1$ and $\dot{y}_2 = \dot{x}_3$, and so in order to ensure that y_1 and y_2 are adjusted to the desired values 3 atm of y_{1_S} and y_{2_S}, the stabilizing controller is designed by linear control theory using the pole-placement strategy [8]. The new control inputs are given by

$$\begin{bmatrix} v_1 \\ v_2 \end{bmatrix} = \begin{bmatrix} \dot{y}_{1\_S} - k_{11}e_1 \\ \dot{y}_{2\_S} - k_{21}e_2 \end{bmatrix} \tag{4.8}$$

where $e_1 = y_1 - y_{1\_S}$ and $e_2 = y_2 - y_{2\_S}$.

Even though the nonlinear system PEMFC is exactly linearized by feedback linearization, there may exist a tracking error in the variation of the parameters, especially when the load changes. To eliminate this tracking error, the integral terms are added in the closed-loop error equation as in [7–8]

$$\begin{bmatrix} v_1 \\ v_2 \end{bmatrix} = \begin{bmatrix} \dot{y}_{1\_S} - k_{11}e_1 - k_{12}\int e_1 dt \\ \dot{y}_{2\_S} - k_{21}e_2 - k_{22}\int e_2 dt \end{bmatrix} \tag{4.9}$$

From Equation 4.9, the error dynamics can be obtained as follows:

$$\begin{aligned} \ddot{e}_1 + k_{11}\dot{e}_1 + k_{12}e_1 &= 0 \\ \ddot{e}_2 + k_{21}\dot{e}_2 + k_{22}e_2 &= 0 \end{aligned} \tag{4.10}$$

By appropriately choosing the roots of the characteristics of $s^2 + k_{11}s + k_{12}$ and $s^2 + k_{21}s + k_{22}$, asymptotic tracking is achieved, so that $P_{H_2} \to y_{1\_S}$ and $P_{O_2} \to y_{2\_S}$ as $t \to \infty$. The overshoots also become small by choosing $k_{11}^2 \gg 4k_{12}$ and $k_{21}^2 \gg 4k_{22}$ [7–8].

As shown in Figure 4.12, the main objective of this control scheme is to design a nonlinear controller by appropriately defining a transformation mapping that transforms the original nonlinear system into a linear and controllable (closed) system, at which point a linear controller can be designed using the pole-placement technique for tracking purposes. However, the control law in Equation 4.4 will be unobservable because the entire dynamics has a 5-order $P_{H_2}, \ldots, P_{H_2O_c}$, whereas only a 2-order P_{H_2} and P_{O_2} are observed in the outputs. So, we may have a problem of internal dynamics. In other words, the

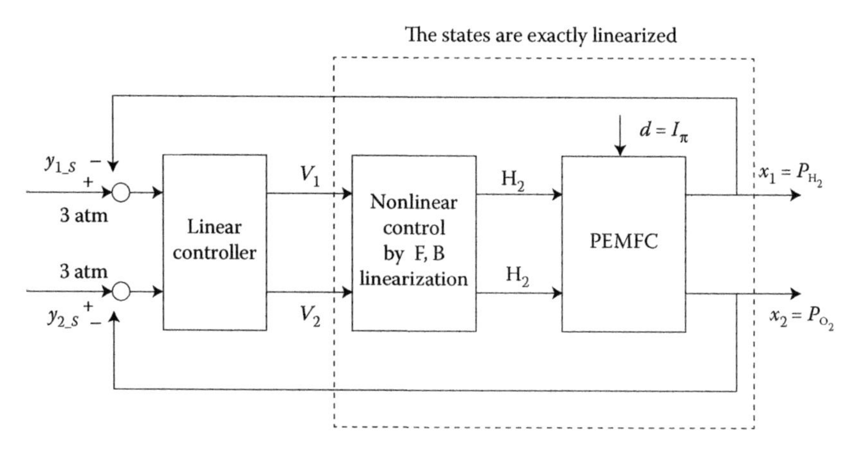

FIGURE 4.12
Overall control block diagram of PEMFC.

internal dynamics of $P_{H_2O_a}$, P_{N_2}, and $P_{H_2O_c}$ must be stable so that the states of the tracking controllers in Equation 4.9 are held in a bounded region during tracking. Otherwise, with external as well as internal dynamics, this control law cannot enhance the overall system performance. However, it is difficult to directly determine the internal dynamics of the system because it is nonlinear, nonautonomous, and coupled to the external closed-loop dynamics, as seen in Equations 4.1 through 4.10. Here, simulation is used to verify whether each state remains within the reasonable bounded area [9]. The comparison between the simulation results and experimental data in Reference 10 was used to verify the performance of the control law.

4.4 Nonlinear Control Design for Interface

For the interface design with the fuel cell nonlinear controller, several main components have to be considered, especially in a fuel cell vehicle system. The following are the main components of a fuel cell vehicle [11]:

- *A fuel cell processor:* gasoline or methanol reformer (in case of using a direct hydrogen, a hydrogen storage tank is required).
- The fuel cell stack, which can produce electricity and includes an air compressor to provide pressurized oxygen to the fuel cell.
- A cooling system, which can maintain the proper operating temperature.
- A water management system, which can manage the humidity and the moisture in the system and keep the fuel cell membrane saturated and at the same time prevent the water from being accumulated at the cathode.
- A DC/DC converter, which can condition the output voltage of the fuel cell stack.
- An inverter system, which can convert the DC to variable voltage and variable frequency to power for the propulsion motor.
- A battery or an ultracapacitor, which can provide supplemental power for a start up of system and store excessive energy during deceleration. For the power management, the bidirectional converter can be used between the DC/DC converter and the battery.

As seen in Figure 4.13, if using methanol or gasoline instead of the direct hydrogen, the fuel storage tank before the reformer is needed. For the direct hydrogen, hydrogen tank is required. Both storage tanks should be fit into the vehicle as small as possible with considering safety hazards.

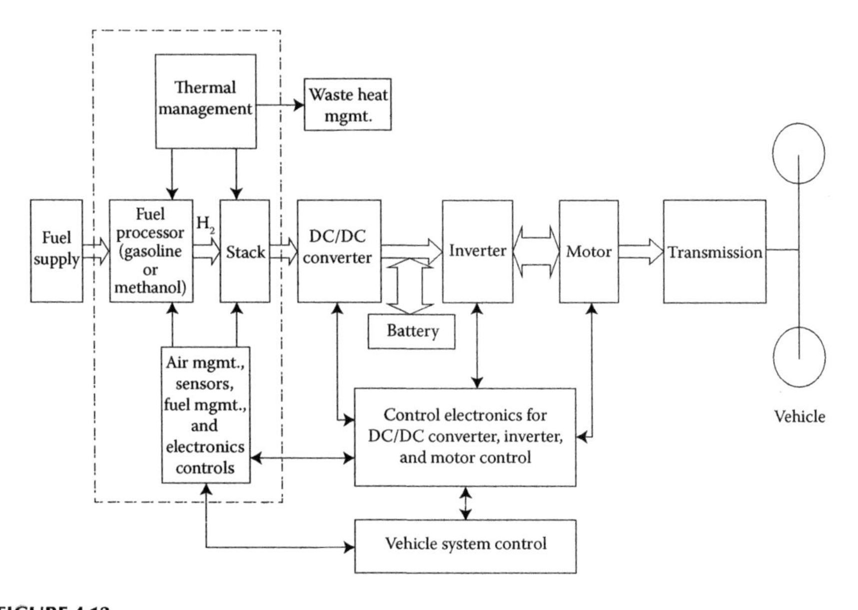

FIGURE 4.13
A typical fuel cell vehicle system. (Adapted from K. Rajashekrara, Propulsion system strategies for fuel cell vehicles, SAE paper 2000-01-0369.)

When high-power demand is required, such as acceleration, the battery will provide the required power. When low power is required, the fuel cell can cover the power and even charge the battery. Figure 4.14 shows the fuel cell converter control system. In Figure 4.14, a DC/DC boost converter is used to boost the fuel cell voltage to the required battery voltage of about 300 V for the propulsion inverter. The DC/DC converter has to be designed based

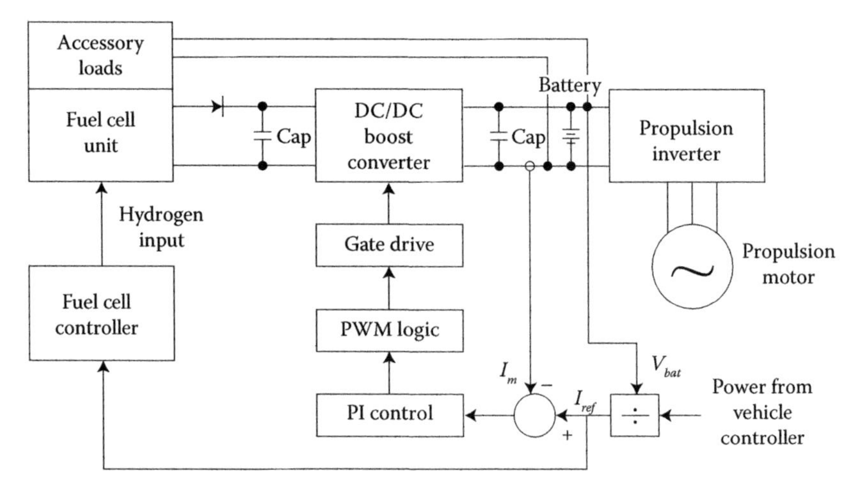

FIGURE 4.14
Fuel cell converter control system.

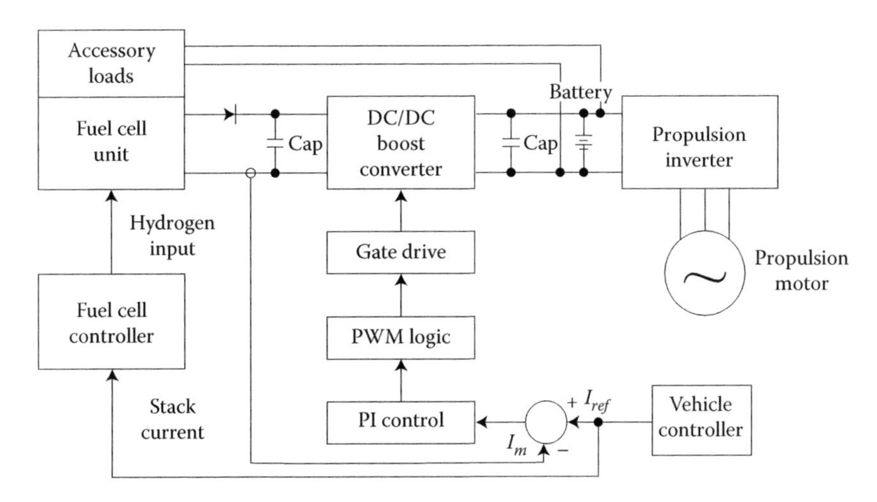

FIGURE 4.15
Modified fuel cell converter control system. (Adapted from K. Rajashekrara, Propulsion system strategies for fuel cell vehicles, SAE paper 2000-01-0369.)

on the maximum power capacity of the fuel cell stack. The blocking diode is needed at the output of the fuel cell stack so as to prevent the reversal current flowing into the stack, which leads to a damage to the fuel cell stack.

The power command from the vehicle controller is divided by the battery voltage and therefore it can generate the current reference, which is used for the DC/DC boost converter and the fuel cell controller. In the DC/DC boost converter, the reference signal is compared with the measured current to derive the duty cycle for controlling the output power of the DC/DC boost converter. Based on the battery voltage change, the fuel output power varies in a wide range. In the fuel cell controller, since the fuel cell output power is proportional to the hydrogen and oxygen flow rates, it has a wide bandwidth to stabilize the system under all operating conditions [11]. Hence, in order to have a constant power output in a wide range, the modified fuel cell converter system is proposed [11]. By directly measuring the fuel cell current in Figure 4.15, this fuel cell current is compared with the reference current, and the error in these currents can derive the duty cycle of the DC/DC converter and produce a constant current and voltage and thus it remains the power constant and makes the system stabilized in a wide range.

4.5 Analysis of Control Design

Several nonlinear control approaches are compared with a proportional integral derivative PID controller, which is used to control the gas pressure loop for both input gases: hydrogen and oxygen. Among them are the

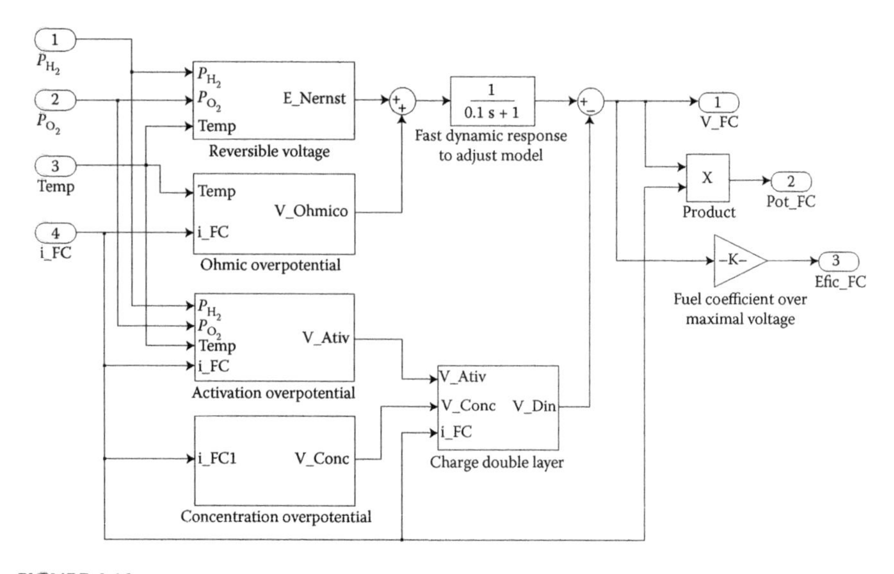

FIGURE 4.16
Dynamic MATLAB–Simulink PEMFC model. (Adapted from P.E.M. Almeida and M.G. Simoes, *IEEE Transactions on Industry Applications*, 41(1), 237–245, 2005.)

feedback linearized-based nonlinear control for both input gases [12] and the neural network-based optimal control (NOC) for the gas pressures. More details about these control designs can be found in Reference 13. Using the electrochemical and thermodynamic relationships in a PEMFC, a dynamic MATLAB–Simulink model of PEMFC was built in Reference 13 by NOC scheme, which is shown in Figure 4.16. In this model, four inputs are fed to the fuel cell. The PID pressure control Simulink model can be found in Figure 4.17 [13]. According to the simulation results in Figures 4.18 and 4.19, NOC can tune the gain in an automated way using following equation:

$$U(t) = \alpha \cdot (V_{ref}(t) - V(t))^2 + \beta \cdot (P_{H_2}(t) - P_{H_2}(t-1))^2 \tag{4.11}$$

This cost criterion $U(t)$ states that the error between the desired state and actual state must be minimized. As seen in Equation 4.11, fine-tuning of parameters α and β is required in order to have the minimum cost.

4.6 Simulation of Nonlinear Control for PEMFC

As to the feedback linearized-based nonlinear control of PEMFC derived in Equation 4.4, the dynamic PEMFC model and control method are tested through simulation in the MATLAB–Simulink environment [9].

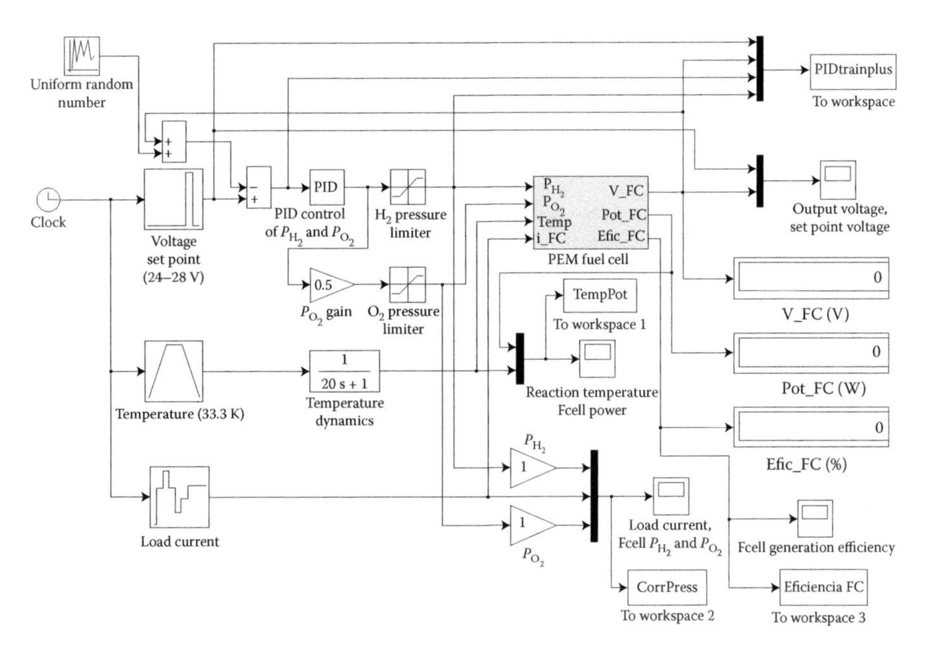

FIGURE 4.17
PID pressure control. (Adapted from P.E.M. Almeida and M.G. Simoes, *IEEE Transactions on Industry Applications*, 41(1), 237–245, 2005.)

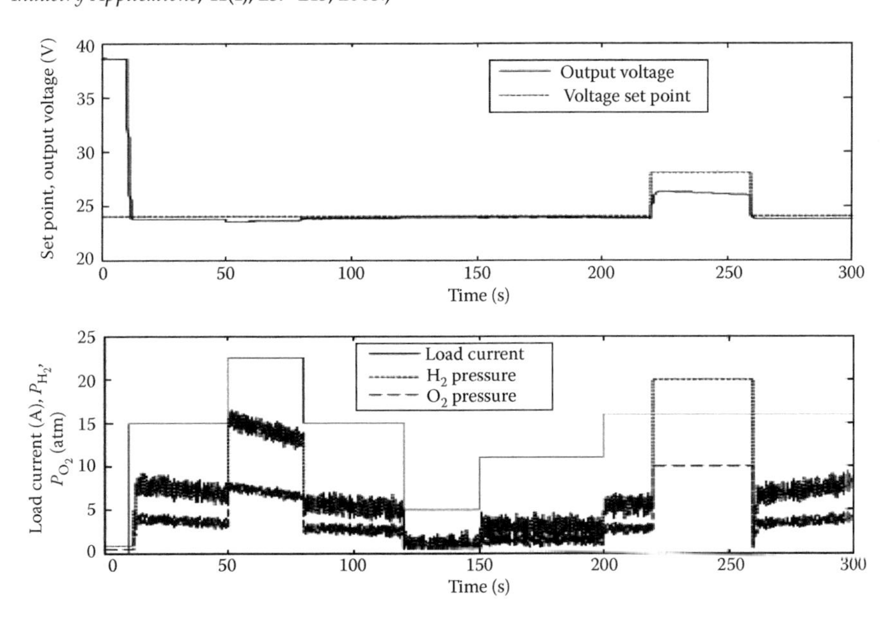

FIGURE 4.18
PEMFC PID control: set point, output voltage, H pressure, O pressure, and load current. (Adapted from P.E.M. Almeida and M.G. Simoes, *IEEE Transactions on Industry Applications*, 41(1), 237–245, 2005.)

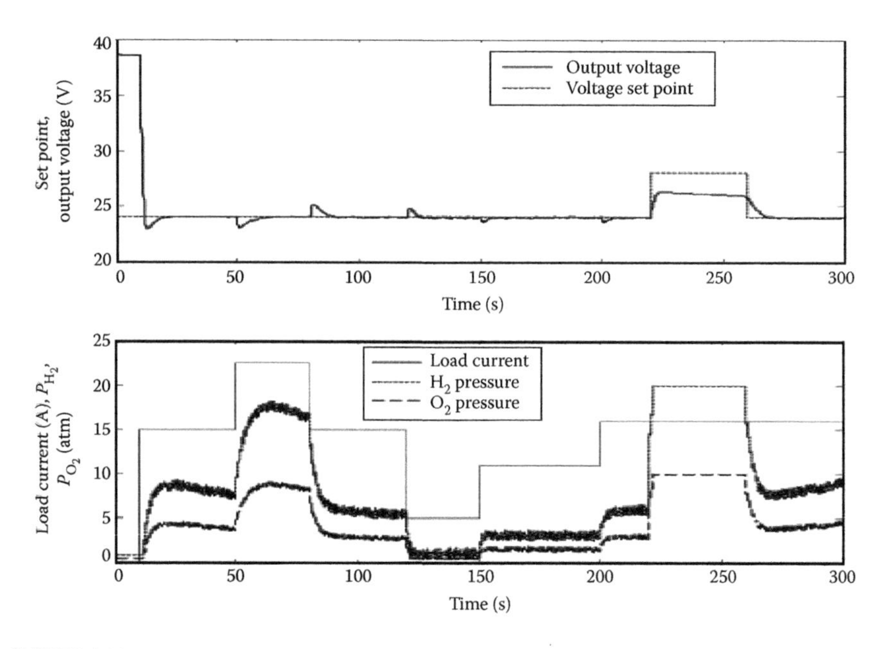

FIGURE 4.19
PEMFC neural optimal control: set point, output voltage, H pressure, O pressure, and load current. (Adapted from P.E.M. Almeida and M.G. Simoes, *IEEE Transactions on Industry Applications*, 41(1), 237–245, 2005.)

For simplicity, the fuel processor, water and heat management, and air compressor models are not considered in the simulation. Experimental data in Reference 10 are used to justify the validity of the proposed dynamic PEMFC model.

The fuel cell system Ballard MK5-E-based PGS-105B system shown in Figure 4.19 is used to test the feedback linearized-based nonlinear controller. This system has a total of 35 cells, connected in series, with a cell surface area of 232 cm^2. The membrane electrode assembly consists of a graphite electrode and a Dow™ membrane. The reactant gases (hydrogen and air) are humidified inside the stack, and the hydrogen is recirculated at the anode side while the air is flowing through the cathode side. The hydrogen pressure is regulated to 3 atm at the anode inlet by a pressure regulator and a back-pressure regulator at the air outlet also maintains 3 atm through the Ballard fuel cell stack. The oxidant flow rate is automatically adjusted to a constant value of 4.5 L/s, based on a programmable load, via a mass flow meter to ensure sufficient water removal at the cathode. Hydrogen is replenished at the same rate as it is consumed. The stack temperature measured at the air outlet is maintained between 72°C and 75°C to produce the maximum power output. The simplified diagram of the Ballard system is depicted in Figure 4.20. The PGS105 system is more likely to be a presetting control system instead of a feedback control system. In the case of outer ranges of settings, the system

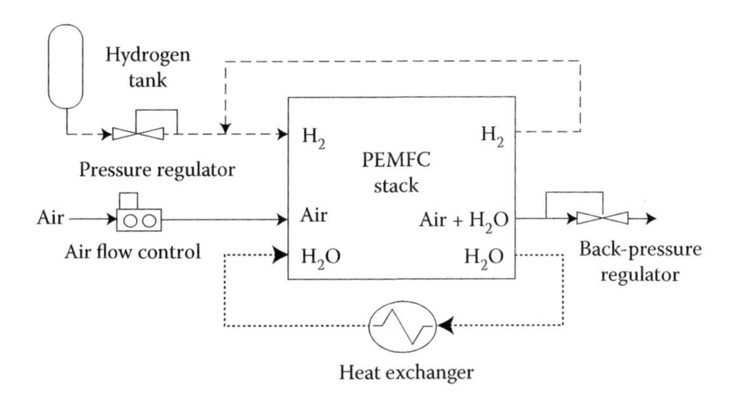

FIGURE 4.20
PEMFC stack based on PGS-105B system. (Adapted from J. Hamelin et al., *International Journal of Hydrogen Energy*, 26, 625–629, 2001.)

automatically shuts down, so a higher-level control system is thus needed. Here, the outline of the experimental setup of the nonlinear controller is explained in Reference 9. To implement this control scheme in the practical system, the sensors to measure the states $P_{H_2},\ldots,P_{H_2O_c}$, the stack current, and the voltage must be installed, and a main controller such as a digital signal processor (DSP) needs to be implemented based on the nonlinear control law obtained using the tool of feedback linearization. With the help of the DSP, the fuel cell can communicate with the electronics rack and the PC for the user interface monitoring shown in Figure 4.16. The safety concerns for the fuel cell system must be solved with high priority before a control scheme is implemented. At least the safety issues related to low cell voltage, stack overload, high temperature, hydrogen leaking, and pressure differences between the anode and the cathode must be considered in the fuel cell control system to prevent a fatal accident and severe damage to the fuel cell system.

To control the entire fuel cell system, the sensors and the temperature and humidity control actuators must be added to the system. In the simulation, since the temperature and the humidity in the system have a very slow response time, a perfectly controlled humidifier and heat exchanger are assumed to be applied (Figure 4.21). In addition, an automatic purge controller for the hydrogen exhaust is required, and with respect to the air and water exhaust, the back-pressure regulator must be coordinated with the air inlet flow-rate input as well as the hydrogen inlet flow-rate input to maintain the pressures for hydrogen and air at the same level. By assuming that the electrical capacitor is 1 F [12] and the sum of the activation and concentration resistance is from 0.2 to 0.3 Ω [13], the time constant of the fuel cell stack is approximately from 0.2 to 0.3 s. With the use of a fast actuator on the cathode side, whose performance setting time is 10%–90% of 50 ms, and another actuator on the anode side whose performance setting time is 10%–90% less than 20 ms, it becomes possible to maintain the anode and cathode pressures

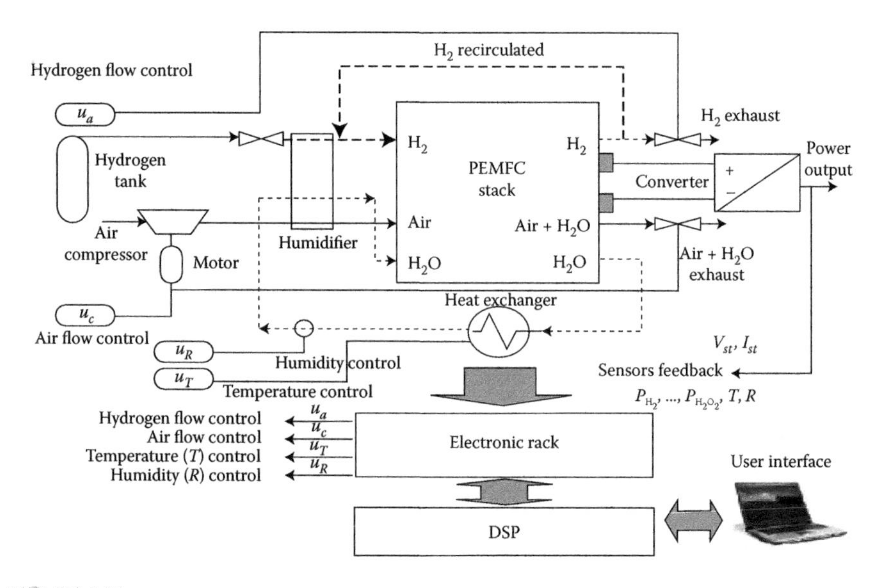

FIGURE 4.21
Simplified experiment setup for the PEMFC nonlinear control. (Adapted from W.K. Na and B. Gou, *IEEE Transaction on Energy Conversion*, 23(1), 179–190, 2008.)

at a certain level. To ensure this performance, the anode pressure controller must be three times faster than that on the cathode side because the anode pressure follows the cathode-side pressure [13]. Currently, these fast actuators are available in the market [14], yet in reality, due to the uncertainties in the fuel cell model parameters and inaccuracy in the measurements, we may encounter a serious obstacle to achieve the desired responses. Moreover, the time delay of actuators, sensors, compressor, and sampling period for the DSP is unavoidable for the design of the nonlinear control of PEMFCs. Even though we could not exactly calculate the time delays, they can still be compensated by adapting a lead compensator in practical systems. However, since the lead compensation is very sensitive to noise, a multiple first-order, low-pass noise filter is recommended [13]. In the simulation in Reference 12, time delays and other problems such as uncertainty and inaccuracy in measurements are not considered.

The nominal values of parameters for the simulation are shown in Table 4.1.

Experimental data from Hamelin et al. [10] were used to justify the validity of the proposed dynamic model of PEM fuel cells. The voltage and the current—the most important variables—were used for this comparison. In Reference 10, a load profile with rapid variations between 0 and 150 A was imposed on the PGS-105B system. The corresponding stack current and voltage transients are plotted in Figure 4.22, where the experimental data [10] are indicated by solid lines.

The details of load profile are shown in Figure 4.23, where the load resistances were changed from 0.145 to 4.123 Ω during the simulation period.

TABLE 4.1

PEMFC Parameters for the Simulation

Parameter	Value and Definition
N	Cell number: 35
V_o	Open cell voltage: 1.032 V
R	Universal gas constant (J/mol K): 8.314 J/mol K
T	Temperature of the fuel cell (K): 353 K
F	Faraday constant (C/mol): 96485 C/mol
α	Charge transfer coefficient: 0.5 [9]
M	Constant in the mass transfer voltage: 2.11×10^{-5} V
N	Constant in the mass transfer voltage: 8×10^{-3} cm²/mA
R	2.45×10^{-4} kΩ cm²
A_{fc}	Fuel cell active area: 232 cm²
V_a	Anode volume: 0.005 m³ [9]
V_c	Cathode volume: 0.01 m³ [9]
k_a	Anode conversion factor: 7.034×10^{-4} mol/s
k_c	Cathode conversion factor: 7.036×10^{-4} mol/s

Figure 4.22 shows a good agreement between the experimental data and the simulated results, except for the time periods 13 s, 15 s and 25 s, 29 s. The main reason of the voltage difference during these periods is the internal resistance variation of PEM fuel cell systems [12]. The rapid current increase can cause an immediate voltage drop across the internal resistor of the fuel cell, which is closely related to the temperature change that adversely affects this resistance. In this simulation, the stack temperature was assumed to be constant at 353 K. A small discrepancy between simulation results and

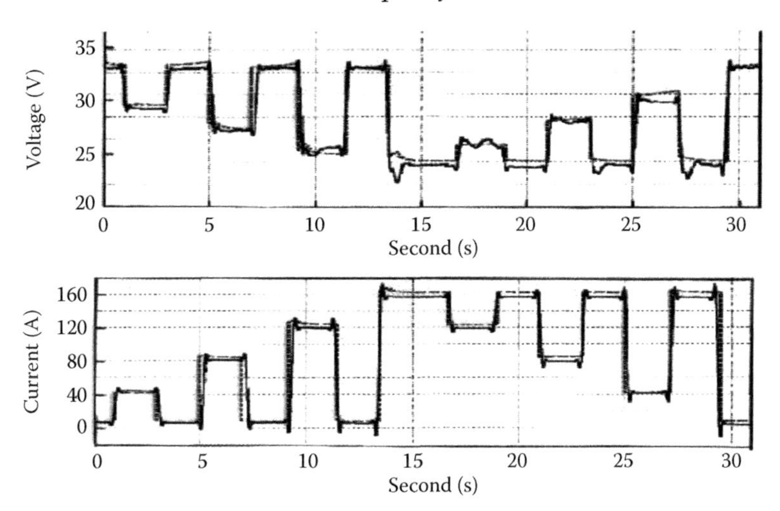

FIGURE 4.22
Voltage and current under load variations.

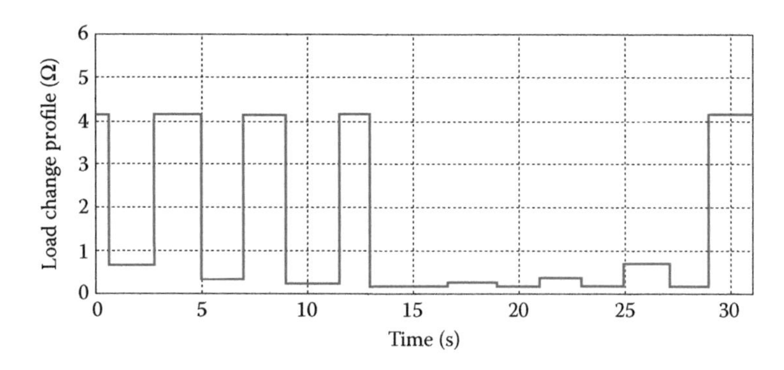

FIGURE 4.23
Load variation profile.

experimental data is inevitable under the rapid current changes. In order to compare the efficiency of the nonlinear controller, the linear controller (PI controller) is used in the fuel cell system.

To achieve fewer overshoot results, the feedback gains k_{11}, k_{21}, k_{12}, and k_{22} in Equation 4.10 have to be tuned to 5 and 1, respectively, which are the optimal values in their feasible ranges [0.1, 10] for k_{11} and k_{21} and [0, 10] for k_{12} and k_{22}, and being beyond the ranges can easily cause a violation of the MATLAB simulation limits.

For the voltage, current, and power shown in Figures 4.22 and 4.24, the discrepancies between the nonlinear control and the linear control are not obvious due to the fast response times, which are less than a few milliseconds. However, other simulation results in Figures 4.25 through 4.29 indicate that better transient performances are observed when using a nonlinear controller because the linear controller is more dependent upon the operating point, while the nonlinear controller is independent of the operating

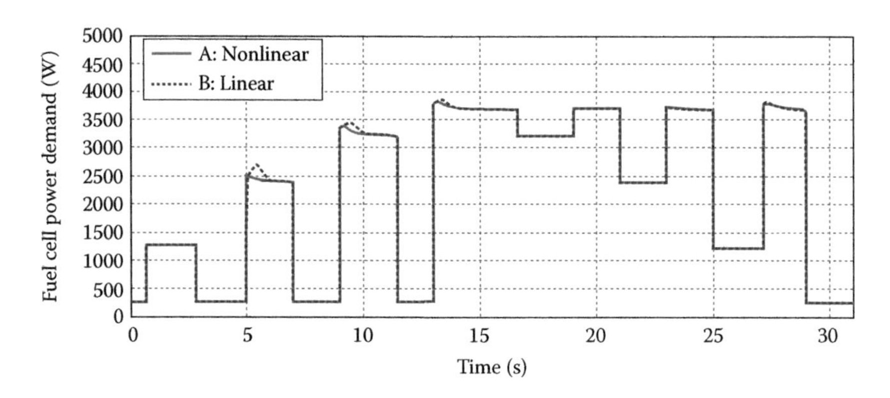

FIGURE 4.24
Fuel cell power demand under load variations.

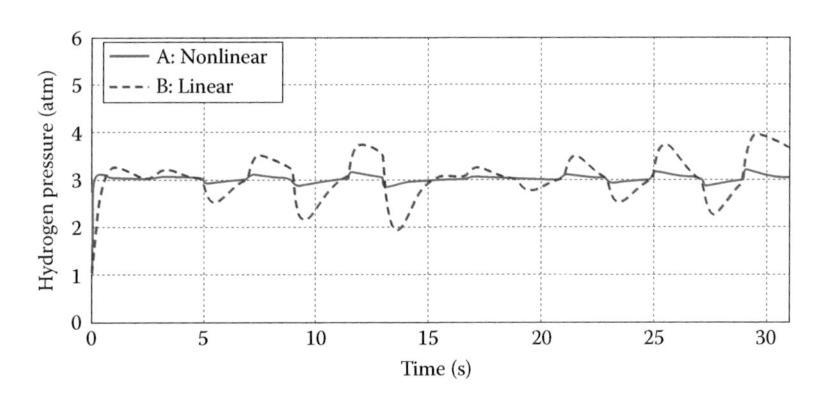

FIGURE 4.25
Variations of hydrogen pressure.

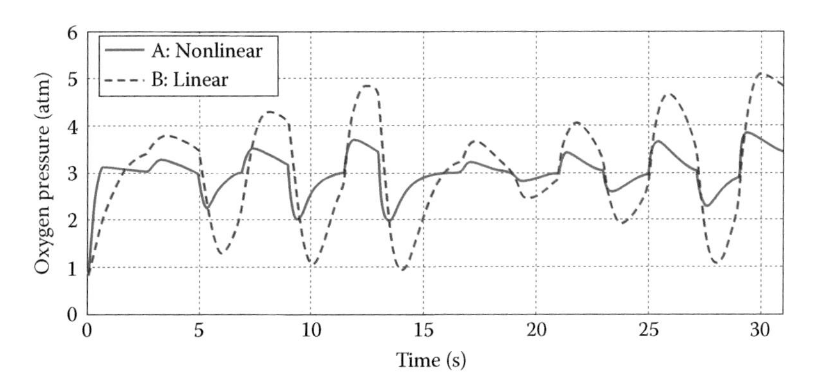

FIGURE 4.26
Variations of oxygen pressure.

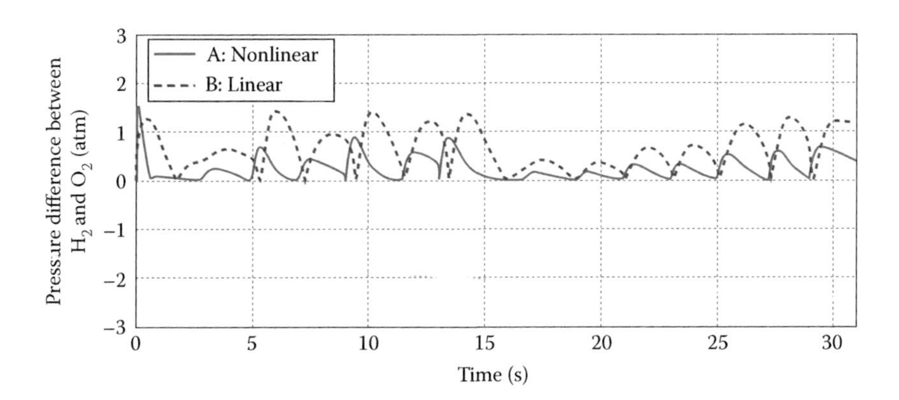

FIGURE 4.27
Variations of pressure difference of H_2 and O_2.

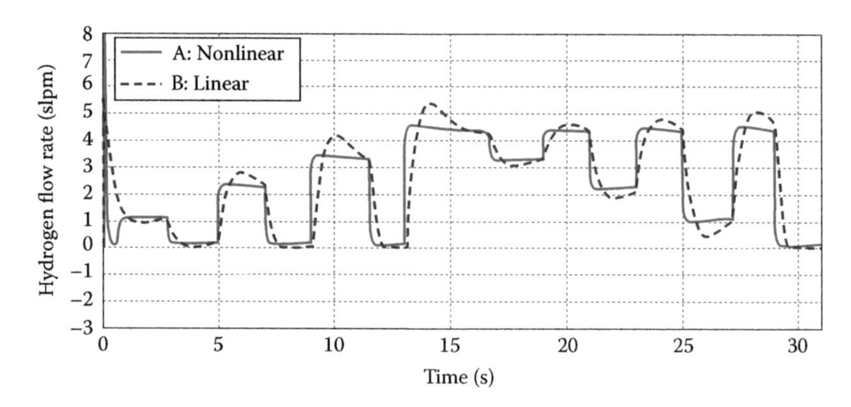

FIGURE 4.28
Variations of hydrogen flow rate.

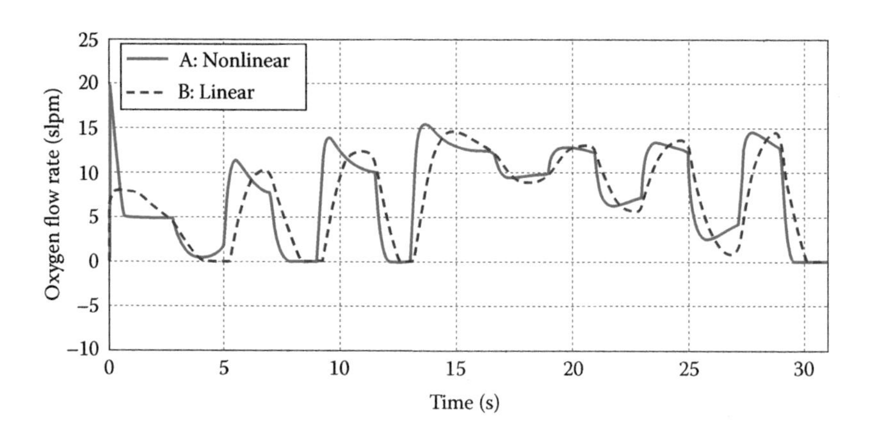

FIGURE 4.29
Variations of oxygen flow rate.

point due to the feedback linearization control design based on the differential geometry [15]. Figure 4.26 shows that the oxygen partial pressure has much bigger overshoot than the hydrogen partial pressure, which implies that the oxygen partial pressure is more sensitive to the load variation than the hydrogen partial pressure. Figure 4.27 displays the absolute value of the difference between the hydrogen and oxygen partial pressures. It is found in Figure 4.27 that the nonlinear controller has a better transient response than the linear controller. Generally, an increase in the stack current causes a decrease in reactant pressures because more fuel consumption is required. However, the flow rates vary with the stack current in the same way and compensate for the increased fuel consumption.

Figures 4.28 and 4.29 give the responses of the hydrogen and the oxygen flow rates under the load variations. The hydrogen flow rate varies between

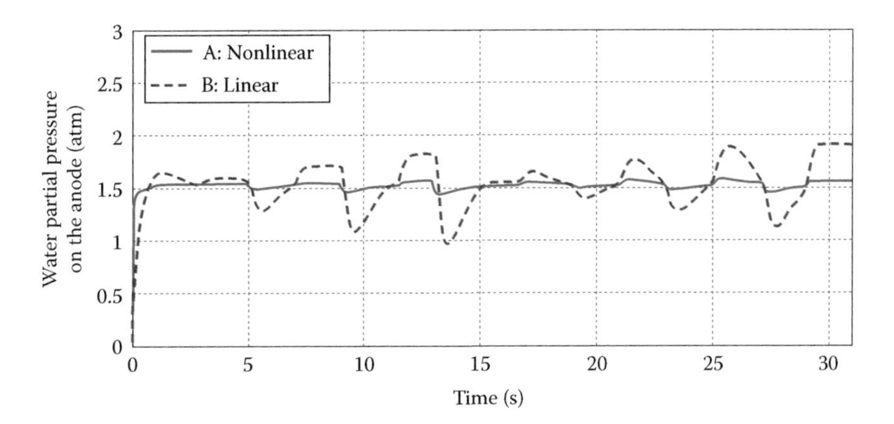

FIGURE 4.30
Variations of water partial pressure on the anode.

0 and 5 slpm, while the oxygen flow rate varies from 0 to 16 slpm. It is observed that the oxygen flow rate has much bigger variations than hydrogen because the oxygen flow rate is more sensitive to the load variation than the hydrogen flow rate, as was seen with the pressure variations.

Although the nonlinear controller has slightly more overshoot in the oxygen flow rate than the linear controller in Figure 4.28, it shows that the response time of the nonlinear controller is faster than the linear controller. Figures 4.20 through 4.32 show that the states $P_{H_2O_a}$, $P_{H_2O_c}$, and P_{N_2} are stable under the load variations.

Figures 4.31 and 4.32 show that the nonlinear controller and the linear controller approximately behave the same way in the water partial pressure on the cathode side and in the nitrogen partial pressure. In addition, even though no tracking controllers are considered in the design of the nonlinear

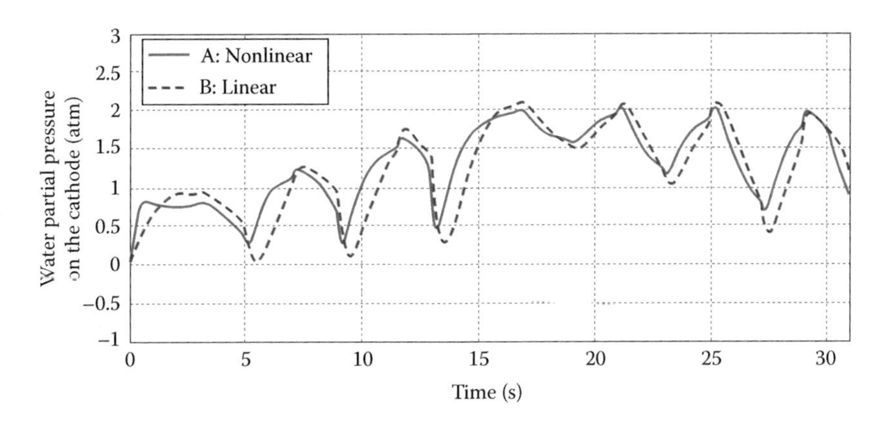

FIGURE 4.31
Variations of water partial pressure on the cathode.

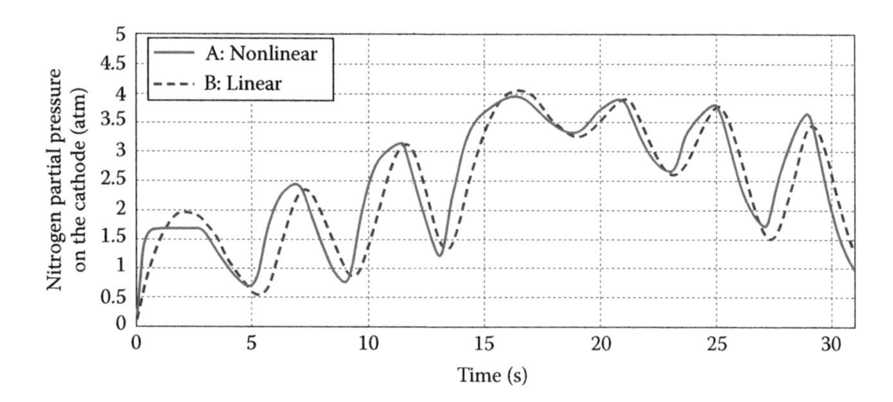

FIGURE 4.32
Variations of nitrogen partial pressure on the cathode.

and linear controls, Figures 4.20 through 4.32 show that $P_{H_2O_a}$, P_{N_2}, and $P_{H_2O_c}$ vary within bounded ranges (0–4 atm) in the load variations, which implies that the internal dynamics problem of the nonlinear tracking controller in our design is not a concern. It is observed that the water partial pressure on the anode side remains a little more stable than the water partial pressure at the cathode and the nitrogen pressure. Furthermore, the water partial pressure on the anode side follows a pattern similar to that of the hydrogen partial pressure because of the high mole fraction of hydrogen, which is about 99%. Therefore, it can be concluded that the load variations have more influence on the cathode side than on the anode side, which implies a more sophisticated control strategy than the one proposed in this chapter needs to be applied on the cathode side.

References

1. J.S. Yi and T.V. Nguyen, An along-the-channel model for proton exchange membrane fuel cells, *Journal of the Electrochemical Society*, 145(4), 1149–1159, 1998.
2. J. Golbert and D. Lewin, Model-based control of fuel cells: (1) Regulatory control, *Journal of Power Sources*, 135, 135–151, 2004.
3. R.N. Methekar, V. Prasad, and R.D. Gudi, Dynamic analysis and linear control strategies for proton exchange membrane fuel cell using a distributed parameter model, *Journal of Power Sources*, 165, 152–170, 2007.
4. W. Yang, B. Bates, N. Fletcher, and R. Pow, Control challenges and methodologies in fuel cell vehicle development, SAE paper 98C054, 1998.
5. J. Purkrushpan, A.G. Stefanopoulou, and H. Peng, Control of fuel cell breathing, *IEEE Control Systems Magazine*, 24(2), 30–46, April 2004.
6. A. Isidori, *Nonlinear Control Systems* (3rd edition), Springer-Verlag, London, 1995.

7. M.A. Henson and D.E. Seborg, Critique of exact linearization strategies for process control, *Journal of Process Control*, 1, 122–139, 1991.
8. J.J.E. Slotine and W. Li, *Applied Nonlinear Control*, Prentice-Hall, Englewood Cliffs, New Jersey, 1991.
9. W.K. Na and B. Gou, Feedback linearization based nonlinear control for PEM fuel cells, *IEEE Transaction on Energy Conversion*, 23(1)179–190, 2008.
10. J. Hamelin, K. Abbossou, A. Laperriere, F. Laurencelle, and T.K. Bose, Dynamic behavior of a PEM fuel cell stack for stationary application, *International Journal of Hydrogen Energy*, 26, 625–629, 2001.
11. K. Rajashekrara, Propulsion system strategies for fuel cell vehicles, SAE paper 2000-01-0369, 2000.
12. J. Larminie and A. Dicks, *Fuel Cell Systems Explained*, Wiley, New York, 2002.
13. C.A. Anderson, M.O. Christensen, A.R. Korsgaad, M. Nielsen, and P. Pederson, Design and control of fuel cells system for transport application, Aalborg University, Project Report, 2002.
14. Burkert, https://www.burkert.com/en/.
15. B.W. Bequette, Nonlinear control of chemical process: A review, *Industrial and Engineering Chemistry* 30, 1391–1413, 1991.
16. P.E.M. Almeida and M.G. Simoes, Neural optimal control of PEM fuel cells with parametric CMAC networks, *IEEE Transactions on Industry Applications*, 41(1), 237–245, 2005.

5

*Simulink Implementation of Fuel Cell Models and Controllers**

5.1 Introduction

The dynamic fuel cell models and controllers were developed using MATLAB–Simulink environment. Simulink is a toolbox extension of the MATLAB program by Mathworks Inc. Simulink is a very powerful tool in modeling and mathematical representation. We can choose a suitable integration method and set up the run-time and initial conditions in the Simulink environment. Systems are drawn on-screen by block diagrams in Simulink. Elements of a block diagram are available, such as transfer functions, a summing junction, etc., as well as virtual input and output devices: function generators and oscilloscopes. Because Simulink provides a graphical user interface, it is easy to build block diagrams, perform simulations, and analyze results. In Simulink, models are hierarchical so that you can view a system at a high level, and details in each block can be viewed by a double click on blocks.

The details of Simulink implementation of the fuel cell model and controllers as well as related elements are described in this chapter.

5.2 Simulink Implementation of the Fuel Cell Models

Two assumptions are made to develop a fuel cell model in Simulink:

- Owing to a slow response time for the stack temperature (about 102 s [1]), the operating stack temperature is assumed to be constant.

* This chapter was mainly prepared by Dr. Woonki Na, California State University, Fresno and Dr. Bei Gou, Smart Electric Grid LLC, USA.

- The fuel cell is well humidified on both the anode and cathode sides.
- For water management, it is assumed that the liquid water does not leave the stack and that it evaporates into the cathode or anode gas if humidity on either side drops below 100% [2].
- Humidifier and temperature controllers are assumed [3].
- The mole fractions of the inlet reactants are assumed to be constant to build the simplified dynamic PEMFC model. In other words, pure hydrogen (99.99%) is fed to the anode, and air that is uniformly mixed with nitrogen and oxygen by a ratio of, say, 21:79 is supplied to the cathode side [3].
- The ideal gas law and mole conservation rule are applied by supposing that all gases are ideal.

Therefore, the water management and temperature control is not a concern in this simulation. PEM fuel cell system composes of four blocks that are an anode model, cathode model, fuel cell voltage model, and control block (Figure 5.1).

Let us first consider the fuel cell voltage model block. The output fuel cell stack voltage V_{st} [1] is defined as a function of the stack current, reactant partial pressures, fuel cell temperature, and membrane humidity:

$$V_{st} = E - V_{activation} - V_{ohmic} - V_{concentration} \tag{5.1}$$

In the above equation,

$E = N_o \cdot \left[V_o + (R^*T/2F)\ln\left(P_{H_2}\sqrt{P_{O_2}}/P_{H_2Oc} \right) \right]$ is the thermodynamic potential of the cell or reversible voltage based on Nernst equation 5.1, $V_{activation}$ the voltage loss due to the rate of reactions on the surface of the electrodes, V_{ohmic} the ohmic voltage drop from the resistances of proton flow in the electrolyte, and $V_{concentration}$ is the voltage loss from the reduction in concentration gases or the transport of mass of oxygen and hydrogen. Their values are given as follows:

$$V_{activation} = N \cdot \frac{R^*T}{2\alpha F} \cdot \ln\left(\frac{I_{fc} + I_n}{I_o} \right) \tag{5.2}$$

$$V_{ohm} = N \cdot I_{fc} \cdot r \tag{5.3}$$

$$V_{concentration} = N \cdot m\exp(n \cdot I_{fc}) \tag{5.4}$$

In Equation 5.1, P_{H_2}, P_{O_2}, and $P_{H_2O_c}$ are the partial pressures of hydrogen, oxygen, and water, respectively. Subscript c means the water partial pressure, which is vented from the cathode side.

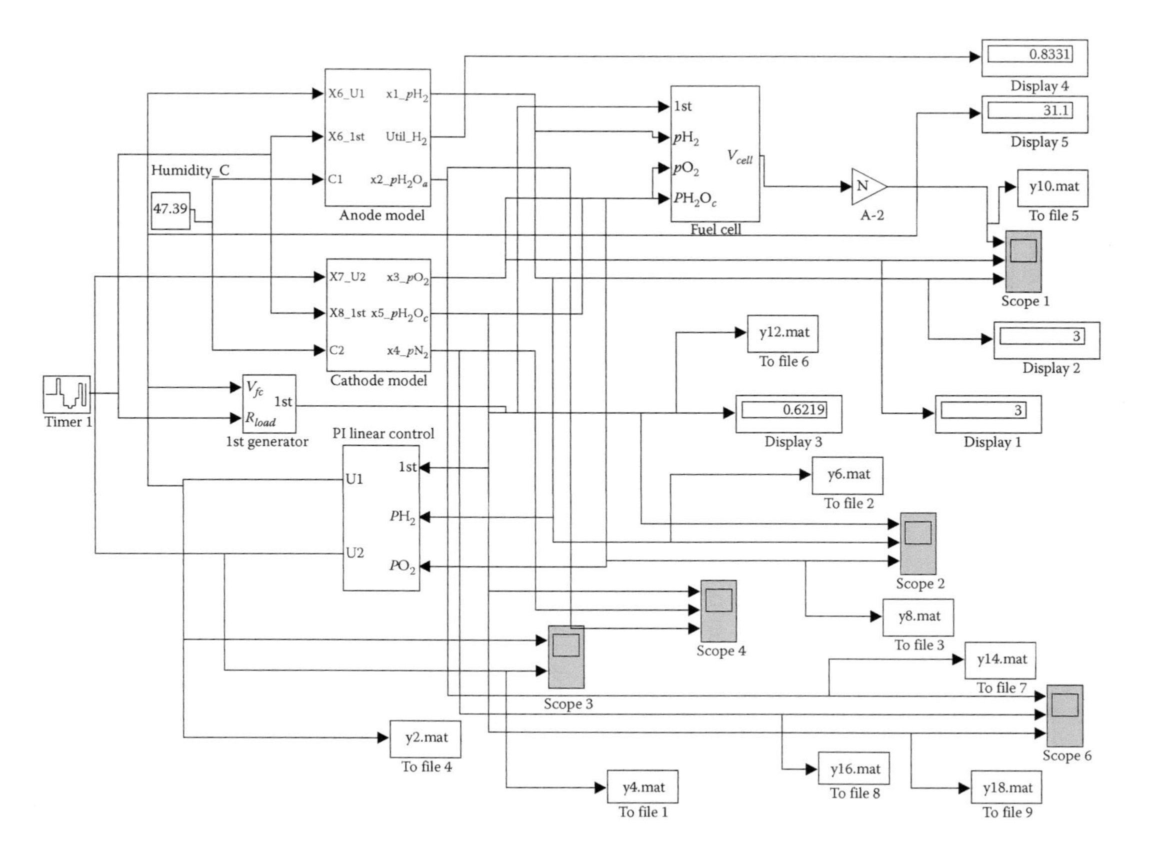

FIGURE 5.1
PEMFC system block diagram.

TABLE 5.1

Cell Voltage Parameters

Parameter	Value and Definition
N	Cell number
V_o	Open-cell voltage (V)
R	Universal gas constant (J/mol K)
T	Temperature of the fuel cell (K)
F	Faraday constant (C/mole)
α	Charge transfer coefficient
I_{fc}	Output current density (A/cm^2)
I_0	Exchange current density (A/cm^2)
I_n	Internal current density (A/cm^2)
m and n	Constants in the mass transfer voltage
r	Area-specific resistance (kΩ cm^2)

Source: Adapted from M.J. Khan and M.T. Labal, *Fuel Cells*, 4, 463–75, 2005.

The cell voltage parameters are seen in Table 5.1.

A detailed explanation of each voltage loss is given in Reference 2, and other voltages are also described in References 1 and 3, where the fuel cell voltage is mainly expressed by the combination of physical and empirical relationships, in which parametric coefficients of the membrane water content, humidity, and temperate, as well as the reactant concentrations are involved. In the book, the general voltage formulation as given in Equation 5.1 will be used because water and temperature factors are assumed to be constant due to their slow response time. According to Equation 5.1, the fuel cell voltage block is built in Figure 5.2.

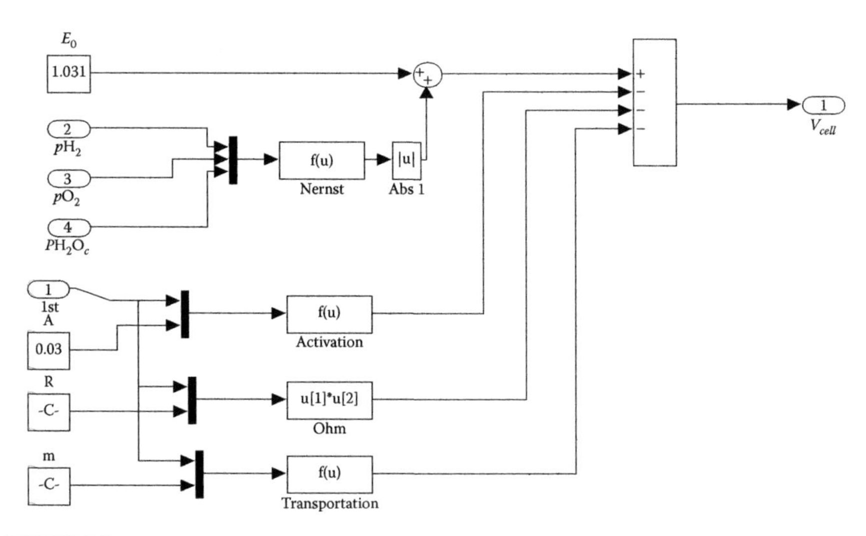

FIGURE 5.2
Fuel cell voltage model block.

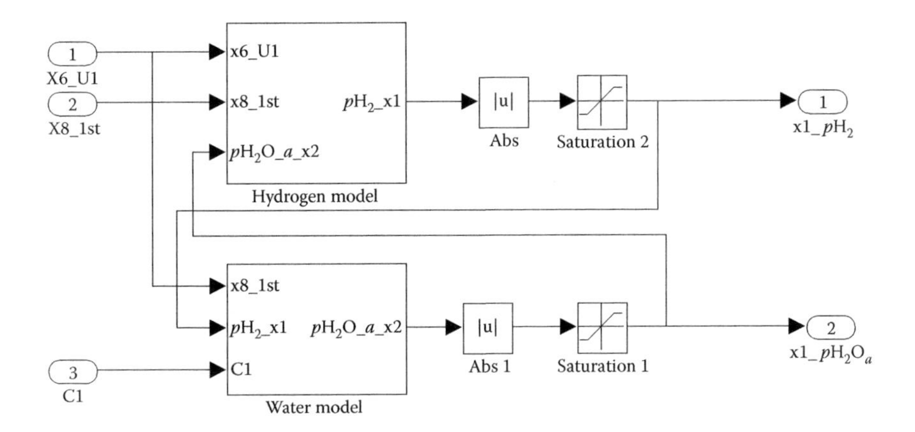

FIGURE 5.3
Anode model.

The reactant flow rates at the anode and cathode sides are determined by the partial pressures and the stack current. Using the ideal gas law, each partial pressure block can be modeled in Figure 5.3. As seen in Figure 5.3, the anode block consists of the hydrogen and water model.

There are many limiters placed in the output of each pressure and even the controller outputs to prevent problems caused by algebraic loops and extreme numerical values [1]. The detailed hydrogen and water models are shown in Figures 5.4 and 5.5, respectively.

The cathode model consists of the oxygen, water, and nitrogen models as seen in Figures 5.6 through 5.9.

In the anode block, the hydrogen inlet flow rate (slpm) is converted into mol/s using a conversion factor (1 slpm: 7.034×10^{-4} mol/s), and the mole fraction of H_2 is assumed to be 99%. In the cathode block, air is uniformly

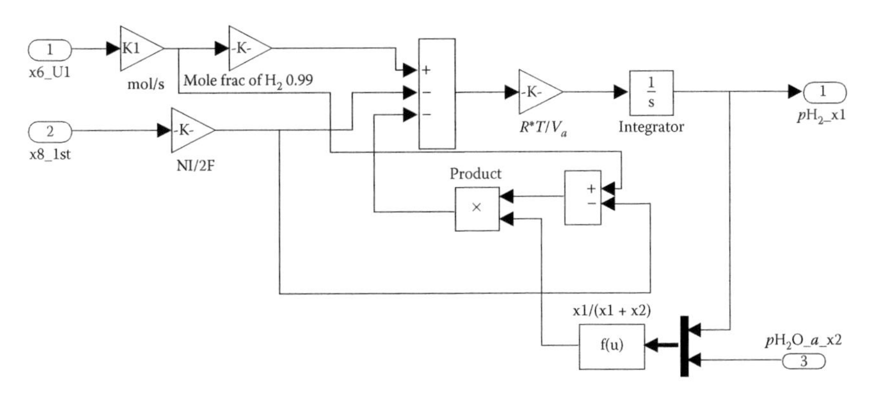

FIGURE 5.4
Hydrogen model.

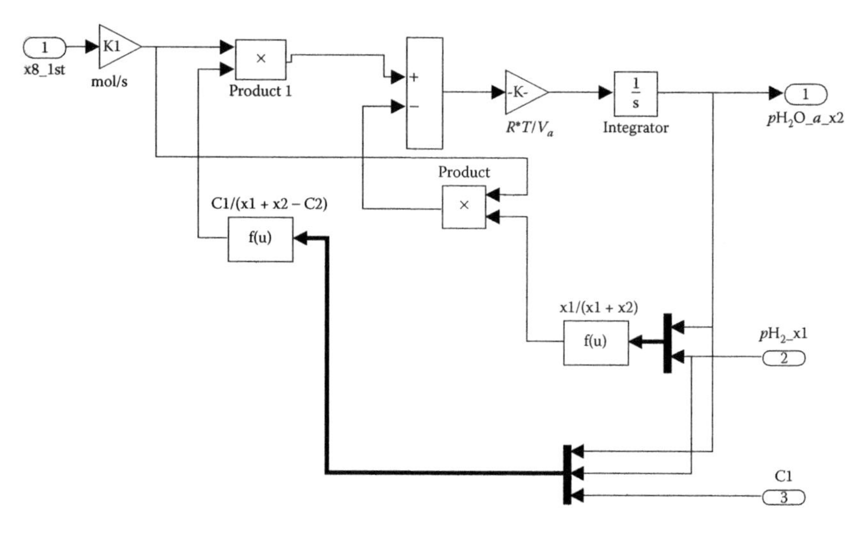

FIGURE 5.5
Water model at the anode side.

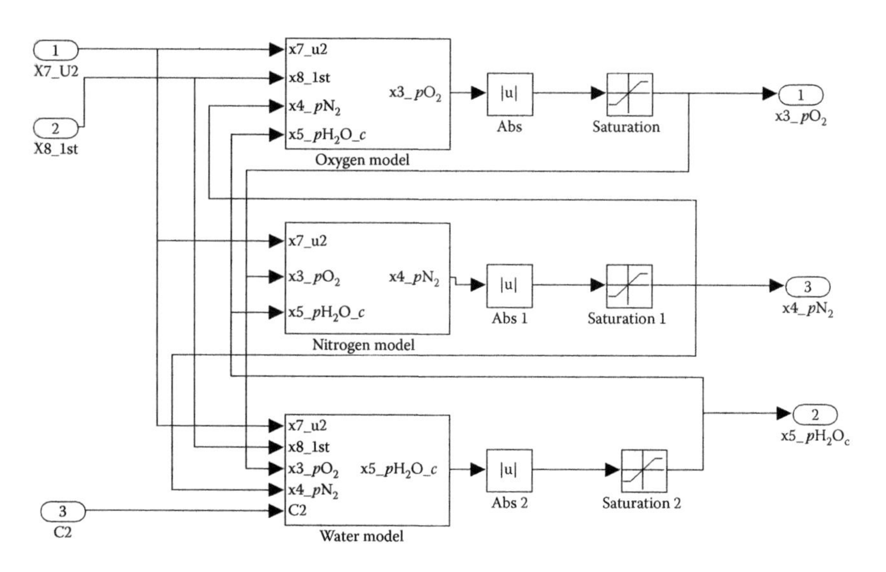

FIGURE 5.6
Cathode model.

mixed with nitrogen and oxygen by a ratio of 21:79 with a conversion factor of 7.034×10^{-4} mol/s [1]. To analyze data in a short time, a fast ordinary differential equation (ODE) solver is used. The ODE45 method is based on Dormand–Prince, which is an explicit, one-step Runge–Kutta that is recommended as a first-try method [4]. All simulation results are finished and

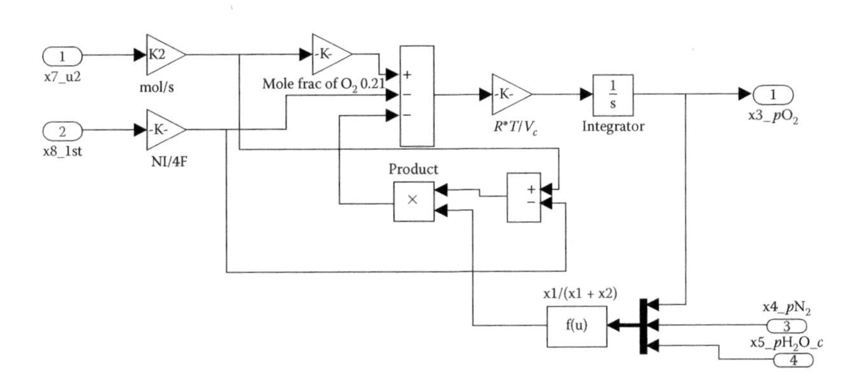

FIGURE 5.7
Oxygen model.

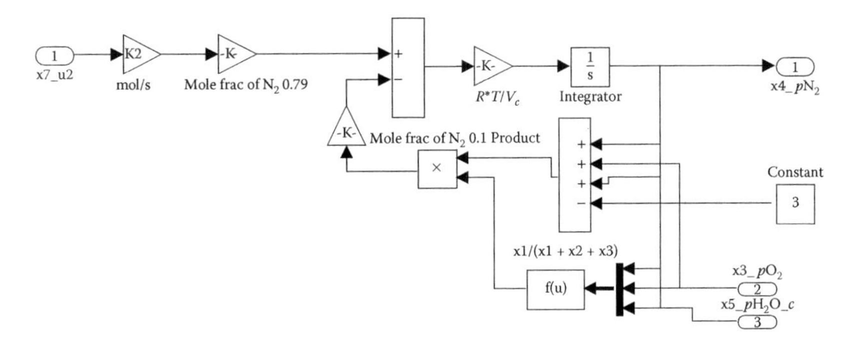

FIGURE 5.8
Nitrogen model.

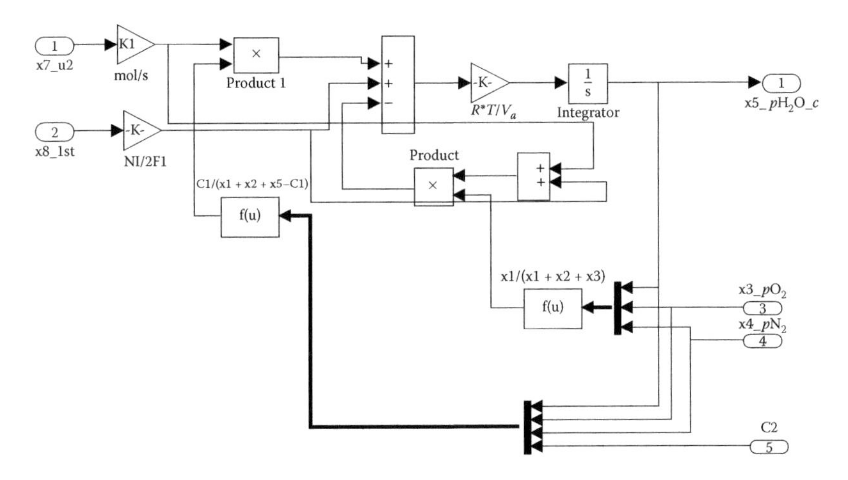

FIGURE 5.9
Water model at the cathode side.

shown within 10 s even though the setup time is 100 s. During the time duration of 0–0.5 s, the unrealistic values of the simulations can be seen due to the inadequacies of initial conditions.

The stack current is generated by the fuel cell voltage, V_{fc} and the load block, R_{load}. The load block is made by the timer, and its time matrix [0, 15.0 s, 20.0 s, 25.0 s, 30.0 s, 35.0 s, 40.0 s, 45.0 s, 50 s, 55 s] corresponds to the amplitude resistor matrix [1 Ω, 0.6 Ω, 0.3 Ω, 0.175 Ω, 0.3 Ω, 1 Ω, 2.5 Ω, 5 Ω, 1 Ω, 5 Ω].

5.3 Simulink Implementation of the Fuel Cell Controllers

For the pressure control of the fuel cell system, the hydrogen and oxygen pressure are sensed and then compared with the reference pressure, 3 atm in the linear PI controllers in Figure 5.10. In case of feed-forward control, the stack current is fed to the controller, but in this simulation, the feed-forward control is not used (Figure 5.11).

Figure 5.12 shows a PI controller for the pressure control. Through trial and error, the optimized PI gains 5 are obtained.

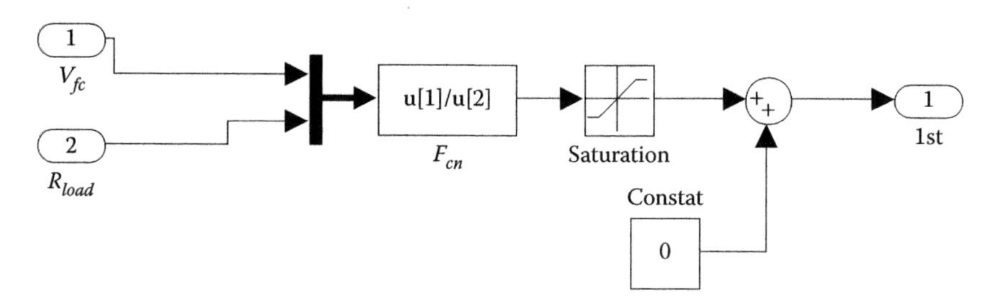

FIGURE 5.10
The load and current block.

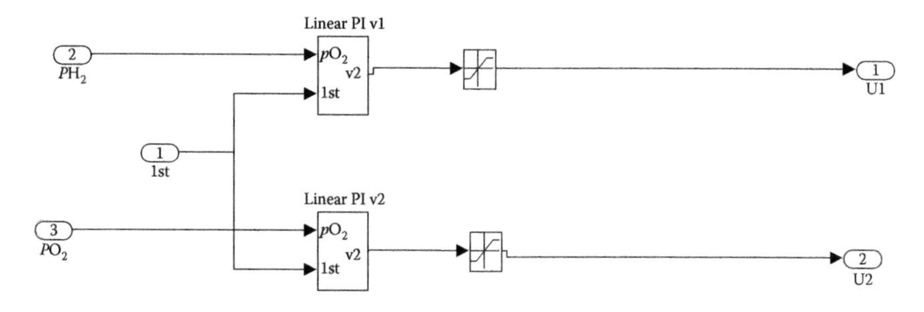

FIGURE 5.11
The linear controller block.

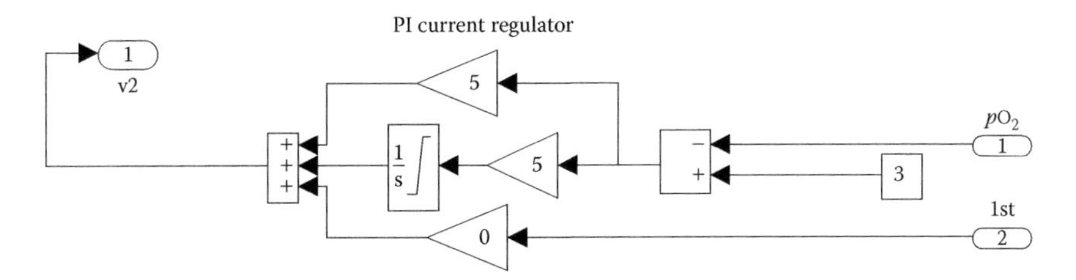

FIGURE 5.12
The linear PI controller block.

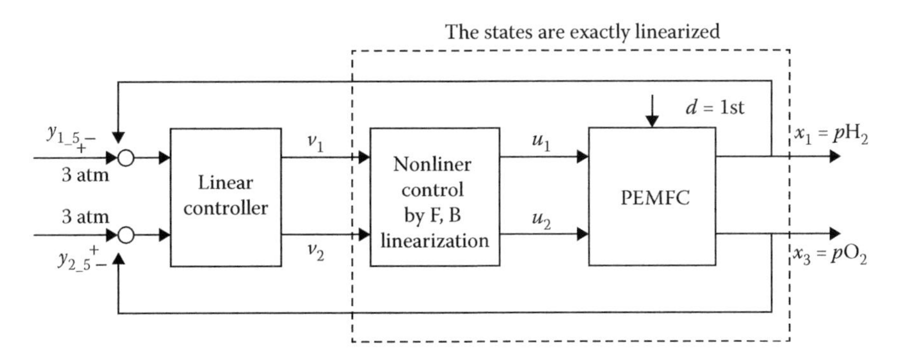

FIGURE 5.13
Nonlinear control block diagram of PEMFC.

For the design of the nonlinear pressure controller, a feedback linearization algorithm is adapted in Figure 5.13. The nonlinear control law is derived by using the decoupling matrix in Section 4.4.

The nonlinear control law vector U is obtained by feedback linearization and the linear controller block is also used in this block to generate a control input V vector in Figure 5.13 (Figure 5.14).

To implement the nonlinear controller law, the multiplexer and other math functions in the Simulink are used. A filter is used before generating the nonlinear control law, which can remove sudden variations in the anode and cathode flow rates.

5.4 Simulation Results

For the simulations, the load profile is generated using a timer in the Simulink. The details of the load profile are shown in Figure 5.15.

In terms of the fuel cell voltage control, the fuel cell voltage will be kept varying based on the load profile as in Figure 5.16, because the fuel cell

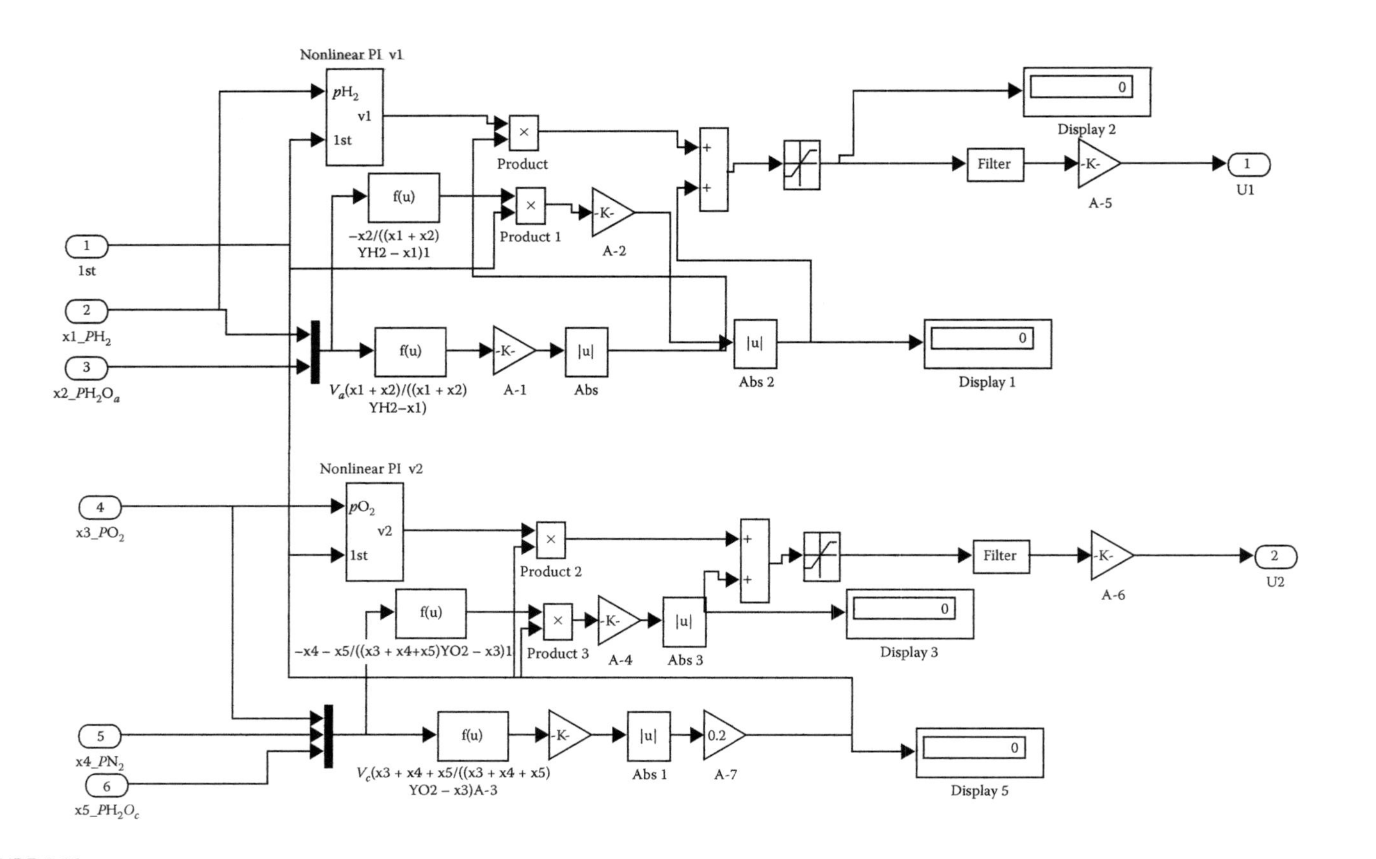

FIGURE 5.14
The nonlinear controller block using feedback linearization.

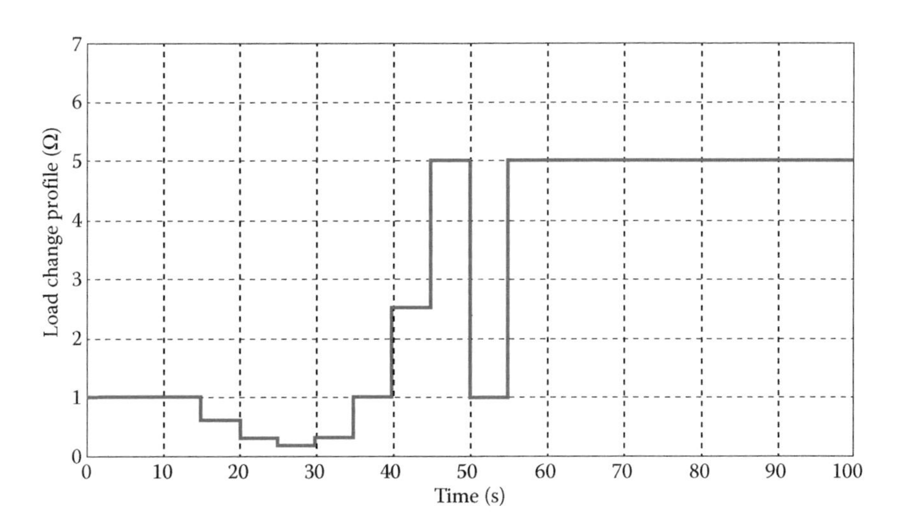

FIGURE 5.15
Load variation profile.

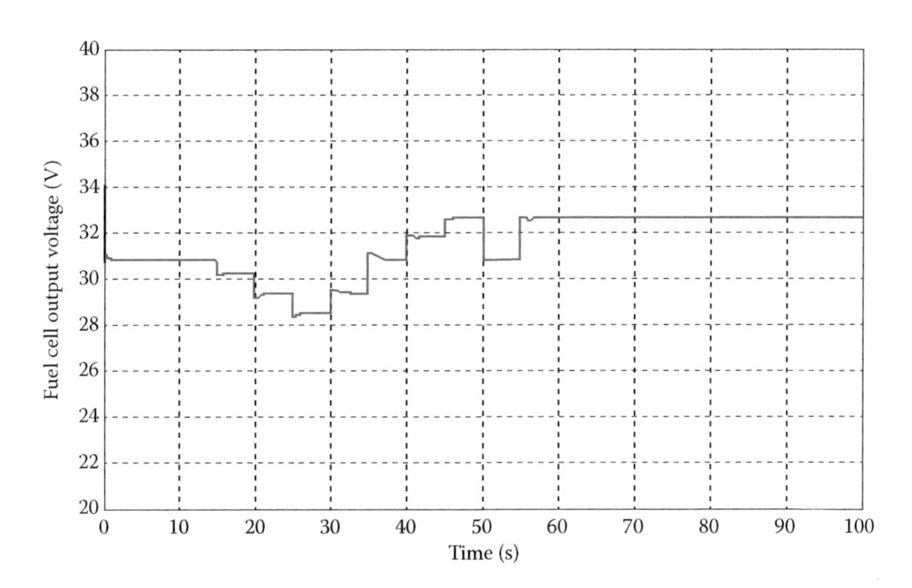

FIGURE 5.16
Fuel cell voltage under load variations.

system cannot compensate for a fast power demand such as an acceleration and deceleration or other fast load changes of the fuel cell vehicle without a secondary power source such as a battery or an ultracapacitor.

On the basis of the load change, the current varies from 6.5 to 160 A and the fuel cell system can generate up to 4.6 kW.

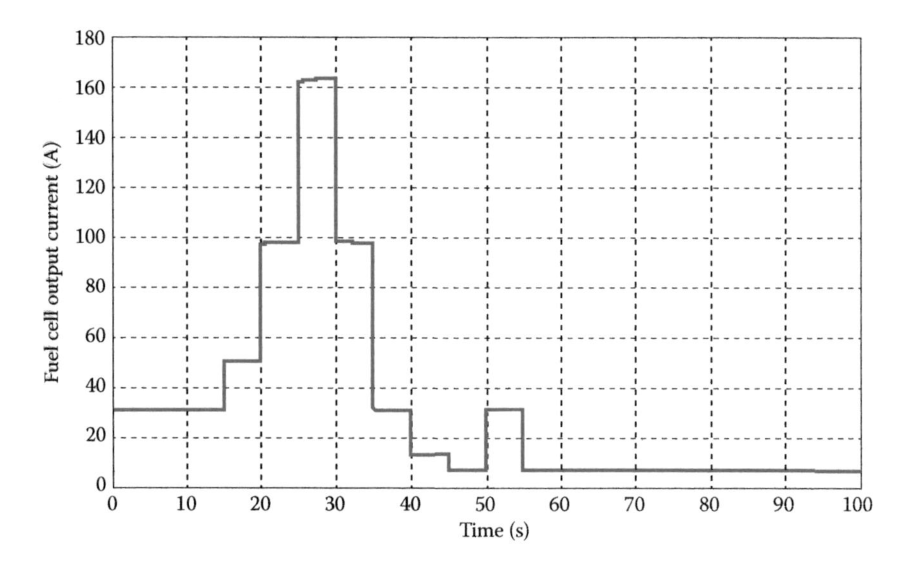

FIGURE 5.17
Fuel cell current under load variations.

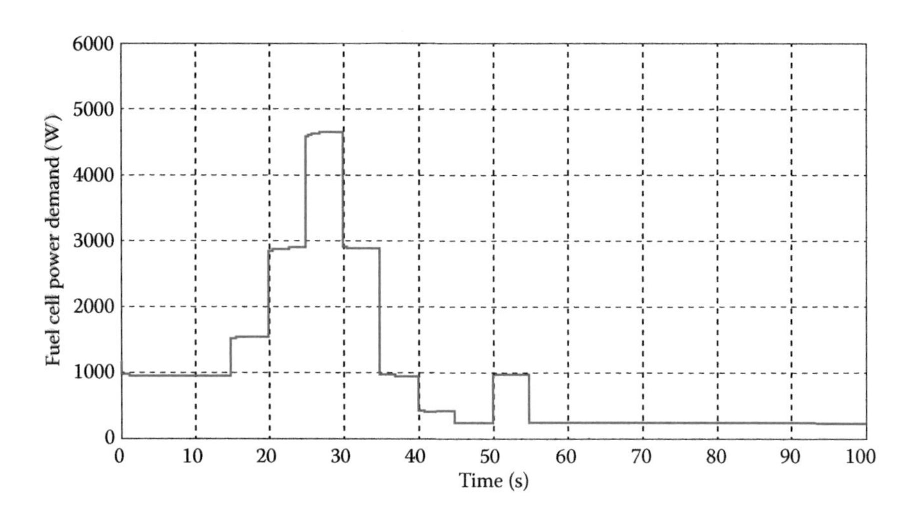

FIGURE 5.18
Fuel cell power under load variations.

For the voltage, current, and power, the discrepancies between the non-linear control and the linear control are not obvious due to the fast response time, which is less than a few milli seconds, as reported in Reference 5 (Figures 5.17 and 5.18).

In Figures 5.19 and 5.20, the hydrogen flow rate varies between 0 and 5 slpm, while the oxygen flow rate varies from 0 to 14 slpm. It can be seen

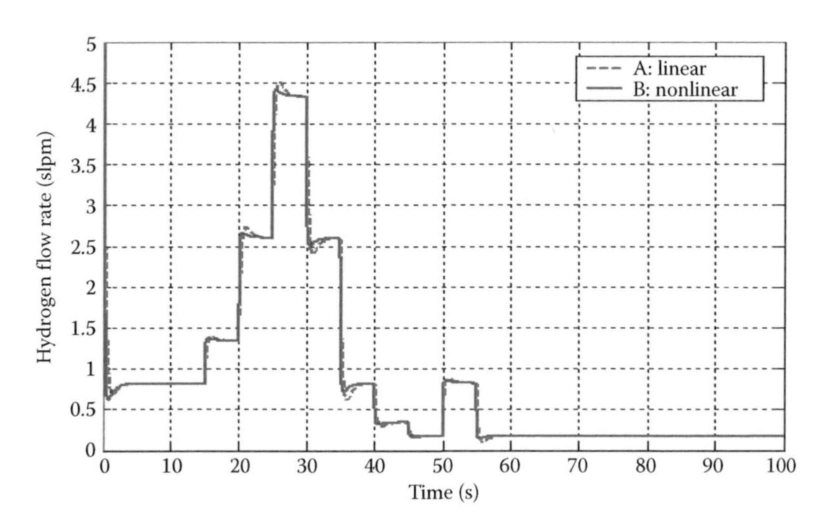

FIGURE 5.19
Variations of hydrogen flow rate.

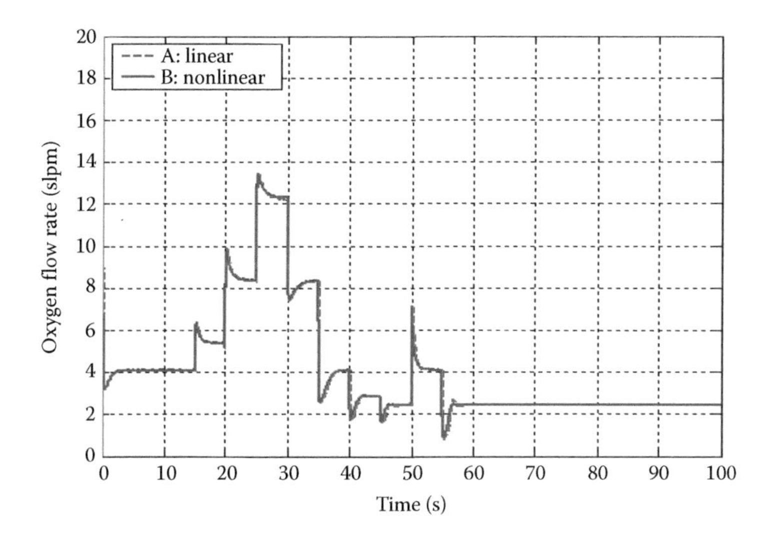

FIGURE 5.20
Variations of oxygen flow rate.

that the oxygen flow rate has much larger variations than those of the hydrogen because the oxygen flow rate is more sensitive to the load variation than the hydrogen flow rate.

According to Figures 5.21 and 5.22, the results show that the linear and nonlinear controller for the pressure of the fuel cell system keep tracking the

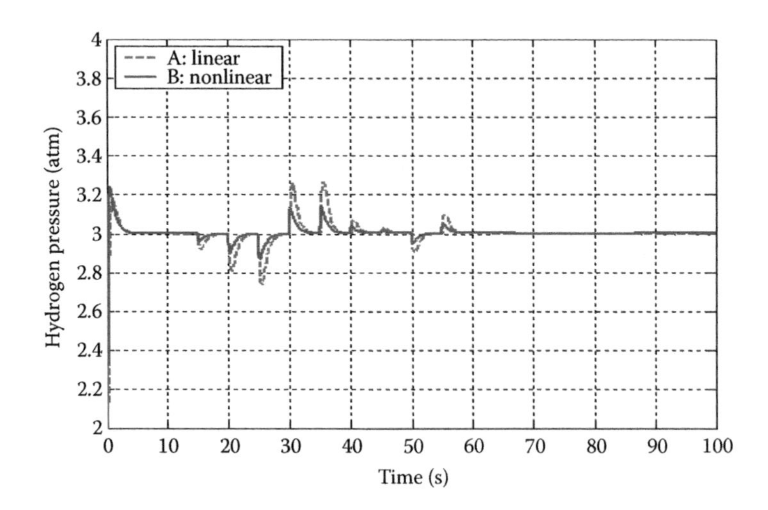

FIGURE 5.21
Variations of hydrogen pressure.

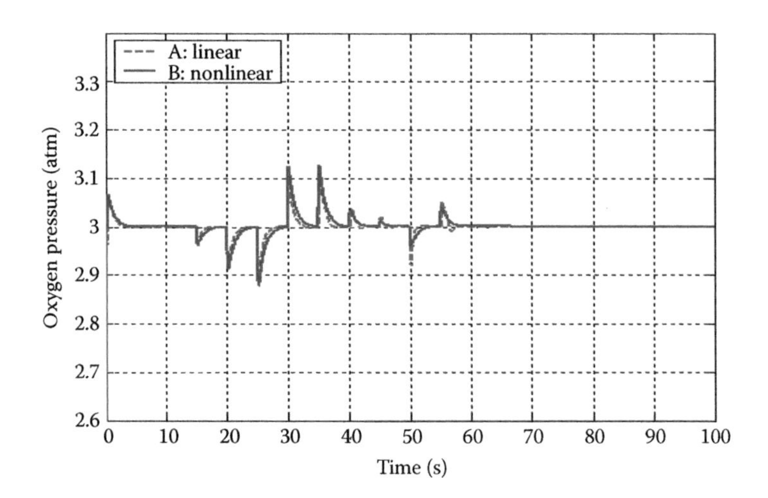

FIGURE 5.22
Variations of oxygen pressure.

reference values at 3 atm. These simulation files can be found in the attached CD. The load profile is likely to be changed and different results can be obtained based on the new load profile.

In this section, 5 kW PEM fuel cell dynamic model has been built in MATLAB–Simulink, and its linear and nonlinear controllers and simulation results have been analyzed and discussed. As seen in Figure 5.16, the fuel cell system should be coordinated with a secondary energy system to avoid a voltage drop due to a load change. To integrate this fuel cell system with a secondary energy system, power electronics devices and electrical components such as a capacitor, resistor, and reactor are needed. Since the switching devices make the simulation slower, it is easier to have a conversion error due to its high-switching frequency over 1 kHz. For example, for 30 s simulation duration, it takes over 5–6 h to finish the simulation, and this book does not deal with the hybrid power fuel cell system simulation.

References

1. M.J. Khan and M.T. Labal, Modeling and analysis of electro chemical, thermal, and reactant flow dynamics for a PEM fuel cell system, *Fuel Cells*, 4, 463–475, 2005.
2. J. Larminie and A. Dicks, *Fuel Cell Systems Explained*, Wiley, New York, 2002.
3. J. Purkrushpan and H. Peng, *Control of Fuel Cell Power Systems: Principle, Modeling, Analysis and Feedback Design*, Springer, Germany, 2004.
4. C-M. Ong, *Dynamic Simulation of Electric Machinery using Matlab/Simulink*, Prentice-Hall, New Jersey, 1998.
5. C.A. Anderson, M.O. Christensen, A.R. Korsgaad, M. Nielsen, and P. Pederson, Design and control of fuel cells system for transport application, Aalborg University, Project Report, 2002.

6

*Applications of Fuel Cells in Vehicles**

6.1 Introduction

Fossil fuels including coal, oil, and gas, which we heavily depend on, can cause air pollution and greenhouse gas problems. A recent study [1] shows that about 18% of CO_2 (carbon dioxide), greenhouse gas, is emitted by motor vehicles. The development of fuel cell powered vehicles (FCVs) is very important to our environment and even our economy, especially for a soaring oil price currently. The fuel cell system is widely regarded as one of the most promising energy sources thanks to its high-energy efficiency, extremely low emission of oxides of nitrogen and sulfur, very low noises, and the cleanness of its energy production. Furthermore, a fuel cell system can operate with other conventional and alternative fuels such as hydrogen, ethanol, methanol, and natural gas. On the basis of the currently used types of electrolytes, fuel cells are classified into PEMFCs, SOFCs, PAFCs, MCFCs, AFCs, DMFCs, ZAFCs, and PCFCs [1].

To date, polymer electrolyte membrane fuel cells (PEMFCs), known as PEMFC, has been considered as the most promising candidate for the fuel cell vehicle (FCV) and small- and mid-size distributed generators because of its high-power density, a solid electrolyte, long stack life, and low corrosion. PEMFCs can operate at low temperatures (50–100°C), which enables a fast start-up. Hence, PEMFCs are particularly attractive for transportation applications that require rapid start-ups and fast dynamic responses over transient times (stopping and running, acceleration, and deceleration).

The fuel cell has a larger efficiency than internal combustion engines (ICEs). It is noted in Reference 2 that the efficiency of an FCV using direct hydrogen from the natural gas is 2 times greater than that of an internal combustion engine (ICE) vehicle. However, for the commercialization of FCVs, their performance, reliability, durability, cost, fuel availability and cost, and public acceptance should be considered [3]. Especially, the performance of the fuel cell systems during transients, the cost of fuel cells, and availability of hydrogen for the success of commercialization of FCVs are the

* This chapter was mainly prepared by Dr. Woonki Na, California State University, Fresno and Dr. Bei Gou, Smart Electric Grid LLC, USA.

critical issues. In this chapter, FCV components are discussed in Section 6.2. A hybrid fuel cell system design and control for electric vehicles are presented in Sections 6.3 and 6.4, respectively. The fault diagnosis of a hybrid fuel cell system is discussed in Section 6.5.

6.2 FCV Components

Fuel cell powered vehicle (FCV) uses the hydrogen fuel as the major source of electric power to drive its electric traction motor through a fuel cell system. There are three major components in the FCV system. The first is the fuel cell and fuel cell subsystem; the second is the hydrogen storage or the fuel processor; and the third is the electric drive system.

6.2.1 Fuel Cell and Fuel Cell Subsystem

Since the principle of fuel cell operation has already been covered in Chapter 2, this section will mainly focus on the fuel cell subsystems—gas flow management, water management, and heat management system.

6.2.1.1 Gas Flow Management Subsystem

Oxygen and hydrogen are fed to the fuel cell system at an appropriate rate according to the current drawn from the load. Oxygen on the cathode side is often supplied with a higher stoichiometric flow rate because the cathode reaction is much slower than the anode reaction. Since the fuel cell voltage is heavily relying on the air stoichiometric flow rate, the role of a compressor is very important to blow more air through the system. Another issue is related to the pressure control between the anode and cathode sides, which is to prevent the membrane from collapsing. With the help of a fast actuator on the cathode side whose performance-setting time is 10%–90% of 50 ms, and another actuator on the anode side whose performance-setting time is 10%–90% of less than 20 ms, it becomes possible to maintain the anode and cathode pressures at a certain level. To ensure this performance, the anode pressure controller must be three times faster than the one on the cathode side because the anode pressure follows the cathode-side pressure [4]. An air compressor with an optimal speed can supply varying amounts of air for the power demand. Since an electric motor driving the compressor produces parasitic power loss, the compressor must be coordinated with the power management system in the fuel cell system to reduce the power loss. For the direct hydrogen fuel cell system, hydrogen needs to be controlled at 100% relative humidity by the humidifier before entering the anode side of the fuel cell stack for the proper operation. The recirculation of hydrogen is needed to improve the utilization of hydrogen. To avoid gas accumulation of inserts and impurities, purging the valve of hydrogen is also required.

6.2.1.2 Water Management Subsystem

To achieve high efficiency of a PEM fuel cell, water management is crucial to the fuel cell system. Water management system has three main functions:

1. Proper hydration of the membrane
2. Removal of the produced water in the stack
3. Cooling of the stack to control the operating temperature

First, the proper hydration of the membrane is related to the membrane conductivity, which greatly affects the stack performance at the high-power-density operation. Second, the proper removal of the produced water in each cell can keep the airflow passage from being blocked by the accumulated water and therefore avoid the degrading of the cell performance. For the cooling of the system, it is closely related to the temperature control, which will be considered in the following section.

6.2.1.3 Heat Management Subsystem

The heat management, one of the important factors for the fuel cell mechanism and the warranty to maintain the stack temperature to the desired level, is directly related to the fuel cell performance [1,5,6].

Figure 6.1 shows that the polarization curve is shifted upward as the temperature increases. A number of control-oriented fuel cell models [4–21] have been developed under the assumption that the operating temperature is constant. Although controlling the temperature is vital for the fuel cell

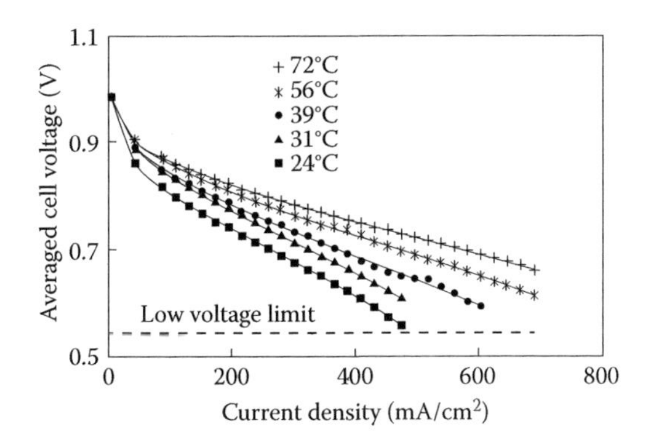

FIGURE 6.1
Polarization curves for different temperatures. (Adapted from J. Hamelin et al., *International Journal of Hydrogen Energy*, 26, 625–629, 2001.)

operation, those models proposed in References 4–21 do not consider the temperature as a state variable due to the complexity of temperature consideration. In Reference 22, the temperature is defined as one of the state variables and a control strategy is developed based on the transient thermal model of PEMFC [11,12,15].

Several thermal models of PEMFC have been reported in References 11, 12, and 15. However, these models are not proposed for the purpose of a control design, but the mathematical analysis and its experimental validation are proposed instead. On the basis of the transient thermal models in References 11, 12, and 15, the control-oriented dynamic thermal model is developed in this section.

A transient energy balance is described by

$$\dot{Q}_{stack} = C_t \frac{dT_s}{dt} = P_{tot} - P_{elec} - \dot{Q}_{cool} - \dot{Q}_{loss} \tag{6.1}$$

where $\dot{Q}_{stack}$ is the rate of heat absorption (J/s) by the stack, C_t the thermal capacitance (J/C), P_{tot} the total power released by a chemical reaction (W), P_{elec} the power consumed by the load (W), $\dot{Q}_{cool}$ the heat flow rate of the cooling system (heat exchanger), and $\dot{Q}_{loss}$ is the heat flow rate through the stack surface.

The total energy can be calculated using the rate of hydrogen consumption

$$P_{tot} = \dot{m}_{H_2\_used} \Delta H = \frac{NI_{fc}}{2F} \Delta H \tag{6.2}$$

where ΔH is the enthalpy change for hydrogen (285.5 kJ mol/s) and $\dot{m}_{H_2\_used}$ is the hydrogen consumption rate.

The electrical power output is given by

$$P_{elec} = V_{st} I_{fc} \tag{6.3}$$

The rate of heat removal by the cooling water is directly related to the water flow in the heat exchanger. The relationship is given as follows [4]:

$$\dot{Q}_{cool} = \dot{m}_{cool\_water} \cdot c_p \cdot \Delta T_s \tag{6.4}$$

where $\dot{m}_{cool\_water}$ is the water pump flow (SLPM), c_p the specific heat coefficient of water (4182 J/kg K), and ΔT_s is the allowable temperature rise (10 K). The water pump flow can be described by the time delay and conversion factor

$$\dot{m}_{cool\_water} = \frac{k_c}{(1+\tau_c s)} u_{cl} \tag{6.5}$$

where τ_c is the time delay constant, 70 s, k_c the conversion factor with a value of 1.5, which means that if the control input u_{cl} for the heat exchanger ranges in 0 ~ 10 (V), $\dot{m}_{cool\_water}$ takes values in the range of 0 and 15 (SLPM) with 70 s delay.

The heat loss by the stack surface is calculated by the following equation:

$$\dot{Q}_{loss} = hA_{stack} \cdot (T_s - T_{amb}) = \frac{T_s - T_{amb}}{R_t} \tag{6.6}$$

where hA_{stack}, the stack heat-transfer coefficient, is 17 W/K [11,12,15], T_{amb} the ambient temperature of 25°C (298.15 K) ± 5%, and R_t is the thermal resistance of the stack, which is the reciprocal of hA_{stack} is 0.0588 K/W.

The thermal time constant of the fuel cell is given by

$$\tau = R_t C_t = \frac{MC}{hA_{stack}} \tag{6.7}$$

where MC is the product of the stack mass and average specific heat, 35 kJ/K, the thermal capacitance C_t is 35 kJ/K, and τ is 2059 s [10].

Equation 6.7 shows that the thermal equivalent circuit model can be developed by using the circuit analogy [17]

$$C\frac{dv}{dt} = i \tag{6.8}$$

where the stack temperature corresponds to voltage, energy flows (P_{tot}, P_{elec}, $\dot{Q}_{cool}$, and $\dot{Q}_{loss}$) correspond to current, and the thermal capacitance C_t corresponds to the capacitance.

In Figure 6.2, load variations cause the changes of the fuel cell stack temperature, and the fuel cell voltage and current are used as external inputs or disturbance, and then the stack temperature can be used as an output when designing the controller. The total power, P_{tot} depends on the hydrogen consumption that is based on the load changes and the electrical power, while

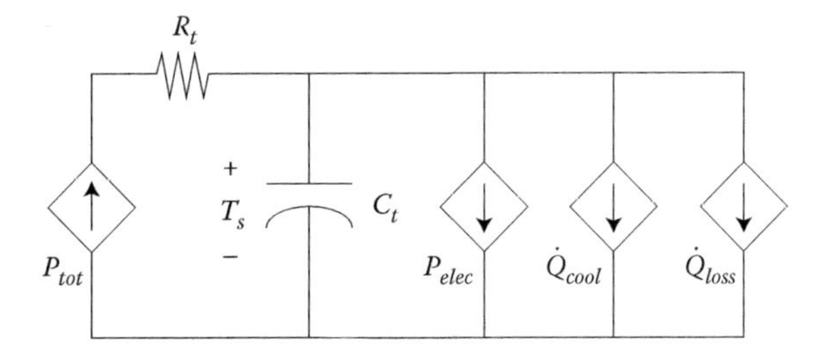

FIGURE 6.2
The thermal equivalent circuit of the fuel cell.

P_{elec} depends on the load current. The cooling power $\dot{Q}_{cool}$ and the heat loss by the stack surface $\dot{Q}_{loss}$ are functions of the fuel cell stack temperature change. For this reason, four dependent energy flow sources have to be considered in the thermal equivalent circuit of the fuel cell shown in Figure 6.2.

6.2.2 Hydrogen Storage and Fuel Processor

The PEMFC needs to be supplied with hydrogen, which is the most abundant resource on Earth. It is a colorless and odorless gas, and an ideal energy source because it has a very high-energy density for its weight compared to an equivalent amount of gasoline, which means hydrogen generally produces much more energy than gasoline. For direct hydrogen fuel cell system, compressed gas or cryogenic liquid-storage tank, or metal hydride energy storage (ES) needs to be installed in the vehicle.

The most common way to store hydrogen is to simply compress it in cylinders with high pressure to increase its density. The major concerns of compressed storage are the required large volume and the weight of gas containers, in which it is normally made from steel alloy. An aluminum body can be possible to make the body of a cylinder, but it is easy to be broken and currently expensive. Typically, storage pressures are between 200 and 450 bars (3000 or 6000 psi) [6]. In practice, storage densities are between 3% and 4% hydrogen. The volume of these storage tanks ranges from 30 to 300 L [6].

At 20 K (temperature) and 0.5 MPa (pressure) vapor pressure, liquid hydrogen can be obtained. Cryogenic liquid hydrogen storage can be used if a large amount of hydrogen is needed. This cryogenic technology has been demonstrated in vehicle applications by BMW, and shows that it can reach a storage efficiency of 14.2%. The main concerns of this technology are to maintain hydrogen at such low temperatures and to minimize hydrogen boil off.

Metal hydride, in which metal atoms bond with hydrogen, is another way to store hydrogen. To release hydrogen from metal hydride, temperature over 100°C is required and therefore, it is not a good option for low-temperature PEM fuel cell application. Even though metal hydride can take a large amount of hydrogen per unit volume, it has the problem of high alloy cost and low gravimetric hydrogen density, and it is heavy and sensitive to gas impurities.

Hydrogen can also be produced from hydrocarbon fuel such as natural gas, gasoline, or methanol through chemical processes: steam reformation, partial oxidation, and autothermal reformation [6]. A fuel processor is required for these chemical processes. Using a fuel processor in vehicle applications causes another problem of slow dynamics due to the slow heat-transfer processes, mass transfer, and mixing delays [23]. To compensate for the slow dynamics, a battery or UC is necessary for the ES during start-up, acceleration, and deceleration in the vehicle operations.

For the example of the practical fuel cell system using a direct hydrogen tank, Ballard MK5-E-based PGS-105B system shown in Figure 6.3 is used to test the proposed nonlinear controller. This Ballard MK-5E system has a

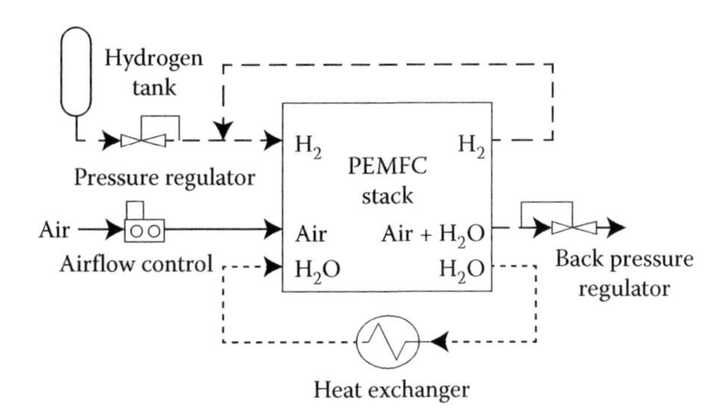

FIGURE 6.3
PEMFC stack based on PGS-105B system. (Adapted from J. Hamelin et al., *International Journal of Hydrogen Energy*, 26, 625–629, 2001.)

total of 35 cells, connected in series with a cell surface area of 232 cm^2. The membrane electrode assembly consists of a graphite electrode and a Dow™ membrane. The reactant gases (hydrogen and air) are humidified inside the stack and the hydrogen is recirculated at the anode while the air is flowing through the cathode. The hydrogen pressure is regulated to 3 atm at the anode inlet by a pressure regulator and the Ballard fuel cell stack keeps at 3 atm through a back-pressure regulator at the air outlet. The oxidant flow rate is automatically adjusted to a constant value of 4.5 L/s for a programmable load, via a mass flow meter to ensure sufficient water removal at the cathode. Hydrogen is replenished at the same rate as it is consumed. The stack temperature measured at the air outlet is maintained at a range of 72–75°C by an internal water to produce the maximum power output. The simplified diagram of the Ballard system is depicted in Figure 6.3. PGS105 system is more likely to be a presetting control system instead of a feedback control system. If it is out of the ranges of setting, the system is automatically shutdown. The safety concern for the fuel cell system and human being operating this system must be handled with a high priority before a control scheme is implemented. For instance, a low cell voltage, stack overload, high temperature, hydrogen-leaking detector, and pressure difference between the anode and the cathode must be considered in the fuel cell control system to prevent a fatal accident and severe damage to the fuel cell system.

6.2.3 Electric Drives Subsystem

An FCV is an "electric drive" vehicle. The rotating torque of the traction motor in the FCV is powered by electricity generated by the fuel cell. An induction motor, permanent magnet (PM) synchronous motor, and switched reluctance motors can be used for FCV applications [23]. More details about the control and integration of the electric drives system in FCV will be explained in the following sections.

6.3 Hybrid Electric Vehicles and Fuel Cell System Design for Electric Vehicles

Today, a combustion engine/hybrid electric vehicles (HEVs) such as the Toyota Prius and Honda Civic Hybrid are very successful. However, HEV still heavily depends on oil. FCV will therefore be the next wave of electric drive vehicles after HEV, even though the issues of hydrogen-refueling infrastructure and high cost of PEMFC have to be solved before commercialization. Before presenting the FCV architecture, it is necessary to describe hybrid vehicle architectures for a better understanding of the FCV technology because, generally, FCV also belongs to HEVs. Since a hybrid vehicle uses an ICE and an electric motor/generator, there are several series and parallel configurations. These hybrid vehicle technologies can increase the efficiency and fuel economy through the use of regenerating energy during braking and storing energy from the ICE coasting.

6.3.1 Series HEVs

In series hybrid vehicles, the ICE mechanical output is first converted into electricity by a generator and the onboard battery charger uses dynamic braking energy from the motor to charge the battery. So, the converted electricity either charges the battery or can bypass the battery to propel the wheels via the same electric motor and mechanical transmission.

In series hybrid vehicles, ICE is operated at an optimal speed and torque according to its speed–torque characteristics because it is fully decoupled from the driven train wheel. Hence, low fuel consumption and high efficiency can be achieved. The disadvantage of the series hybrid vehicle is that it has a low efficiency compared to other parallel and combination configurations because it has two energy conversion stages (ICE/generator and generator/motor) (Figure 6.4).

There are six possible different operation modes in a series HEV [25]:

1. *Battery-alone mode:* The engine is off, and the vehicle is powered only by the battery
2. *Engine-alone mode:* Power from ICE/G
3. *Combined mode:* Both the ICE/G set and battery provide power to the traction motor
4. *Power-split mode:* ICE/G power split to drive the vehicle and charge the battery
5. Stationary-charging mode
6. Regenerative-braking mode

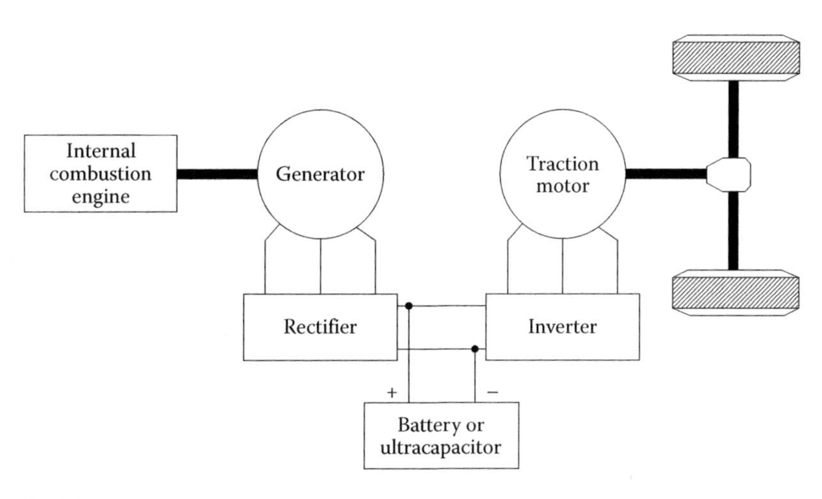

FIGURE 6.4
Series hybrid vehicle configuration. (Adapted from A. Emadi, S.S. Williamson, and A. Khaligh, *IEEE Transactions on Power Electronics*, 21(3), 567–577, 2006.)

6.3.2 Parallel HEVs

As shown in Figure 6.5, ICE and the motor are connected in parallel to drive the wheels in the parallel HEV. It allows to choose three options: (1) both ICE and the motor, (2) ICE alone, and (3) motor alone.

Normally at low speeds, the motor-alone drive strategy is preferred because it has a high efficiency than the ICE and uses the ICE at a high speed.

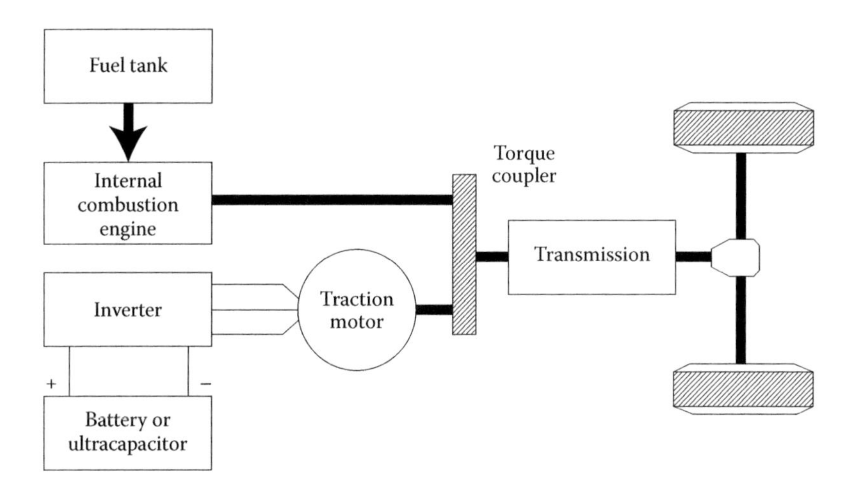

FIGURE 6.5
Parallel hybrid vehicle configuration. (Adapted from A. Emadi, S.S. Williamson, and A. Khaligh, *IEEE Transactions on Power Electronics*, 21(3), 567–577, 2006.)

Similar to a series HEV, during regenerative braking, the electric motor can be used as a generator to charge the battery and this battery can also be absorbed power from the ICE when the output power is greater than the required power.

The advantage of the parallel HEV is that it has fewer energy conversion stages than the series HEV, which leads to less power loss in the parallel HEV. Another advantage over the series HEV is that a smaller ICE and a smaller motor can be possible as long as enough battery energy is provided. The following are the possible different operation modes of parallel HEVs [25]:

1. *Motor-alone mode:* The engine is off, and the vehicle is powered only by the motor
2. *Engine-alone mode:* The vehicle is propelled only by the engine
3. *Combined mode:* Both the ICE and motor provide power to drive the vehicle
4. *Power-split mode:* ICE power is split to drive the vehicle and charge the battery (the motor becomes a generator)
5. Stationary-charging mode

6.3.3 Series–Parallel HEVs

To achieve both advantages of the series and parallel HEVs, their combination is considered in Figures 6.6 and 6.7. An additional mechanical link between the generator and the electric motor can be seen in Figure 6.6. Although the

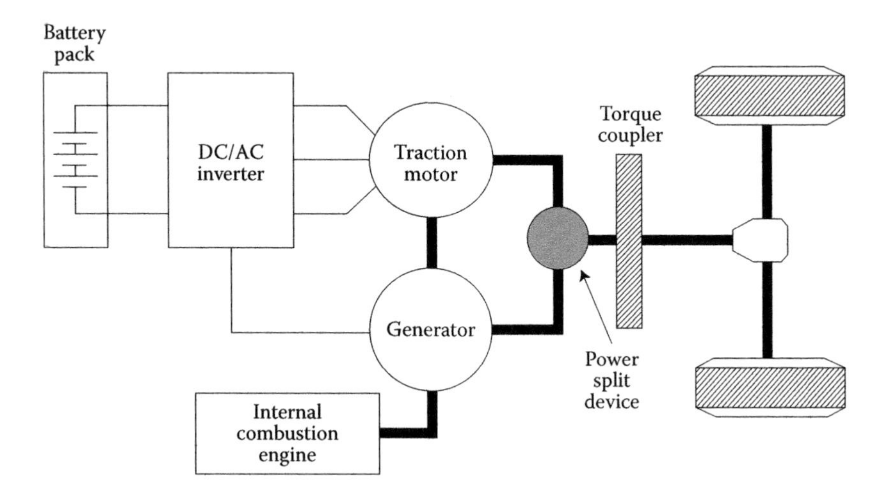

FIGURE 6.6
Series–parallel hybrid vehicle configuration. (Adapted from A. Emadi, S.S. Williamson, and A. Khaligh, *IEEE Transactions on Power Electronics*, 21(3), 567–577, 2006.)

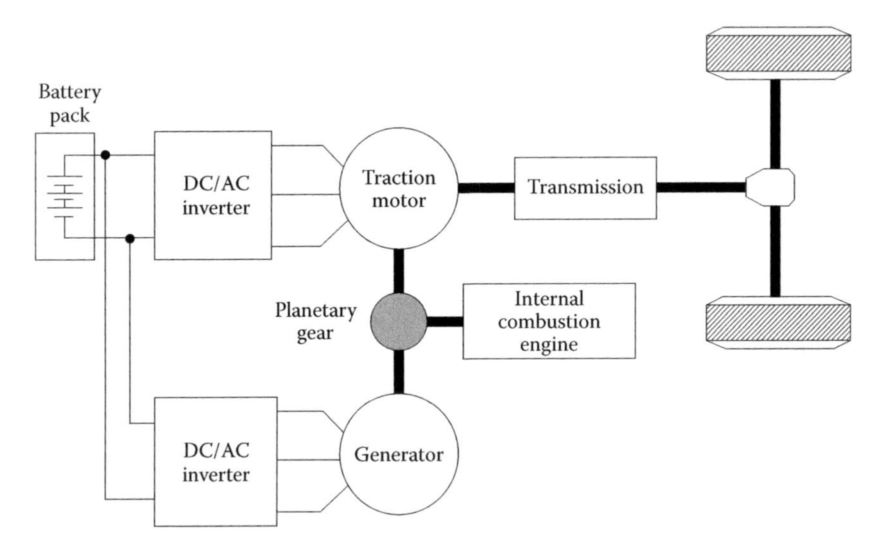

FIGURE 6.7
Complex hybrid vehicle configuration. (Adapted from A. Emadi, S.S. Williamson, and A. Khaligh, *IEEE Transactions on Power Electronics*, 21(3), 567–577, 2006.)

complex hybrid in Figure 6.7 is similar to the series–parallel HEV, the main difference is that the complex hybrid has a bidirectional power flow of the generator and the motor, but the series and parallel HEV has a unidirectional power of the generator.

Even though these drive strategies suffer from a higher complexity and cost problems, many hybrid vehicles still use the combination configurations to adopt the advantages of series and parallel HEVs.

The following are the possible different operation modes of complex HEVs:

1. *Motor-alone mode:* The engine is off, and the vehicle is powered only by the motor
2. *Engine-alone mode:* The vehicle is propelled only by the engine
3. *Combined mode:* Both the ICE and motor provide power to drive the vehicle
4. *Power-split mode:* ICE power is split to drive the vehicle and charge the battery (the motor becomes a generator)
5. Generating mode (if a battery is necessary to be charged, the generator is in the active mode)

6.3.4 Fuel Cell Vehicle

As seen in Figures 6.8 and 6.9, FCV can be considered as a series-type hybrid vehicle. A typical hybrid FCV is shown in Figures 6.8 and 6.9 when the direct

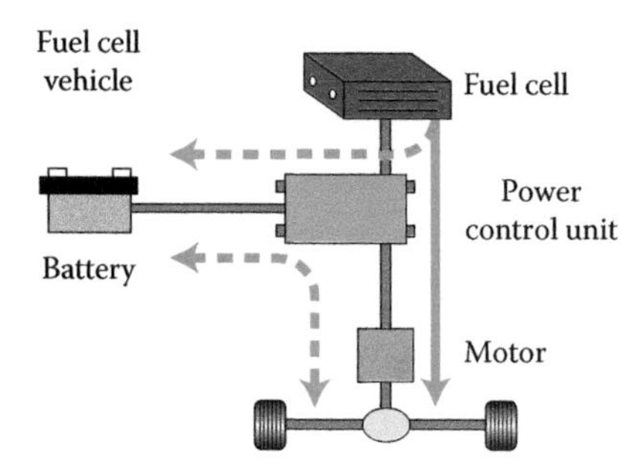

FIGURE 6.8
Hybrid fuel cell vehicle configuration (Toyota).

hydrogen supply to a fuel cell stack is employed. The hydrogen gas that feeds the fuel cell stack is stored in high-pressure tanks in the vehicle. The flow rates of hydrogen and oxygen vary according to the current drawn in the stack. The fuel cell output is fed to the power control unit (power converters) to control the traction motor and the distribution of electric power from the fuel cell secondary ES (a battery or an UC). Secondary ES not only provides additional power to the motor when it is accelerating, but also stores the power generated by braking. The traction motor generates the force to propel the vehicle.

Power conditioners in the power control unit must have minimal losses and higher efficiency up to 80%. This section discusses how to design and

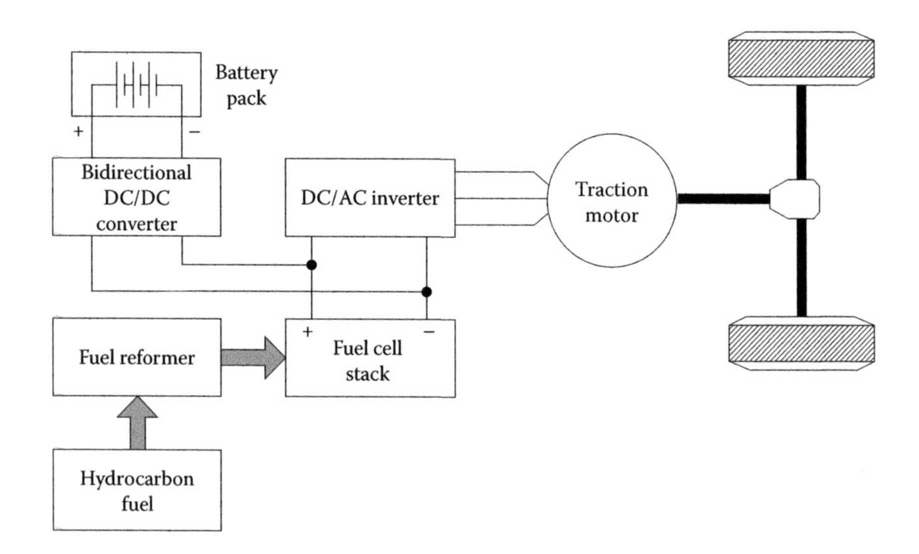

FIGURE 6.9
Typical hybrid fuel cell vehicle configuration. (Adapted from A. Emadi, S.S. Williamson, and A. Khaligh, *IEEE Transactions on Power Electronics*, 21(3), 567–577, 2006.)

coordinate the energy management systems including ES, electric motors, electric motor controller/inverter, and auxiliaries for FCV.

6.3.4.1 Energy Management Systems for FCVs

A detailed schematic diagram of the energy management system of the FCV is shown in Figure 6.10.

The fuel cell voltage is transferred to a higher level using a DC–DC boost converter. The output of the boost converter is fed to a three-phase DC–AC inverter. Because the traction motor needs AC voltage, a DC–AC inverter converts the voltage from DC to AC with three-phase variable voltage/frequency. The fuel cell power is also provided to the load of the electric-driven components in the BOP systems such as pumps, fans, blowers, actuators, and so on through another DC–DC converter in Figure 6.10. The fuel cell system needs to be equipped with a battery system to start-up or to supply peak load demand. The battery or an UC, serving as the secondary power source, can be used for both load leveling and regenerative brake energy capture. This secondary power source can be charged during the steady states from the fuel cell system and be discharged during the transient period when the fuel cell system responds to sudden load changes. The charging and discharging processes are carried out by the battery discharge/charge unit through the

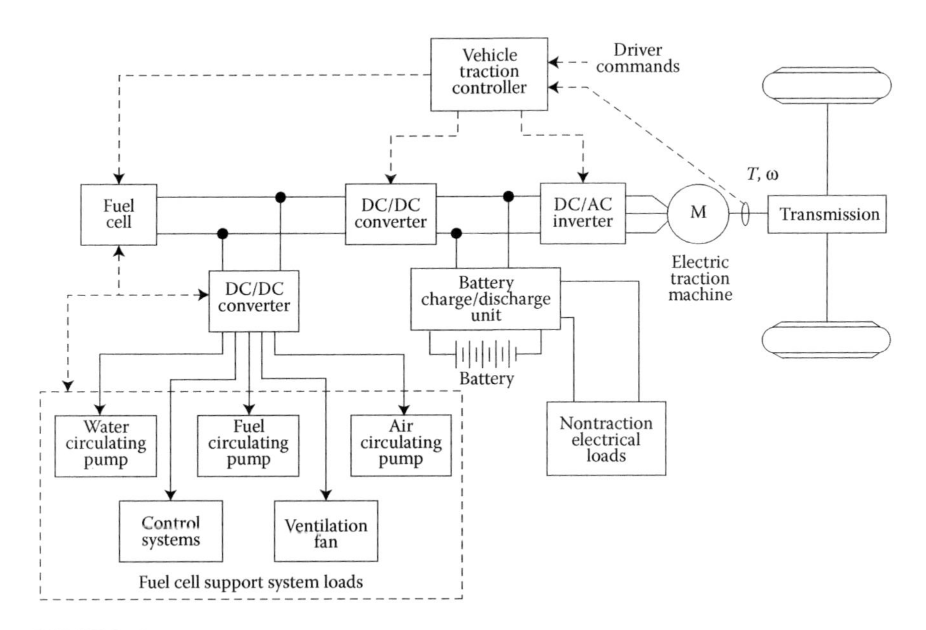

FIGURE 6.10
Detailed schematic diagram of the energy management system of the fuel cell vehicle. (Adapted from A. Emadi, S.S. Williamson, and A. Khaligh, *IEEE Transactions on Power Electronics*, 21(3), 567–577, 2006.)

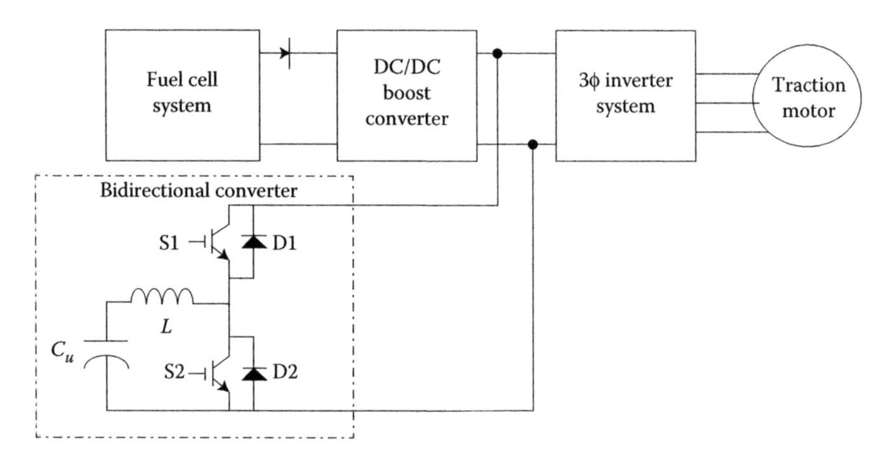

FIGURE 6.11
Energy management configuration in the FCV.

bidirectional DC–DC converter. For simplicity of the control design for the bidirectional converter, a simple buck/boost converter is used in Figure 6.11.

In the boost mode, the switch S2 and the diode D1 are on and then the UC or battery can release the energy from the ES; in the buck mode, the switch S1 and the diode D2 are on and the energy from the braking can be stored to the energy-storage unit.

6.3.4.2 Electric Motors and Motor Controller/Inverter for FCV

An FCV consists of high-voltage components at a range of 300–400 V, such as a traction motor, inverter drives systems, battery, DC to DC converter, and so on.

The power demand of an FCV's electric motor ranges from 75 to 120 kW. With great progress in power electronics and microcontroller-based controllers, compact, cost-effective, and highly efficient inverters make it possible for an AC induction motor and brushless permanent magnet (BPM) motor to be able to be utilized in FCVs. In general, both of these motors (an AC and BPM motor) provide high efficiency over a wide range of operation, but these motors require complicated control schemes such as a space vector pulse with modulation (SVPWM) to produce the desired torque through processing the feedback signals of the current and rotor position compared to DC brush motors. A traction motor plays an important role in the FCV. The main characteristics of a traction motor for vehicle applications are [26]

- High torque density and power density.
- High torque for starting, at low speeds and hill climbing, and high power for high-speed cruising.

- Wide-speed range, with a constant power operating range of around 3–4 times the base speed, being a good compromise between the peak torque requirement of the machine and the volt–ampere rating of the inverter.
- High efficiency over wide-speed and torque ranges, including a low torque operation.
- Intermittent overload capability, typically twice the rated torque for short durations.
- High reliability and robustness appropriate to the vehicle environment.
- Acceptable cost.

PM synchronous or PM brushless motors, induction motors, and switched reluctance motors in Figure 6.12 are used for traction applications.

Torque–power–speed characteristics required for traction applications are shown in Figure 6.13. An idealized torque–power–speed characteristics for this application are illustrated in Figure 6.14.

In Figure 6.14, three torque–power regions are observed. In the constant torque region I, the maximum torque capability, especially at low-speed ranges, is determined by the current rating of the inverter and the ratio of the magnitude of the inverter output voltage to the inverter frequency is adjusted to maintain air-gap flux approximately constant in this region [27]. As the maximum available voltage of the inverter is reached, the constant torque region I is switched to the constant power region II. In this region, stator voltage is held at its rated value and stator current is regulated to obtain constant power [28]. The motor operates in flux weakening due to the inverter voltage and current limits. As the frequency continuously keeps increasing in this mode, it operates with a reduced air-gap flux by increasing the slip to maintain the stator current at its limit. When the slip reaches a value corresponding to the pull-out torque, region II switches to region III. In region III, the stator voltage is held at its rated value and slip speed is regulated just below

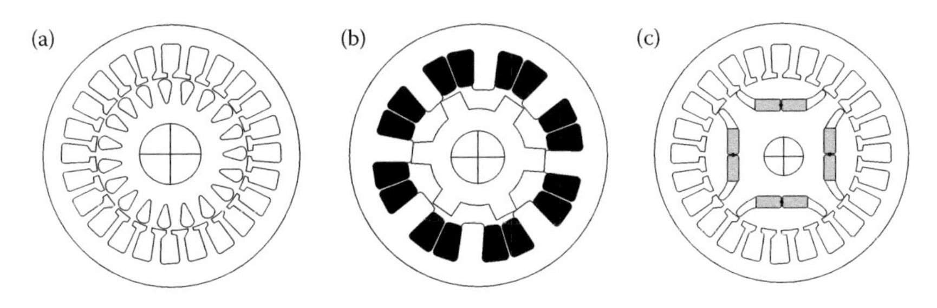

FIGURE 6.12
Main traction machine technologies. (Adapted from Z. Q. Zhu and D. Howe, *Proceedings of the IEEE* 95(4), 746–765, 2007.) (a) IM—induction machine. (b) SRM—switched reluctance machine. (c) PMM—permanent magnet machine.

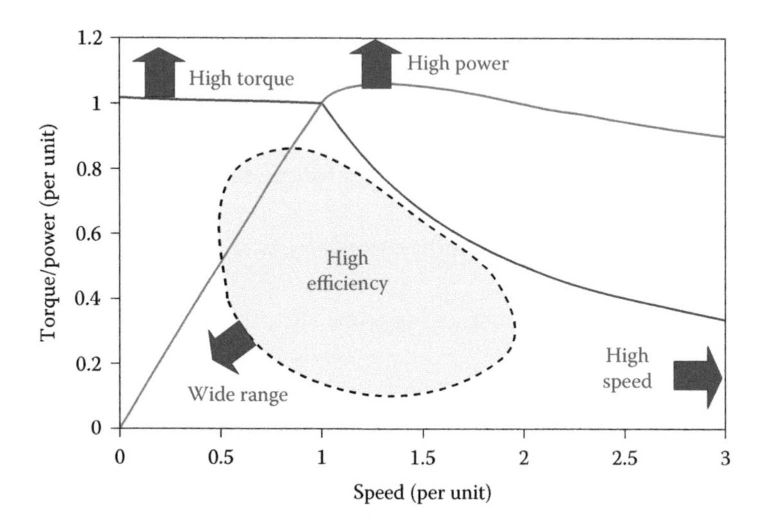

FIGURE 6.13
Torque–power requirements for traction machines. (Adapted from Z. Q. Zhu and D. Howe, *Proceedings of the IEEE* 95(4), 746–765, 2007.)

its pull-out torque value. The slip will be kept just under its pull-out torque value. The torque and the power reduce due to the increasing influence of the back-electromotive force (back-EMF). In traction applications, a wide-range speed control and control beyond a constant power range are also needed.

Since induction motors are simple and robust as well as having a wide-speed range, they are desirable for traction application. A good dynamic torque control performance can be achieved by field-oriented control

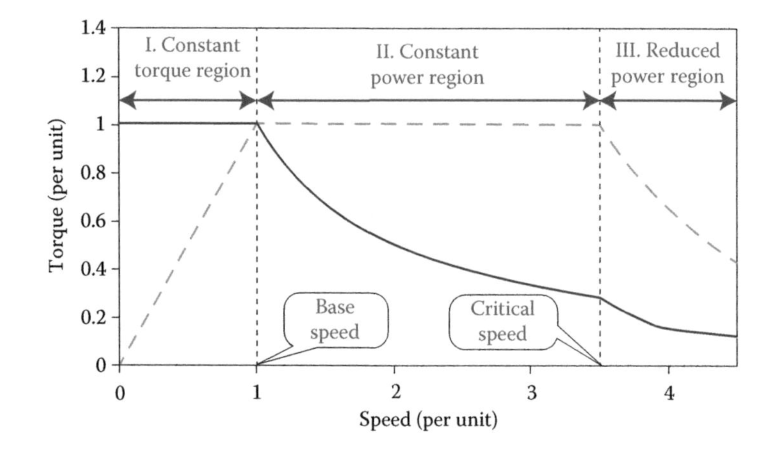

FIGURE 6.14
Idealized torque–power–speed characteristics for traction applications. (Adapted from Z. Q. Zhu and D. Howe, *Proceedings of the IEEE* 95(4), 746–765, 2007.)

(vector control) or direct torque control because an induction machine can be modeled with a d–q model so as to behave just like a DC machine [25]. However, the efficiency of an induction motor is lower than a PM motor due to the rotor loss and its size is bigger than the same power and speed-rated PM motor. A PM motor has a high efficiency, high torque, and high-power density, but a PM motor has a short constant range that can limit its field-weakening capability compared with an induction motor. The switch reluctance motor (SRM) is also a good candidate for traction application due to its simple and rugged construction, simple control, the ability of an extremely high-speed operation, and hazard-free operation [25]. But, SRM has not been widely used and is much expensive than other motors. The motor controller–inverter system converts the power of the battery or the fuel cell DC power into AC power to control AC and BPM motors by insulated-gate bipolar transistors (IGBT) serving as the high-power-switching devices.

6.3.4.3 *Auxiliary Systems in FCVs*

Auxiliary systems in FCVs such as pumps, air conditioning, various sensors, and gauges in the electric drive systems can work at the voltages of 12 V, 140 V, 300 V, or the standard voltage of 42 V. The auxiliary systems as well as other systems must maintain a high efficiency, low electromagnetic interface, compliance (EMI/EMC), and low cost. The design change/optimization of FCV varies from manufacturers and vehicle models. The basic design rules for a hybrid FCV are to consider how to construct the energy management systems, electric motors, motor controllers, and auxiliary systems as mentioned above. By downsizing the related electric drives systems and components, cost reduction is possible.

6.4 Control of Hybrid Fuel Cell System for Electric Vehicles

According to the FCV components described in Section 6.2, there are three major control modules of FCVs: drive-train control including power control, motor/inverter control, and fuel cell control including a fuel processor. Using a fuel processor may cause the system to be very complex. The detailed control descriptions of a fuel cell processor are out of the scope of this book; however, a simplified control of a fuel cell processor will be described.

6.4.1 Drive-Train Control

Drive-train control manages all driver's signals and system-operating condition signals such as key states (start-up or shutdown), accelerator pedal position, select gear or brake position, master cylinder pressure, temperature, displays of speed, torque, current, mileage, and so on. On the basis of the power demand of the electric motor on the road condition and vehicle capacity, the desired current is calculated through the main microprocessor, and then this calculated current signal is fed to the fuel cell control and power control modules. Since the drive-train control is closely related to the power control in FCV, the following section will be helpful for the understanding of the drive-train control.

6.4.2 Power Control

6.4.2.1 Power Control

Power control module controls the power flow between the fuel cell system, the ES (battery or UC), and the drive train based on the motor power command and the energy level of the ES. Three power control modes are given as follows [24]:

1. *Standstill mode:* It is an idle mode. Neither the fuel cell system nor the ES is inactive.
2. *Braking mode:* The fuel cell system is inactive, but the regenerative-braking energy is absorbed to the ES until the ES is fully charged.
3. *Traction mode:* There are two cases in the traction model. One is the hybrid mode and the other is the sole mode. In the hybrid mode, if the command motor input power P_{comm} is greater than the fuel cell total rated power P_{fc}, the hybridization mode is active. So, the fuel cell system produces the rated power within its optimal operating region and the remaining power ($P_{comm} - P_{fc}$) is covered by the ES system. In the sole mode, if the command motor input power is less than the fuel cell rated power, the energy difference between the fuel cell and the motor is absorbed by the ES system until its maximum charging level. In this case, the fuel cell system is the only source to drive the vehicle.

Once being fully charged, the ES system alone drives the vehicle until the energy level of the ES system reaches its minimum level.

6.4.2.2 Inverter Control

Nowadays, the real-time microprocessor makes the advanced inverter-switching method possible such as SVPWM and vector control for AC induction motors. SVPWM is a special technique to determine the switching sequence of the upper-level power of the three power transistors in a three-phase voltage source inverter. It could not only generate less harmonic distortion in the output voltages or current in the winding of the motor load, but also provide a more efficient use of the DC input voltage, compared with the sinusoidal PWM. For the vector control of an AC induction machine, an

AC induction machine has to be modeled in detail. In Appendices C and E, induction machine modeling, vector control, and SVPWM for HEV applications are described in detail. Also, Appendix D is explained for the coordinate transformation for induction machine modeling and vector control. Here, indirect and direct field-oriented control schemes are briefly introduced. Since the time of the introduction to the field-oriented control, the indirect field-oriented control method becomes more popular, due to its feasibility, for achieving a reliable and accurate speed and torque control than the direct method. The direct scheme electrically determines the rotor flux position using field angel sensors, while the indirect scheme determines the rotor flux from the calculation of the slip speed of the rotor. This field-oriented control is based on three major factors: (1) the machine current and voltage space vectors, (2) the transformation of a three-phase speed and time-dependent system into a two-coordinate time-invariant system, and (3) an effective pulse width modulation (PWM) pattern generation [28]. Because of these factors, an AC machine can be modeled like a DC machine and can achieve a good transient and steady-state control performance. Field-oriented controlled machines have two components: the torque component (aligned with the q coordinate) and the flux component (aligned with the d coordinate). In a conventional indirect vector control mode shown in Figure 6.15, the d-axis current producing the flux is controlled to be constant and the q-axis current producing the torque is controlled to have the machine to operate at the speed at which the machine supplies the maximum output power. The torque can be controlled by regulating i_{qs}^e and slip speed $\omega_e - \omega_r$. The rotor flux can be controlled by i_{ds}^e [27]. The main disadvantage of the

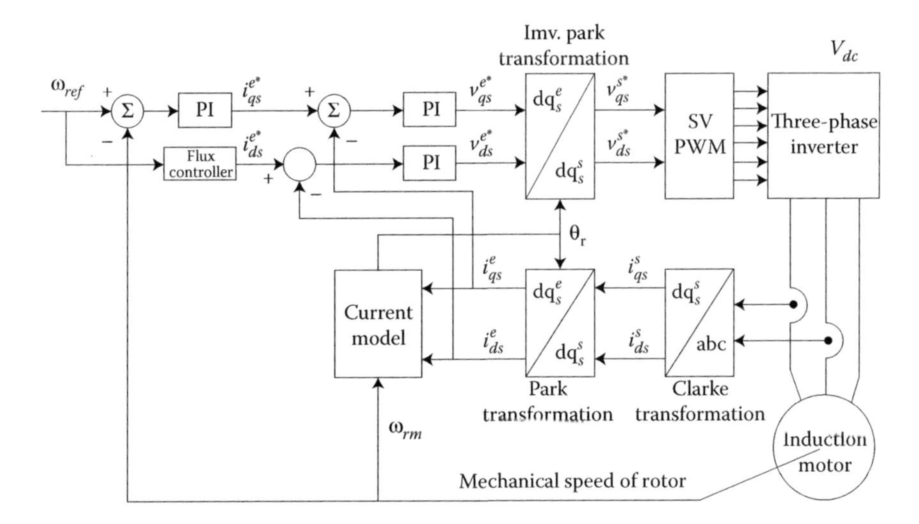

FIGURE 6.15
Indirect vector control scheme for an induction machine. (Adapted from H. Toluyat and S. Campbell, *DSP-Based Electromechanical Motion Control*. CRC Press, 2006.)

indirect vector control is that it is too sensitive to motor parameters, and therefore, the accuracy of the control gain is highly required.

6.4.3 Fuel Cell Control

First of all, a fuel cell has a continuous supply of reactants adjusted by the load current. Hence, internal states (humidity, temperature, and pressure) and external inputs (hydrogen and oxygen) have to be properly controlled to prevent from reactant starvation. The temperature of a fuel cell system reaches steady states through the humidity controller and the temperature controller. The energy for the cold start-up of the fuel cell can be supplied by the backup battery.

Using a microprocessor or DSP, internal states and an internal input can be controlled. The control scheme is shown in Figure 6.16. In Figure 6.16, control commands are issued for opening and closing the hydrogen solenoid valve and purge valve of the system. The speed of the air compressor varies according to the current demand. The speed of the cooling fan is controlled to regulate the fuel cell stack temperature through an air compressor. The fuel cell operating temperature is maintained at 65°C by varying the speed of the cooling fan. In addition, the pressure difference between the anode and cathode can be minimized to prolong the stack life by controlling the actuator at both anode and cathode sides.

6.4.4 Fuel Processor or Reformer

In Reference 30, a simple model of a reformer that generates hydrogen through reforming methanol was introduced. Using a second-order transfer

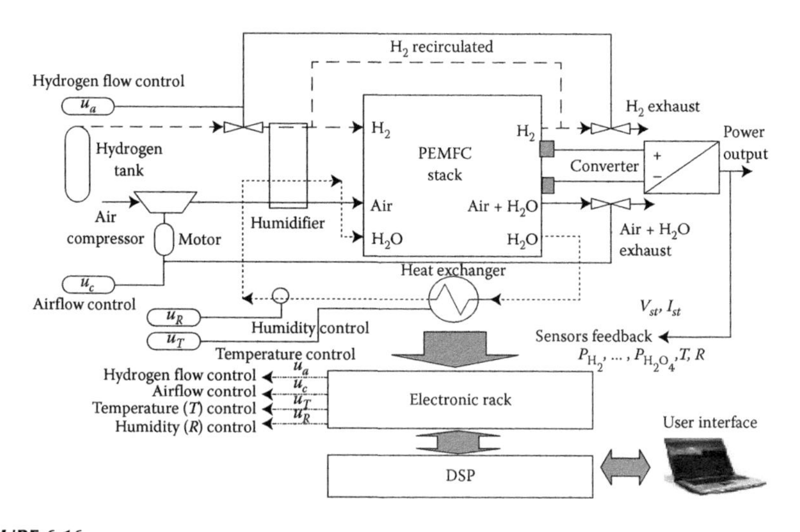

FIGURE 6.16
Simplified experimental setup for the PEMFC control. (Adapted from W. Na and B. Gou, *IEEE Transactions on Energy Conversion*, 23(1), 179–190, 2008.)

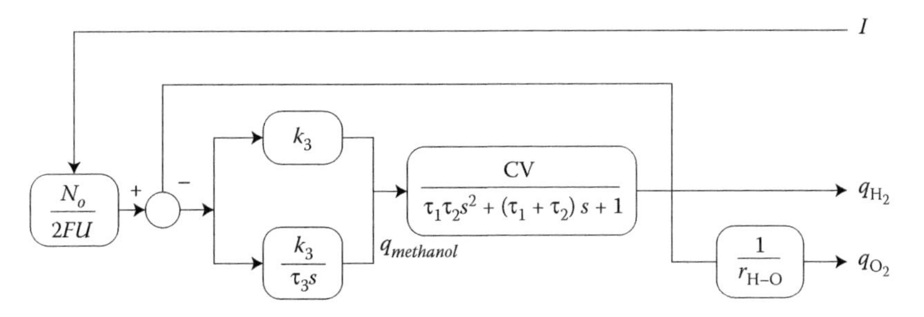

FIGURE 6.17
Reformer and reformer controller model. I, stack current (A); N_o, number of series of a fuel cell in the stack; F, Faraday's constant (C/kmol); U, fuel utilization factor; k_3, PI, controller gain; τ_3, PI controller time constant; q_{O_2}, oxygen flow rate (mol/s); r_{H-O}, hydrogen–oxygen flow ratio. (Adapted from M.Y. El-Sharkh et al., *IEEE Transactions on Power Delivery*, 19(4), 2022–2028, 2004.)

function model, the mathematical form of the fuel flow response to an input step of methanol flow is given as follows:

$$\frac{q_{H_2}}{q_{methanol}} = \frac{CV}{\tau_1\tau_2 s^2 + (\tau_1 + \tau_2)s + 1} \tag{6.9}$$

where q_{H_2} hydrogen flow rate (mol/s), $q_{methanol}$ methanol flow rate (mol/s), CV conversion factor (kmol of hydrogen per moles of methanol), τ_1, τ_2 reformer time constant (s).

A proportional–integral (PI) controller determines the methanol flow into the reformer by controlling hydrogen flow according to the current shown in Figure 6.17. Oxygen flow is determined by the hydrogen–oxygen flow ratio r_{H-O}.

6.5 Fault Diagnosis of a Hybrid Fuel Cell System

Fault diagnosis is to detect, isolate, and identify an impending or incipient failure condition—the affected component in the hybrid fuel cell system is still operational even though at a degraded mode. In this section, fault-diagnostic issues and procedures will be described by using a practical system called the Nexa fuel cell stack, which is of 1.2 kW [31].

6.5.1 Fuel Cell Stack

The 1.2 kW Nexa fuel cell stack output voltage varies with power, ranging from ~43 V without load to ~26 V with full load. During operations, the fuel cell stack voltage is monitored for fault diagnosis, control, and safety purposes. A cell voltage checker (CVC) system monitors the performance of individual cell pairs and detects the presence of a poor cell. The Nexa unit will shutdown if a cell failure or a potentially unsafe condition is detected in the fuel cell stack.

6.5.2 Hydrogen Supply System

The fuel-supply system monitors and regulates the hydrogen supply to the fuel cell stack. The main concern of the fuel (hydrogen) supply system is the safety of hydrogen supplying for the fuel cell. The fuel-supply operation is taken into account in the following components:

- A pressure transducer monitors fuel delivery conditions to ensure an adequate fuel supply is present for Nexa system operation.
- A pressure relief valve protects downstream components from overpressure conditions.
- A solenoid valve provides isolation from the fuel supply when it is not in operation.
- A pressure regulator maintains an appropriate hydrogen supply pressure to the fuel cell.
- A hydrogen leak detector monitors hydrogen levels near the fuel delivery subassembly. Warning and shutdown alarms are implemented for product safety.
- The accumulation of nitrogen and water in the anode side degrades the fuel cell performance, monitored by the purge cell voltage, a hydrogen purge valve at the stack outlet is periodically opened to flush out inert constituents in the anode and to restore performance.

Only a small amount of hydrogen purges from the system, less than 1% of the overall fuel consumption rate. Purged hydrogen is discharged into the cooling air stream before it leaves the Nexa system. Hydrogen quickly diffuses into the cooling air stream and is diluted to a level much less than the lower flammability limit. The hydrogen leak detector, situated in the cooling air exhaust, ensures that flammable limits are not reached.

6.5.3 Air, Humidifier, and Water Management System

Like the anode side, the cathode side needs two actuators to control both the pressure and the airflow. These two actuators can minimize the pressure difference between the anode and the cathode side, and control airflow through the compressor because the fuel cell output voltage is largely dependent on the airflow.

Oxidant air has to be humidified for the proper operation of fuel cells. Excess product water may cause flooding, which results in loss of the cell potential. So, excess water needs to be passively evaporated into the surrounding environment, or alternatively, product water can be drained and collected through the water management system. However, a small amount of water also causes membrane dying, which is one of the reasons to reduce the cell potential. By monitoring the cell potential during the operation, flooding and drying will be alarmed.

6.5.4 Hydrogen Diffusion and Cooling System

The hydrogen quickly diffuses into the cooling air and is diluted to levels that are far below the lower flammability limit (LFL) of hydrogen. For safety purposes, a hydrogen sensor is located within the cooling air outlet stream and provides feedback to the control system. The control system generates warning and alarm signals if the hydrogen concentration approaches 25% of the LFL. The LFL of hydrogen is defined to be the smallest amount of hydrogen that will support a self-propagating flame when it is mixed with air and ignited. At concentrations less than the LFL, there is insufficient fuel present to support combustion. The LFL of hydrogen is 4% by volume.

6.5.5 Safety Electronics System

Unusual or unsafe operating conditions result in either a warning or alarm, and may cause an automatic shutdown, depending on the severity. During a warning, the Nexa power module continues to operate and the controller attempts to remedy the condition. During an alarm, the controller initiates a controlled shutdown sequence. There are two kinds of error levels [4].

First-level errors:

- *Lower cell voltage:* Alarm if one cell voltage drops below 0.3 V
- *Stack overload:* It varies on the stack power capacity. If 125%–150% rate current is drawn, an alarm signal is activated
- *High temperature:* The temperature exceeds 87.4°C

Second-level errors:

These cases enforce the system to disconnect all power supply from the system components and all valves are set to their initial conditions.

- *High-level hydrogen:* According to the hydrogen sensor, 15% over the LFL, an early alarm is triggered, and 25% over the LFL, an error signal is turned on.
- *The pressure difference protection:* If the pressure difference between the anode and cathode sides is over 0.5 bar, an error signal is initiated.

References

1. J. Larminie and A. Dicks, *Fuel Cell Systems Explained*. Wiley, New York, 2002.
2. K. Rajashekrara, Propulsion system strategies for fuel cell vehicles, SAE paper 2000-01-0369, 2000.
3. F. Barbir and T. Gomez, Efficiency and economics of PEM fuel cells, *International Journal of Hydrogen Energy*, 22(10/11), 1027–1037, 1997.
4. C.A. Anderson, M.O. Christensen, A.R. Korsgaad, M. Nielsen, and P. Pederson, Design and control of fuel cells system for transport application, Aalborg University, Project Report, 2002.
5. J. Purkrushpan and H. Peng, *Control of Fuel Cell Power Systems: Principles, Modeling, Analysis and Feedback Design*. Springer, Germany, 2004.
6. F. Barbir, *PEM Fuel Cells: Theory and Practice*. Elsevier Academic Press, London, 2005.
7. J. Hamelin, K. Abbossou, A. Laperriere, F. Laurencelle, and T.K. Bose, Dynamic behavior of a PEM fuel cell stack for stationary application, *International Journal of Hydrogen Energy*, 26, 625–629, 2001.
8. W. Na and B. Gou, Nonlinear control of PEM fuel cells by feedback linearization, *IEEE Transactions on Energy Conversion*, 23(1), 179–190, 2008.
9. J. Purkrushpan, A.G. Stefanopoulou, and H. Peng, Control of fuel cell breathing, *IEEE Control Systems Magazine*, 24(2), pp. 30–46, April 2004.
10. J. Purkrushpan, A.G. Stefanopoulou, and H. Peng, Modeling and control for PEM fuel cell stack system, *Proceedings of the American Control Conference*, Anchorage, Alaska, pp. 3117–3122, 2002.
11. J.C. Amphlett, R.M. Baumert, R.F. Mann, B.A. Peppy, P.R. Roberge, and A. Rodrigues, Parametric modeling of the performance of a 5-kW proton exchange membrane fuel cell stack, *Journal of Power Sources*, 49, 349–356, 1994.
12. J.C. Amphlett, R.M. Baumert, R.F. Mann, B.A. Peppy, and P.R. Roberge, Performance modeling of the Ballard Mark IV solid polymer electrolyte fuel cell, *Journal of Electrochemical Society*, 142(1), 9–15, 1995.
13. R.F. Mann, J.C. Amphlett, M.A. Hooper, H.M. Jesen, B.A. Peppy, and P.R. Roberge, Development and application of a generalized steady-state electrochemical model for a PEM fuel cell, *Journal of Power Sources*, 86, 173–180, 2000.
14. M.J. Khan and M.T. Labal, Dynamic modeling and simulation of a fuel cell generator, *Fuel Cells*, 1, 97–104, 2005.
15. M.J. Khan and M.T. Labal, Modeling and analysis of electro chemical, thermal, and reactant flow dynamics for a PEM fuel cell system, *Fuel Cells*, 4, 463–475, 2005.
16. M.Y. El-Sharkh, A. Rahman, M.S. Alamm, A.A. Sakla, P.C. Byrne, and T. Thomas, Analysis of active and reactive power control of a stand-alone PEM fuel cell power plant, *IEEE Transactions on Power Delivery*, 19(4), 2022–2028, 2004.
17. P. Famouri and R.S. Gemmen, Electrochemical circuit model of a PEM fuel cell, *IEEE Power Engineering Society General Meeting*, 3, 13–17, 2003.
18. A. Sakhare, A. Davari, and A. Feliachi, Fuzzy logic control of fuel cell for standalone and grid connection, *Journal of Power Sources*, 135(1–2), 165–176, 2004.
19. P. Almeida and M. Godoy, Neural optimal control of PEM fuel cells with parametric CMAC network, *IEEE Transactions on Industry Applications*, 41(1), 237–245, 2005.

20. L.Y. Chiu, B. Diong, and R.S. Gemmen, An improved small-signal mode of the dynamic behavior of PEM fuel cells, *IEEE Transactions on Industry Applications*, 40(4), 970–977, 2004.
21. J. Sun and V. Kolmannovsky, Load governor for fuel cell oxygen starvation protection: A robust nonlinear reference governor approach, *IEEET Transactions on Control Systems Technology*, 3(6), 911–913, 2005.
22. W. Na and B. Gou, A thermal equivalent circuit for PEM fuel cell temperature control design, *IEEE International Symposium on Circuits and Systems*, Seattle, Washington, May 2008, pp. 2825–2828, May 2008.
23. W. Yang, B. Bates, N. Fletcher, and R. Pow, Control challenges and methodologies in fuel cell vehicle development, SAE paper 98C054, 1998.
24. A. Emadi, S.S. Williamson, and A. Khaligh, Power electronics intensive solutions for advanced electric, hybrid electric, and fuel cell vehicular power systems, *IEEE Transactions on Power Electronics*, 21(3), 567–577, 2006.
25. C.C. Chan, The state of the art of electric, hybrid, and fuel cell vehicles, *Proceedings of the IEEE*, 95(4), 704–718, 2007.
26. Z. Q. Zhu and D. Howe, Electrical machines and drives for electric, hybrid, and fuel cell vehicles, *Proceedings of the IEEE* 95(4), 746–765, 2007.
27. C.-M. Ong, *Dynamic Simulation of Electric Machinery*. Prentice-Hall, Upper Saddle River, NJ, 1997.
28. Texas Instruments Literature Number BPRA073, Field Oriented Control of 3-Phase AC motors, February 1998.
29. H. Toluyat and S. Campbell, *DSP-Based Electromechanical Motion Control*. CRC Press, Boca Raton, FL, 2006.
30. K. H. Hauer, Analysis tool for fuel cell vehicle hardware software (control) with an application to fuel economy comparisons of alternative system designs, PhD dissertation, Department of Transportation Technology and Policy, University of California, Davis, California, 2001.
31. *Ballard Nexa Fuel Cell Stack 1.2 kW User Manual*. www.heliocentris.com

7

*Application of Fuel Cells in Utility Power Systems and Stand-Alone Systems**

7.1 Introduction

Fuel cells ranging from sub-kilowatt portable power units to multi-megawatt stationary power plants are emerging to deliver clean and efficient power. This new technology is suitable for producing heat and power for residential, commercial, and industrial customers. Because of high fuel conversion efficiency, combined heat and power generation flexibility, friendly siting characteristics, negligible environmental emissions, and lower carbon dioxide emissions, fuel cells are considered at the top of the desirable technologies for a broad spectrum of power generation applications. Among the available fuel cells, the PEM fuel cell is seen as the system of choice for portable, vehicular, and residential applications.

In the United States, the Federal Energy Regulatory Commission has issued several rules and Notices of Proposed Rulemaking to set the road map for the utility deregulation. The California crisis has drawn great attention and sparked intense discussion within the utility industry. One possible solution is to rejuvenate the idea of integrated resource planning and promote the distributed generation (DG) via traditional or renewable generation facilities for the deregulated utility systems. The technology challenges of PEM fuel cells relating to the low-temperature operation include water management, heat removal, and anode poisoning by trace amounts of CO present in hydrocarbon-derived fuels.

Fuel cell is the most promising renewable generation technology for the residential and small commercial users. Fuel cells are static energy conversion devices directly converting the chemical energy of fuel into electrical energy. Compared with conventional power generation systems, they have many advantages, such as high efficiency, zero or low emission (of pollutant gases), and flexible modular structure. Fuel cell is a promising energy form that is expected to play an important role in future DG applications.

* This chapter was mainly prepared by Dr. Bei Gou, Smart Electric Grid LLC, USA and Dr. Woonki Na, California State University, Fresno, USA.

It is desirable for these renewable generation facilities to be interconnected with the utility grid to perform peak shaving, demand reduction, and to serve as emergency and standby power supply. However, the emerging integration problems, including control strategy design, energy management, mismatch between the utility tie protection and the equipment protection, etc., need to be studied and solved. Fuel cell DGs (FCDGs) can either be connected to a utility power system for network reinforcement or installed in a remote area to supply stand-alone power. This chapter is aimed to introduce the applications of fuel cells to the utility power systems, stand-alone applications, power electronics interface design for stand-alone fuel cell and ultracapacitor based power system.

7.2 Utility Power Systems and Residential Applications

One of the most important applications of fuel cells is utility power systems. The frequent occurrence of faults or big disturbances in power distribution systems causes the problem of loss of power supply; the energy transfer through the transmission network from generation plants in remote areas increases the energy cost. It is desirable to install renewable generation facilities near the customers to avoid the frequent occurrence of loss of power supply and the high-energy cost. This section will discuss the technologies of fuel cell installation in the distribution systems.

7.2.1 Modeling and Control of PEM Fuel Cell DG System

Wang et al. proposed a typical configuration of fuel cell DG system [1], shown in Figure 7.1. The system configuration ratings and parameters are given in

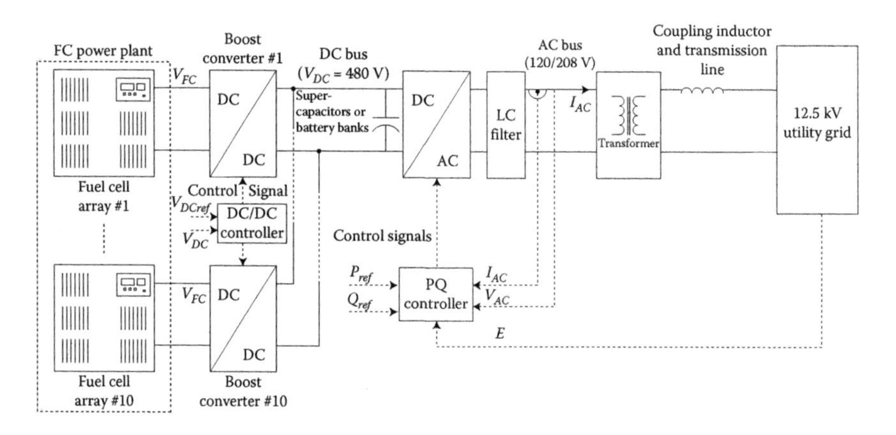

FIGURE 7.1
Block diagram of a fuel cell distributed generation system. (Adapted from C. Wang, M.H. Nehrir, and H. Gao, *IEEE Trans. Energy Convers.*, 21, 586–595, 2006.)

TABLE 7.1

Configuration Parameters of the Proposed System

PEMFC power plant	216 V/480 kW ten 48-kW FC arrays are connected in parallel
PEMFC array	216 V/48 kW, consisting of 8 (series) × 12 (parallel) 500-W fuel cell stacks
Boost DC/DC converter	200 V/480 V, 50 kW each 10 units connected in parallel
Three-phase DC/AC inverter	480 V DC/208 V AC, 500 kW
LC filter	$L_f = 0.15$ mH, $C_f = 306.5$ uF
Step-up transformer	$V_n = 208$ V/12.5 kV, $S_n = 500$ kW $R_1 = R_2 = 0.005$ p.u., $X_1 = X_2 = 0.025$ p.u.
Coupling inductor	$X_C = 50\ \Omega$
Transmission line	0.5 km ACSR 6/0 $R = 2.149\ \Omega$/km, $X = 0.5085\ \Omega$/km
DC bus voltage	480 V
AC bus voltage	120 V/208 V

Table 7.1. The PEMFC power plant consists of 10 parallel-connected fuel cell arrays. Each array is rated at 48 kW, for a total of 480 kW. A boost converter is used to adapt the output voltage of each fuel cell array to the DC bus voltage. In addition, a three-phase six-switch inverter is used to convert the power available at the DC bus to AC power. Following the inverter, an LC filter is applied, which is followed by a transformer to increase the AC voltage from 208 V to 12.5 kV, the voltage level of the utility grid.

The parameters used in this structure are given in Table 7.1.

7.2.1.1 Modeling of PEM Fuel Cells

To simplify the analysis, the following assumptions are made [2]:

1. One-dimensional treatment.
2. Ideal and uniformly distributed gases.
3. Constant pressures in the fuel cell gas flow channels.
4. The fuel is humidified and the oxidant is humidified air. Assume the effective anode water vapor pressure is 50% of the saturated vapor pressure while the effective cathode water pressure is 100%.
5. The fuel cell works under 100°C and the reaction product is in liquid phase.
6. Thermodynamic properties are evaluated at the average stack temperature, temperature variations across the stack are neglected, and the overall specific heat capacity of the stack is assumed to be a constant.
7. Parameters for individual cells can be lumped together to represent a fuel cell stack.

7.2.1.2 Equivalent Electrical Circuit

According to assumptions, the output voltage of fuel cell is given as

$$V_{out} = E - V_c - V_{act1} - V_{ohm} \tag{7.1}$$

where

$$E = E_0 + \frac{R*T}{2F}\ln\left(p_{H_2}^* \cdot \sqrt{p_{O_2}^*}\right) - E_d$$

$$V_c = -\frac{R*T}{zF}\ln\left(1 - \frac{I}{I_{limit}}\right)$$

$$V_{act1} = \eta_0 + a(T - 298)$$

$$V_{ohm} = IR_{ohm}$$

where R is the gas constant, 8.3143 J/mol K; E_0 the standard reference potential at standard state, 298 K and 1 atm pressure; T the temperature (K); $p_{H_2}^*$ the effective value of partial pressure of hydrogen; $p_{O_2}^*$ the effective value of partial pressure of oxygen; I the fuel cell current (A); I_{limit} the limitation current (A); z the number of participating electrons; F the Faraday constant (96487 C/mol); R_{ohm} the conducting resistance between the membrane and electrodes; η_0 the temperature invariant part of V_{act} (V); and a the constant terms in Tafel equation (V/K).

From Equation 7.1, we can get the equivalent circuit of the fuel cell voltage shown in Figure 7.2.

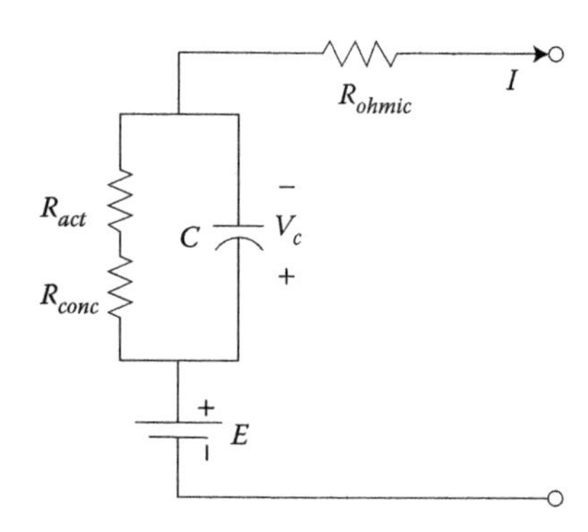

FIGURE 7.2
Equivalent electrical circuit of the double-layer charging effect inside the PEM fuel cell.

7.2.1.3 Energy Balance of the Thermodynamics

The net heat that causes the temperature of the fuel cell to vary, which is generated by the chemical reaction inside the fuel cell, can be expressed as follows:

$$\dot{p}_{net} = \dot{p}_{chem} - \dot{p}_{elec} - \dot{p}_{sens+latent} - \dot{p}_{loss} \tag{7.2}$$

where $\dot{p}_{net}, \dot{p}_{chem}, \dot{p}_{elec}, \dot{p}_{sens+latent}$, and $\dot{p}_{loss}$ are net energy of fuel cell, chemical energy, electrical energy, sensible and latent heat, and the heat loss, respectively. Their values are given in Reference 2.

7.2.1.4 Control Design for PEM Fuel Cells

Two controllers are designed for the boost DC/DC converter and the three-phase voltage source inverter.

7.2.1.4.1 Controller Design for the Boost DC/DC Converter

To design the controller for the boost DC/DC converter, we need to first build its state-space model.

The equivalent circuit of the boost DC/DC converter is given in Figure 7.3.

The small signal state-space model for the boost DC/DC converter is given as follows [1]:

$$\begin{cases} \begin{bmatrix} \dot{i}_{L_{dd}} \\ \dot{v}_{C_{dd}} \end{bmatrix} = \begin{bmatrix} 0 & \dfrac{-(1-D)}{L_{dd}} \\ \dfrac{1-D}{C_{dd}} & \dfrac{-1}{RC_{dd}} \end{bmatrix} x + \begin{bmatrix} \dfrac{X_2}{L_{dd}} \\ \dfrac{-X_1}{C_{dd}} \end{bmatrix} u \\ V_{dd\_out} = [0\ 1] \begin{bmatrix} i_{L_{dd}} \\ v_{C_{dd}} \end{bmatrix} \end{cases} \tag{7.3}$$

where X_1 and X_2 are the steady-state values of $i_{L_{dd}}$ and $v_{C_{dd}}$, respectively, and D is the pulse duty ratio at the rated operating point. Other variables are given in Figure 7.3.

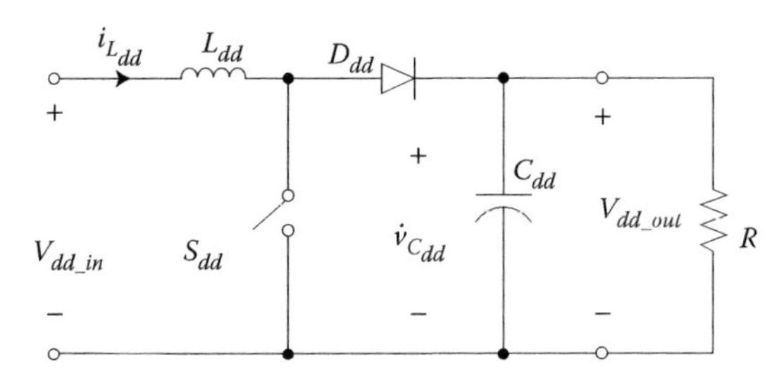

FIGURE 7.3
Boost DC/DC converter.

TABLE 7.2

Parameters of the Boost DC/DC Converter

Parameter	Values
L_{dd}	1.2 mH
C_{dd}	2500 μF
D_N	0.5833
R_{load}	4.608 Ω
X_1	250 A
X_2	480 V
k_{di}	20
k_{dp}	0.02

A PI controller is designed based on Equation 7.3, whose parameters are given in Table 7.2.

7.2.1.4.2 Controller Design for the Three-Phase Voltage Source Inverter

A three-phase PWM controller is designed for the inverter to satisfy voltage regulation as well as to achieve real and reactive power control. A voltage regulator is used in this design to take the error signals between the actual output voltage in dq frame ($V_{d,q}$) and the reference voltage ($V_{d,q(\mathrm{ref})}$) and generates the current reference signals ($I_{d,q(\mathrm{ref})}$) for the current control loop. The overall control design is given in Figure 7.4 [1].

7.2.2 Operation Strategies

In this section, we will introduce a concept of cogeneration of fuel cells for residential applications, proposed in Reference 3. Residential fuel cells have been launched recently. However, studies regarding hydrogen networks, hydrogen energy society, use of recovered heat from FCs in combined heat and power mode, and production and delivery of hydrogen are required. Neither a clear blueprint nor a clear path exists for achieving a future hydrogen energy society [4,5]. A cooperative and coordinated manner of operation and a hierarchical control of DG are necessary to avoid disorder and to sustain the reliability of power systems when a large number of DG systems are employed.

The concept of cogeneration via networked fuel cells is presented in Figure 7.5 [3]. Five homes are connected to the energy network, which provides the electricity interchangeable between the existing grid and fuel cells. Hot water piping is installed, and three homes are equipped with FC stacks. Hydrogen is interchangeable between them via hydrogen piping. Two of these homes are also equipped with fuel processors that are

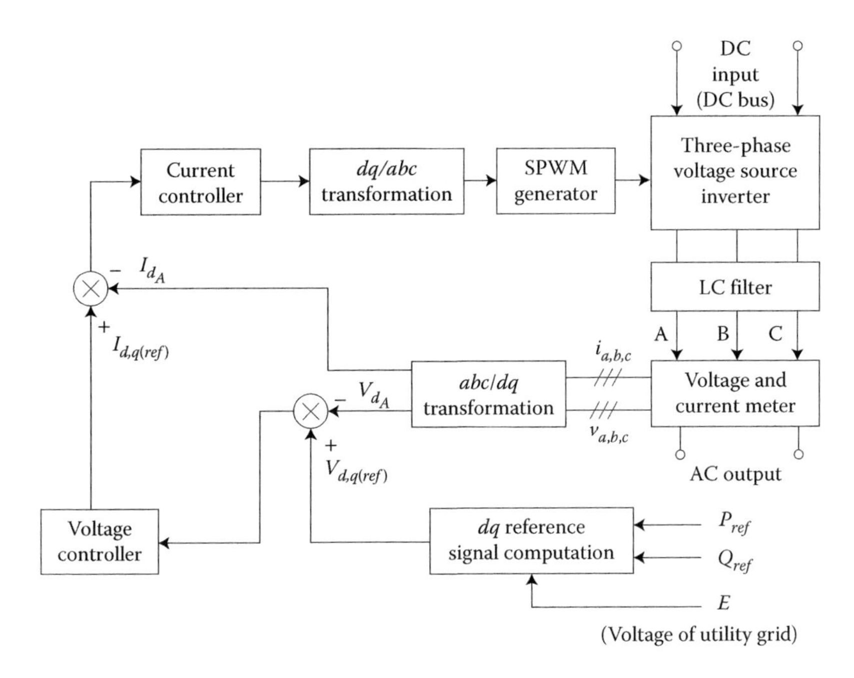

FIGURE 7.4
Overall control design for the three-phase voltage source inverter.

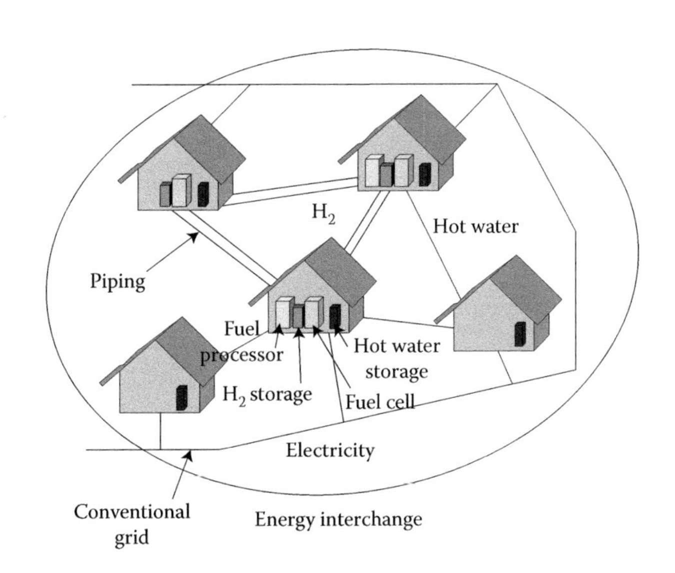

FIGURE 7.5
Example of proposed energy network for residential homes.

constantly operated at their rated load without any interruption while producing hydrogen efficiently. The FC stacks are operated depending on the load of the homes.

Based on the structure of the cogeneration network, four operation strategies (OS) rules were proposed and studied in Reference 3.

OS Rule 0: The electricity dispatch of each FC is generated to match the electricity demand of the home in which the FC is installed. Electricity interchange is not available between them.

OS Rule 1: All the FCs equally share the total electricity demand of the homes.

OS Rule 2: A logic similar to that used for internal combustion engines and unit controls is applied. The FCs are switched on one by one as the electricity demand increases and vice versa.

OS Rule 3: The FCs are switched on one by one as the electricity demand increases, similar to Strategy 2; however, all the FCs, which are switched on, equally share the total electricity demand. In this strategy, Strategies 1 and 2 are combined.

Rule 0 is proposed as a baseline case of the analysis in this structure. Rule 1 deals with partial load operations, which provide higher electricity generation efficiencies and FCs can be operated for longer periods than in other rules. Thus, Rule 1 provides efficient electricity supply rather than hot water supply. In Rule 2, FCs are operated at partial load while the others at the rated load, which provides the highest heat recovery efficiency. This rule attempts to recover hot water (heat) from the FCs in the most efficient manner. Rule 3 is the intermediate one between Rules 1 and 2.

Details of this cogeneration concept and the simulation results can be found in Reference 3.

7.3 Stand-Alone Application

The available power of a fuel cell (FC) power plant may not be able to meet load demand capacity, especially during peak demand or transient events encountered in stationary power plant applications. Normally, FCs need to work together with other sources to meet high load demand capacity. An UC bank is commonly chosen for this purpose, which can supply a large burst of power, but it cannot store a significant amount of energy. Uzunoglu and

Alam [6] proposed the modeling and control strategies for a combination of a fuel cell and an UC bank.

7.3.1 Dynamic Modeling of Fuel Cells and UC Bank

7.3.1.1 Modeling of Fuel Cell

A dynamic model was proposed in Figure 7.6 [6]. In this model, the relationship between the molar flow of any gas (hydrogen) through the valve and its partial pressure inside the channel, and three significant factors—hydrogen input flow, hydrogen output flow, and hydrogen flow during the reaction—are considered.

Assuming constant temperature and oxygen concentration, the FC output voltage can be expressed as follows:

$$V_{cell} = E + \eta_{act} + \eta_{ohmic} \tag{7.4}$$

where

$$\eta_{act} = -B\ln(CI'_{FC})$$

$$\eta_{ohmic} = -R^{\text{int}} I'_{FC}$$

and

$$E = N_0\left[E_0 + \frac{R^*T}{2F}\log\left[\frac{pH_2\sqrt{P_{O_2}}}{pH_2O}\right]\right]$$

The cell voltage is a sum of three voltages, which can be seen in Figure 7.6.

7.3.1.2 Modeling of UC Bank

The classical equivalent model of UC is given in Figure 7.7. It consists of an equivalent series resistance, an equivalent parallel resistance, and a capacitor. Owing to the limited capacitor of an UC, a bank of ultracapacitors (UCs) needs to be used via series and parallel connections.

7.3.2 Control Design of Combined Fuel Cell and UC Bank

In this structure, the fuel cell system and the ultracapitor bank are connected through a power diode, which is shown in Figure 7.8.

The FC system, which decides the voltages of the UC bank and the load, prevents the power capability of the UC bank from being fully utilized.

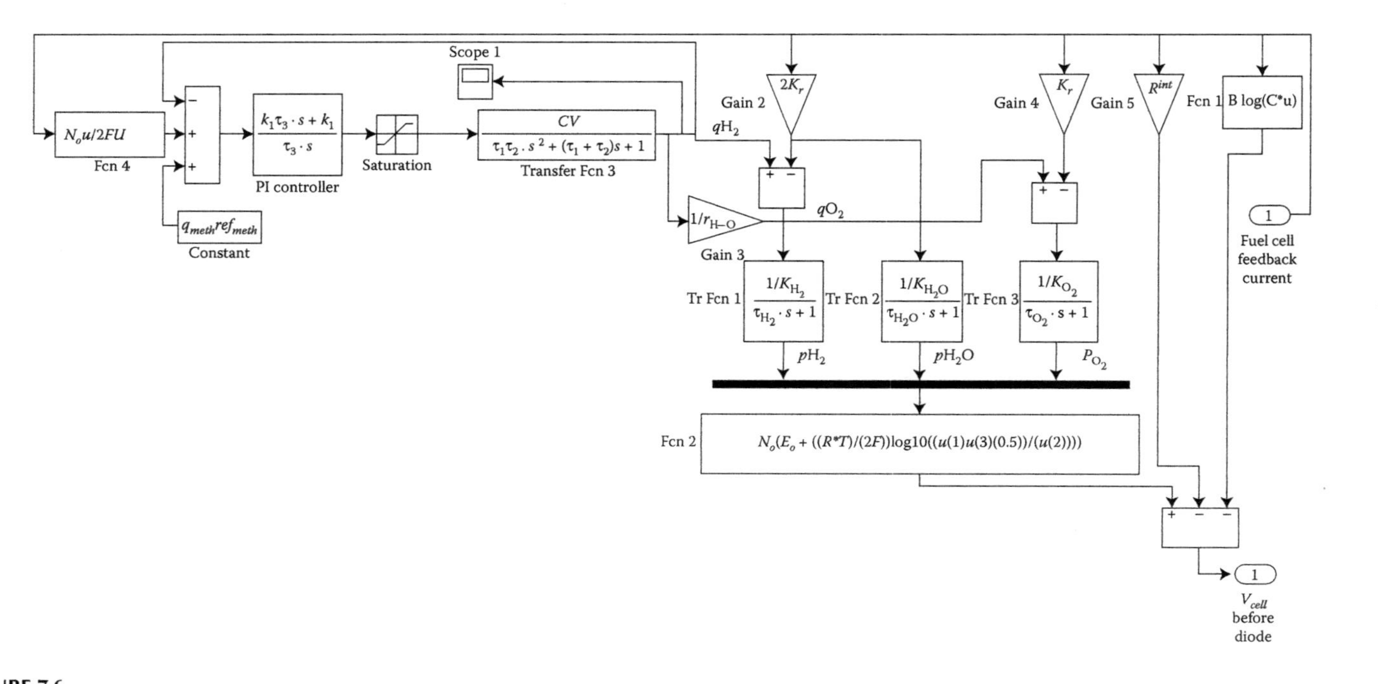

FIGURE 7.6
Dynamic model of fuel cell system.

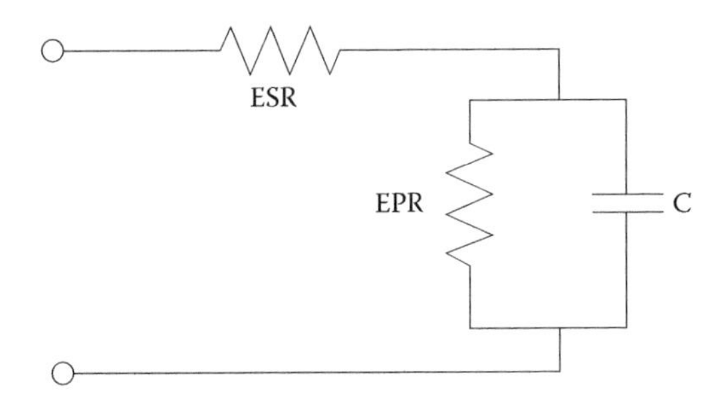

FIGURE 7.7
Classical equivalent model of UC.

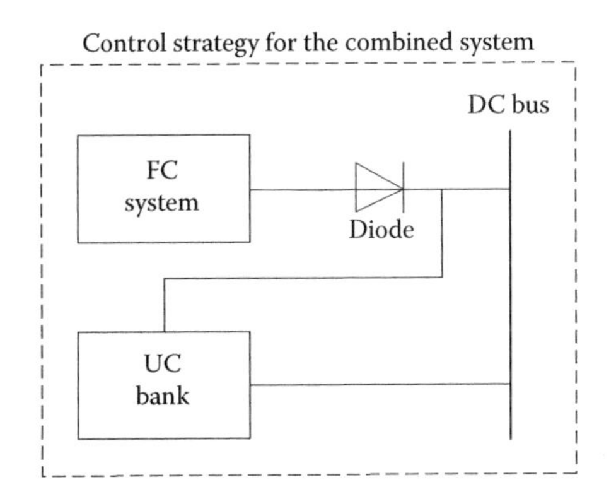

FIGURE 7.8
Combination of FC system and UC bank.

The power sharing between the FC system and UC bank is determined by the total resistance between these two systems. The main control strategy for the combined system can be summarized as follows [6]:

1. During low-power-demand periods (<5 kW), the FC system generates up to its load limit, and the excess power is used to charge the UC. The charging or discharging of the UC bank occurs according to the terminal voltage of the overall load requirements.
2. During high-power-demand periods (≥5 kW), the FC system generates the rated power and the UC is discharged to meet the extra power requirements that cannot be supplied by the FC system.
3. Short-time power interruptions in the FC system can only be supplied by the UC bank.
4. The UC bank is designed to avoid overcharge or undercharge conditions.
5. About 75% of the initial energy stored in the UC bank can be utilized if the terminal load voltage is allowed to decrease to 50% of its initial value.

To realize the control system of the above-mentioned combined system, PI controllers, ideal switching elements, and current and voltage sensors are used in the simulation model.

7.3.3 Active and Reactive Control for Stand-Alone PEM Fuel Cell System

Based on the real power and reactive power of synchronized generators in power grids, a similar model for real power and reactive power control was proposed in Reference 7.

Assume a lossless inverter and a small phase angle $\sin(\delta) \cong \delta$; using the output voltage and the output power and electrochemical relationships, the relationship between output voltage phase angle δ and hydrogen flow qH_2 is obtained as $\delta = (2FUX/mV_sN_o)qH_2$.

As we know, for synchronized generators, to increase the amount of steam input to the turbine can increase the output power, which is used to control the output power of synchronized generators. A similar scenario can be adopted to control the output power from the FC system, the only difference is that, there is no speed change, which can affect the frequency.

In the FC system, the frequency is fixed at 60 Hz by the inverter. Figure 7.9 shows that the amount of hydrogen flow can be controlled manually or automatically according to the methane reference signal and the current feedback signal, respectively, and the current feedback signal is proportional to the terminal load. A limit switch is added to control the amount of hydrogen, which simulates the actual switch setting of the FC system. Using the methane reference signal $q_{meth_{ref}}$, the operator can manually set the power output from the FC system to a value less than the maximum value set by the limit switch.

For a synchronized generator, controlling its reactive power is achieved by controlling the excitation voltage, which controls the generator output voltage and the amount of reactive power flowing in or out of the generator. In the FC system, the reactive power can be controlled by the modulation index m. The modulation index controls the terminal voltage value, which also controls the reactive power output from the FC system. The modulation index m can be set manually using the voltage reference signal V_r.

7.4 Power Interface Design for Fuel Cell and Ultracapacitor (UC) Hybrid Systems

A fuel cell voltage varies under the load changes. For instance, the fuel cell voltage reaches to the maximum when no load is applied, while it drops as the load current connected to the fuel cells increases. Particularly for high currents, significant voltage drop can be seen due to the activation overvoltage, and ohmic resistance losses in the membrane [8,9]. Thus, a secondary energy source is

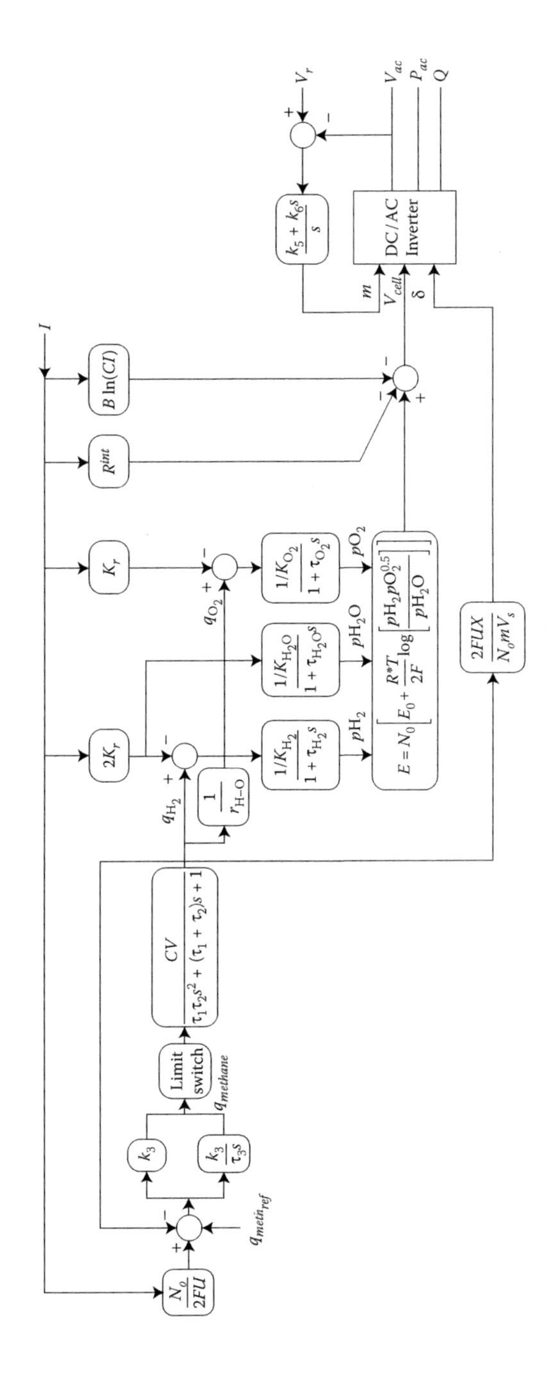

FIGURE 7.9
Real power and reactive power control for FC system. (Adapted from M.Y. El-Sharkh et al., *IEEE Trans. Power Syst.*, 19, 4, 2004.)

needed to satisfy the load demands. Presently, many studies regarding using the secondary power sources for the fuel cell system have been performed [6,10–12]. Most studies have been conducted for transportation applications [10–12].

Lately, Uzunoglu and Alam [6] used an UC in parallel to fuel cell systems without using a bidirectional converter for residential applications. This will create a problem when optimally charging, and discharging the UC during the transients especially for transportation applications. In this section, a fuel cell-based hybrid power system in which the UC is connected to the fuel cell terminal voltage through the bidirectional converter is explained for stationary power applications.

The controller for the bidirectional converter is designed so that any substantial voltage drop of the fuel cell can be prevented during transient energy deliveries and peak power-demand periods. The control analysis is performed in boost (charging), and buck (discharging) mode separately. A 5 kW PEM fuel cell system is considered for the residential system, and the nominal voltage of UC is set to 74 V [11]. For the fast load changing, the fuel cell current slope is limited to $4As^{-1}$[12] in order to prevent the fuel cell damage because a frequent fast step up and down load may apply an unnecessary stress to the fuel cell control systems such as pumps, valves, compressors, etc.

Although, during the peak power period, the total maximum power was assumed to be 7.4 kW in Reference 6, in the simulation, the total power demand can be up to 10 kW. So, the additional power exceeding the fuel cell power can be compensated by the UC through the bidirectional converter during that period. In the normal load (less 5 kW) and steady state conditions, the UC also can be charged by controlling the bidirectional converter. This fuel cell based hybrid system was tested using Matlab–Simulink.

7.4.1 Small Signal Transfer Function of Bidirectional Converter

The configuration of the bidirectional converter is shown in Figure 7.10. The UC bank C_u is considered as energy storage because it offers not only a higher energy density than conventional capacitors and batteries, but also minimal maintenances [13]. It is actually a buck-boost structure. It can be analyzed in two different modes (Boost and Buck) separately. In the boost mode, if the switch S_2 and the diode D_1 are in active mode, then the UC can absorb the energy from the fuel cell system. In the buck mode, if the switch S_1 and the diode D_2 are in active mode, then the energy from the capacitor can be released to the fuel cell terminal. The main objective of this system is to quickly provide the transient energy to the load from the UC and to transfer the steady-state energy to UC. Let us consider the boost mode first.

7.4.1.1 Boost Mode

Since the fuel cell voltage varies from 36 to 24 V between the light load and the heavy load, the nominal voltage of UC is set to 74 V. The voltage of the

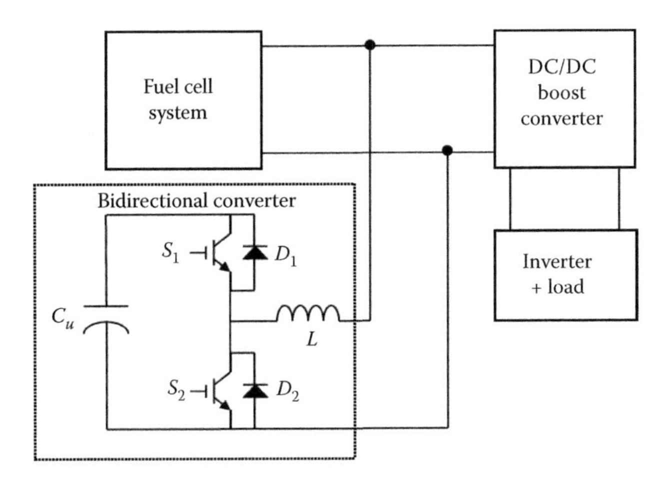

FIGURE 7.10
Functional diagram of the system.

UC can be charged when the fuel cell system operates in steady state by the boost operation of the converter. In this mode, the average small signal circuit model [14] is often derived as shown in Figure 7.11 to build up the system transfer functions.

where
009 is the fuel cell terminal voltage variation;
00 the inductor current variation;
$\hat{d}$ the duty cycle variation;
D' the DC values of $1 - d$;
L, V_{fc}, I_L, R_s, R_P, and C_u are the DC values; and
Z_{Leq} the load impedance.

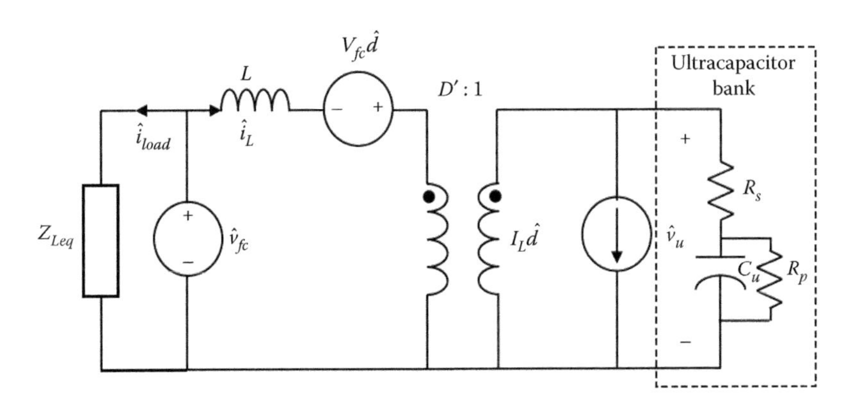

FIGURE 7.11
Average small signal circuit model of boost converter.

The UC bank in Figure 7.11 is comprised of the equivalent series resistance R_s and the parallel resistance R_p and the capacitance C_u [15]. In the boost mode, the fuel cell voltage works like a constant voltage source, while in the buck mode, the fuel cell system is modeled with an open voltage and RC model [16]. In this section, the parallel resistance R_p is not considered in the transfer functions because R_p is significantly larger than R_s, which is set to be infinite in most of the analyses [15]. The transfer functions from the duty cycle to the inductor current, and to the UC voltage can be obtained as shown in Figure 7.11.

$$\frac{\hat{i}_L}{\hat{d}} = G_{di1}(s) = V_{fc} \cdot D' \frac{s/L}{s^2 + (R_s(D')^2/L)s + ((D')^2/LC_u)} \tag{7.5}$$

$$\frac{\hat{v}_u}{\hat{v}_{fc}} = G_{vv1}(s) = (D') \frac{R_s Cs + 1}{LCs^2 + R_s C \cdot (D')^2 s + (D')^2} \tag{7.6}$$

The transfer functions from the fuel cell voltage to UC voltage is

$$\frac{\hat{v}_u}{\hat{d}} = G_{dv1}(s) = V_u \cdot D' \frac{(R_s/L)s + (1/LC_u)}{s^2 + (R_s(D')^2/L)s + ((D')^2/LC_u)} \tag{7.7}$$

The transfer function from the UC voltage to the inductor current is as shown below

$$\frac{\hat{i}_L}{\hat{v}_u} = G_{vi1}(s) = \frac{Cs}{D'} \tag{7.8}$$

Each transfer function from Equations 7.5 to 7.8 is very useful to construct the control system block diagram in Section 7.4.3 and analyze frequency responses using the Bode plot.

7.4.1.2 Buck Mode

When the load current is increased, the UC is discharged through the buck mode converter to stabilize the fuel cell terminal voltage. Like the boost mode, an average small signal circuit model is derived as below.

Where $\hat{v}_u$ is the capacitor voltage variation; $\hat{d}$ the duty cycle variation; and D the DC values of duty cycle d. In the fuel cell model shown in Figure 2.3; $\hat{i}_{fc}$ is the fuel cell current variation; E_o is the open circuit voltage; R_a is the sum of the activation and concentration resistance; C_a is the capacitive constant; and R_r is the ohmic resistance.

By assuming that the impedance of E_o in the fuel cell model be less than the internal impedances in other fuel cells, the total internal fuel cell impedance [16] can be calculated by

$$Z_{fc}(s) = \left(R_a + \frac{R_a}{R_a \cdot Cs + 1} \right) \tag{7.9}$$

And the fuel cell terminal voltage can be defined as

$$V_{fc} = E_o - Z_{fc}(s) \cdot I_{fc}(s) \tag{7.10}$$

The transfer functions from duty cycle to inductor current and to fuel cell voltage can be obtained based on Figure 7.12.

$$\frac{\hat{i}_L}{\hat{d}} = G_{di2}(s) = \frac{V_u}{D} \cdot \frac{Z_{Leq}Z_{fc}Cs + Z_{Leq} + Z_{fc}}{Z_{Leq}Z_{fc}LCs^2 + (Z_{Leq} + Z_{fc})Ls + Z_{Leq}Z_{fc}} \tag{7.11}$$

$$\frac{\hat{v}_{fc}}{\hat{d}} = G_{dv2}(s) = \frac{V_u}{D} \cdot \frac{Z_{Leq}Z_{fc}}{Z_{Leq}Z_{fc}LCs^2 + (Z_{Leq} + Z_{fc})Ls + Z_{Leq}Z_{fc}} \tag{7.12}$$

The transfer function from the UC voltage to the fuel cell voltage is given as

$$\frac{\hat{v}_{uc}}{\hat{v}_{fc}} = G_{vv}(s) = D \cdot \frac{Z_{Leq}Z_{fc}}{Z_{Leq}Z_{fc}LCs^2 + (Z_{Leq} + Z_{fc})Ls + Z_{Leq}Z_{fc}} \tag{7.13}$$

As seen in Figure 7.12, although the inductor current is coupled with the fuel cell current, the direct relationship between the inductor current and the fuel cell voltage can be derived by Z_{fc}.

$$\frac{\hat{i}_L}{\hat{v}_{fc}} = G_{iv}(s) = \frac{Z_{Leq}Z_{fc}Cs + Z_{Leq} + Z_{fc}}{Z_{Leq}Z_{fc}LCs^2 + (Z_{Leq} + Z_{fc})Ls + Z_{Leq}Z_{fc}} \tag{7.14}$$

Specific control architectures of boost and buck modes are described using the derived transfer functions in the following section.

7.4.2 Design of the Controller and Bandpass Filter

There are two voltage controllers: one is the UC voltage controller, and the other is the fuel cell voltage controller. And the current controller controls the inductor current in both converters. The control architecture will be different in each mode because the input and output are exchanged vice versa for charging and discharging modes.

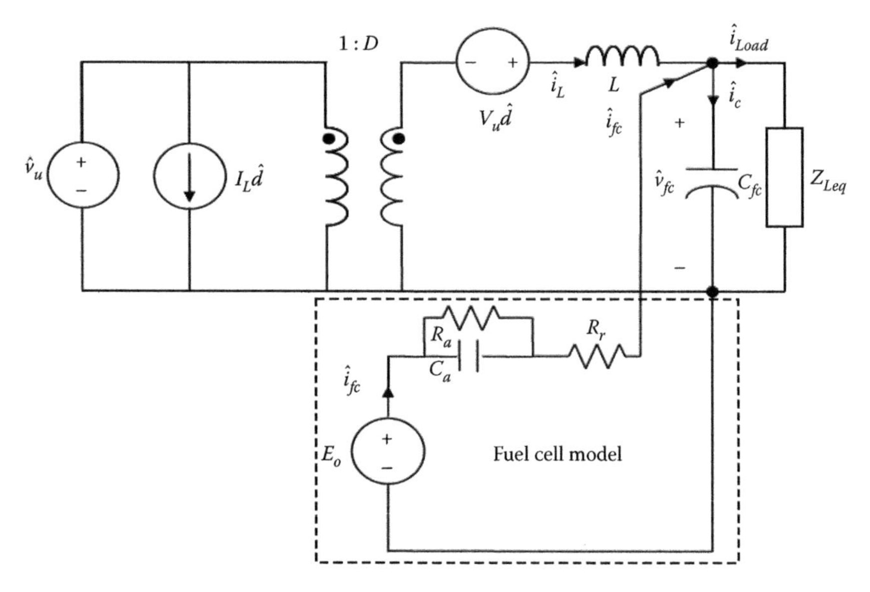

FIGURE 7.12
Average small signal circuit model of buck converter and fuel cell.

7.4.2.1 The Controller

7.4.2.1.1 Boost Mode

If the load variations keep lasting over 5 s and their capacities are less than the fuel cell stack capacity, 5 kW, the boost mode (charging mode) is activated. In steady states of fuel cell systems, since the fuel cell voltage controller charges the UC voltage, the fuel cell voltage controller is placed in the outermost loop, which generates the reference UC voltage through the BPF1, bandpass filter1, as shown in Figure 7.13. To generate the reference UC voltage 74 V, the initial duty ratio is set to $D = 0.527$ based on the reference fuel cell voltage being assumed to be 35 V because the output voltage V_o of the boost converter is calculated by the duty ratio and the input voltage V_{in} shown in Equation 7.16 during the continuous conduction mode [17].

$$\frac{V_o}{V_{in}} = \frac{1}{(1-D)} \tag{7.15}$$

The UC voltage controller in the inner loop also generates the reference inductor current through the BPF2, bandpass filter2. These two bandpass filters, BPF1 and BPF2, can determine the frequency range in each control loop. For the UC voltage controller in the inner loop, the low cutoff frequency of BPF2 is inferred to be $f_1 = 100$ Hz because at any given load the change is slower than f_1 during the boost mode, and the UC would respond to it. The high cutoff frequency of BPF2 is limited to be $f_2 = 10$ kHz to avoid the correlation between the switching frequency of the power converter.

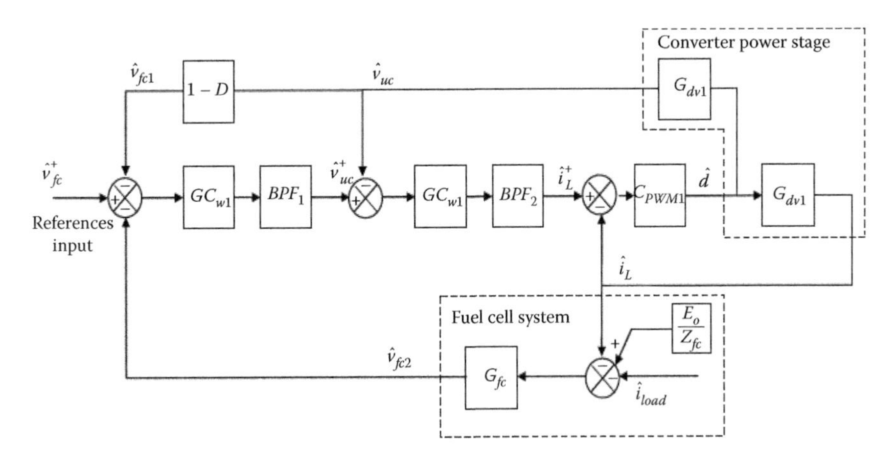

FIGURE 7.13
Control block diagram in the boost mode.

The outermost loop cutoff frequencies of BPF1 are set to be in the range ($0.1f_1$, $0.1f_2$) because it is expected to respond much slower than the inner loop controller. However, in the buck mode, as the UC voltage will stabilize the fuel cell voltage, the UC voltage controller must be placed in the outermost loop.

7.4.2.1.2 *Buck Mode*

If the load variations last for 5 s, and their power capacities exceed 5 kW, the buck mode is activated. As seen in Figure 7.14, during buck (discharging) mode, the UC voltage controller generates the reference fuel cell voltage, and the fuel cell voltage controller generates the reference inductor current of the converter. To generate the reference fuel cell voltage of 35 V, the initial duty

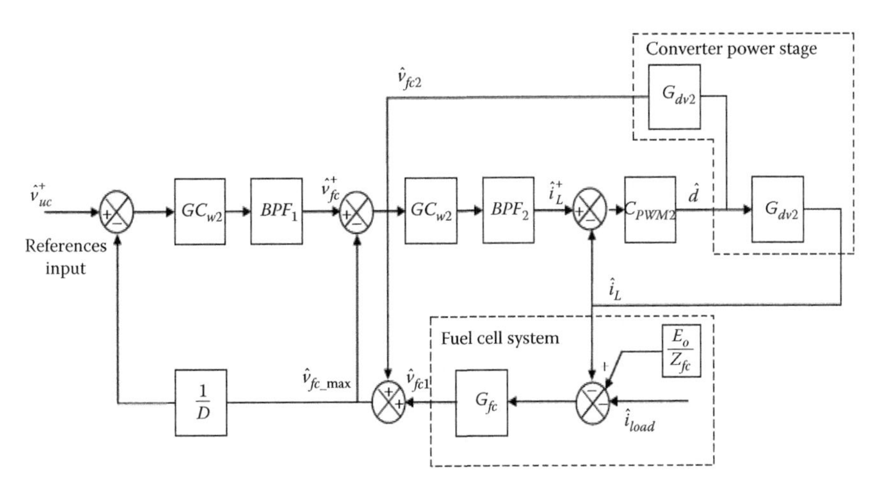

FIGURE 7.14
Control block diagram in the buck mode.

ratio is set to $D = 35/74 = 0.473$ based on the reference, UC voltage is assumed to be 74 V because the output voltage V_o of the buck converter is calculated by the duty ratio and the input voltage V_{in} is given in Equation 7.16 during the continuous conduction mode [15].

$$\frac{V_o}{V_{in}} = D \tag{7.16}$$

In Figures 7.13 and 7.14, the GC means the combined transfer function including the controllers. In terms of the fuel cell voltage feedback, the maximum fuel cell voltage output that is the sum of the fuel cell voltage from the buck converter, and the fuel cell system are considered because the fuel cell terminal voltage is supplied by these two sources, the fuel cell and the UC. The functions of bandpass filters are the same as in the boost mode.

7.4.2.2 Bandpass Filters Design

There are two bandpass filters, BPF1 and BPF2, in the control block diagram.

BPF1 is characterized by the frequency band $10\text{ Hz} < \omega < 1\text{ kHz}$ and BPF2's bandpass $100\text{ Hz} < \omega < 10\text{ kHz}$ is because the BPF1 actually has a slower response than that of BPF2. Especially, in the buck mode, for any load change which is lower than the low cutoff frequency 100 Hz, the fuel cell system can respond to it without the help of UC [18]. And the high cutoff frequency of BPF2 must be low enough so that it does not interact with the switching frequency 10 kHz of the bidirectional converter. Hence, the high cutoff frequency of BPF2 is limited to be 10 kHz, and the second-order wide bandpass filter [19] is used as in Equation 7.17 by letting $s \rightarrow j\omega$.

$$H(j\omega) = H_0 \frac{j\omega/\omega_L}{(1 + j\omega/\omega_L)(1 + j\omega/\omega_H)} \tag{7.17}$$

where H_o is the mid-frequency gain, ω_L, the low cutoff frequency, and ω_H, the high cutoff frequency.

On the basis of the characterized cutoff frequencies in the BPF1 and BPF2, the Bode plot of the bandpass filters is shown in Figure 7.15.

7.4.3 Simulation Results and Analysis

For the simulation, the 2500-ft^2 house load profile illustrated in Figure 7.16 [6] is used as a reference to calculate the transient power of the UC.

Figure 7.17 shows that the peak loads vary from 7 to 7.4 kW and the sampling interval of the load profile is 15 s. The maximum period of the peak demand is 75 s when the load demand exceeds 5 kW, which is the maximum available power from the fuel cell system. By using the system shown in

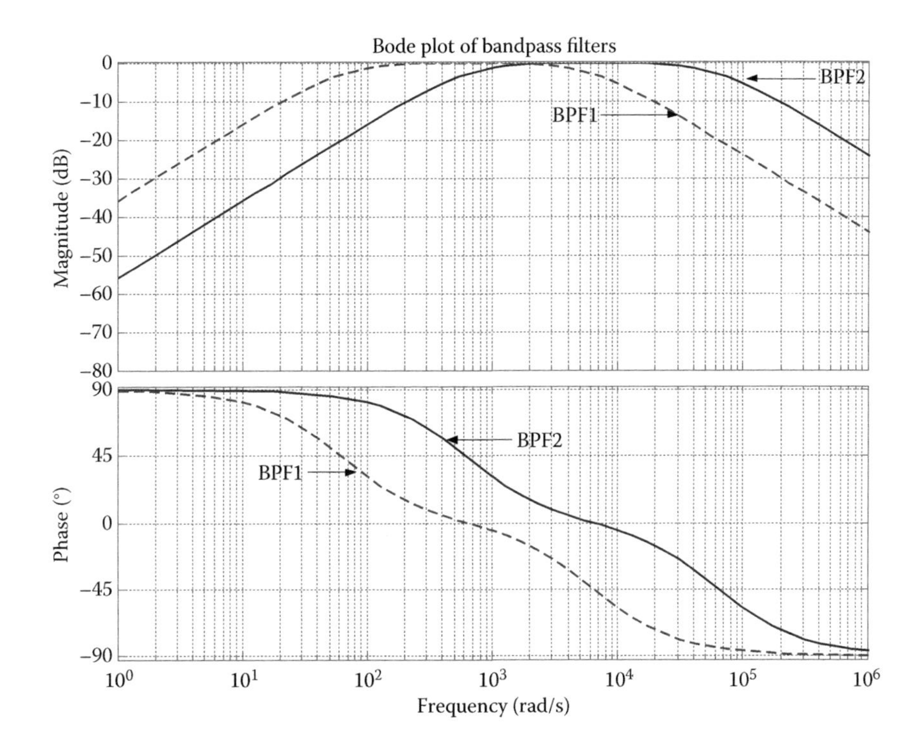

FIGURE 7.15
Bode plot of the bandpass filters BPF1 and BPF2.

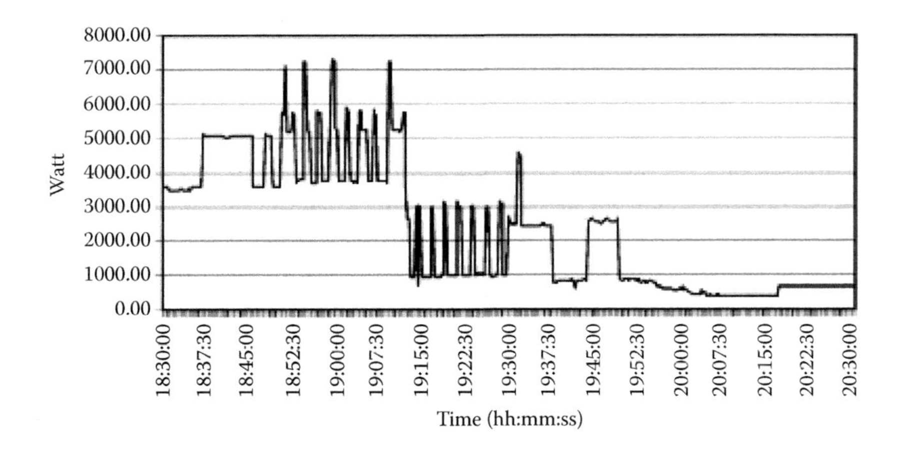

FIGURE 7.16
2500-ft^2 house load profile (Real power). (Adapted from M. Uzunoglu and Alam, M.S., *IEEE Transactions on Energy Conversation*, 21(3), 2006.)

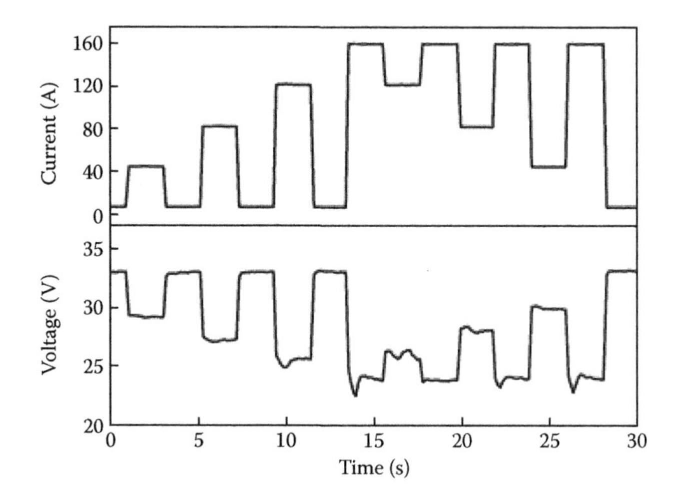

FIGURE 7.17
Fuel cell current and voltage under load variations. (Adapted from J. Hamelin et al., *International Journal of Hydrogen Energy*, 26, 625–29, 2001.)

Figure 7.10, the UC is able to support the extra load of 2.4 kW for 75 s, which is 50 Wh of the energy and its minimum energy capacity rating of the UC is 50/0.75 = 66.67 Wh [6].

The Maxwell boostcap PC2500 UC [20] is selected for the simulation. Its nominal voltage is 2.5 V, and the capacitance value is 2700 F. Thus, 30 units of UC in series (the total capacitance is 2700 F/30 = 90 F) is required to keep charging the reference voltage to 74 V through the bidirectional converter because 74 V/2.5 V = 29.6. The calculated energy storage can be $0.5 \times 90 \times 74^2/3600 = 68.46$ Wh, which can sustain the minimum energy rating of the UC of 66.67 Wh. However, if we directly connect the fuel cell system to the UC without the bidirectional converter, 15 UCs (total 2700/15 = 180 F) are needed because 36 V/2.5 V = 14.4. And to meet the extra load demand of 66.67 Wh, at least two 15 UCs in parallel and some more UCs are expected because 15 UCs in series can store the energy calculated by $0.5 \times 180 \times 36^2/3600 = 32.4$ Wh, and 30 UCs can support 64.8 Wh in this case. It is obvious that using the bidirectional converter is more beneficial, especially in terms of saving the capacitor, rather than just coordination with the UC without using the converter.

The simulation is carried out by a Ballard 5 kW PEMFC fuel cell stack model MK5-E composed of 36 cells; each cell has a 232 cm^2 active area, graphite electrodes, and a Dow membrane [21]. Experimental data from Hamelin et al. [21] were used to compare the validity of the proposed system. The voltage and the current, the most important variables, were used for this comparison. In Reference 21, a load profile with rapid variations between 0 and 150 A was imposed on the PGS-105B system [9]. The corresponding stack current and voltage transients are plotted in Figure 5.8, where the experimental data [21] are indicated by solid lines. The details of load profile are

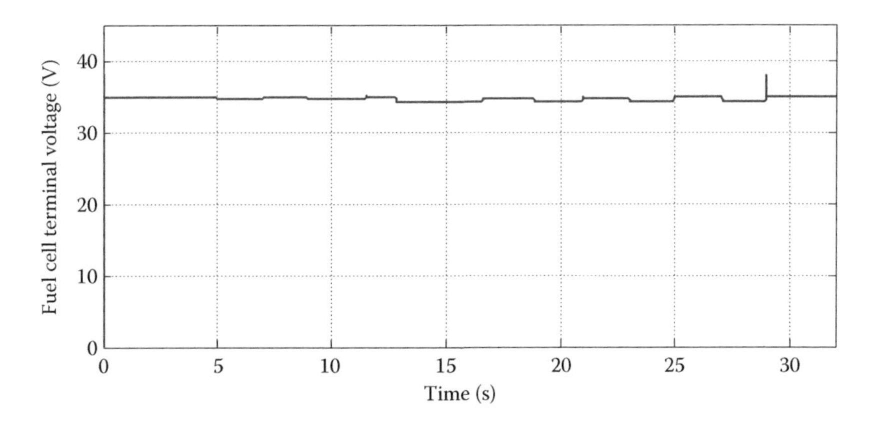

FIGURE 7.18
Fuel cell terminal voltage (discharging mode).

shown in Figure 7.17, where the load resistances were changed from 0.119 to 4.15 Ω during the simulation period.

As shown in Figure 7.17, the fuel cell terminal voltage drops below 25 V. The rapid current increase can cause an immediate voltage drop across the internal resistors (activation and concentration) of the fuel cell. Based on the frequent load changes seen in Figure 7.18, normally the buck mode (discharging) is imposed to the system.

The value of the inductor of the bidirectional converter is set to 1 mH to operate the boost, and the buck mode properly without a high surge voltage. With the help of the UC and bidirectional converter, the fuel cell terminal voltage does not drop below 30 V. But when the load resistance is changed from the smallest value of 0.119 Ω to the biggest value 4.15 Ω, a 2–3 V voltage spike can be observed.

Thus, in case of a sudden load change with a big variation, the surge protection should be mounted. In the simulation, the surge protection arrestor which has the maximum limit of 50 V and 200 A, is used.

Figures 7.19 and 7.20 show the load current and load power demand, respectively. Owing to the UC and bidirectional converter, the load current lifts up to 290 A, and the load power demand can reach up to 10 kW. To utilize the total load power demand of 10 kW, an additional UC bank 90 F is needed because this and the fuel cell stack can support the total power of 7.4 kW. The UC voltage gradually decreases, and a voltage drop can be observed until 68.7 V due to sudden load changes as shown in Figure 7.21.

In case of a light load change in which the total load power demand is less than 5 kW and the load change duration is beyond 5 s, the boost mode (charging mode) is imposed.

The fuel cell voltage varies between 35 V and ±0.3 V because of the light load as seen in Figure 7.22. The load power varies less than 5 kW and the load current is changed between 18 and 58 A in Figures 7.23 and 7.24.

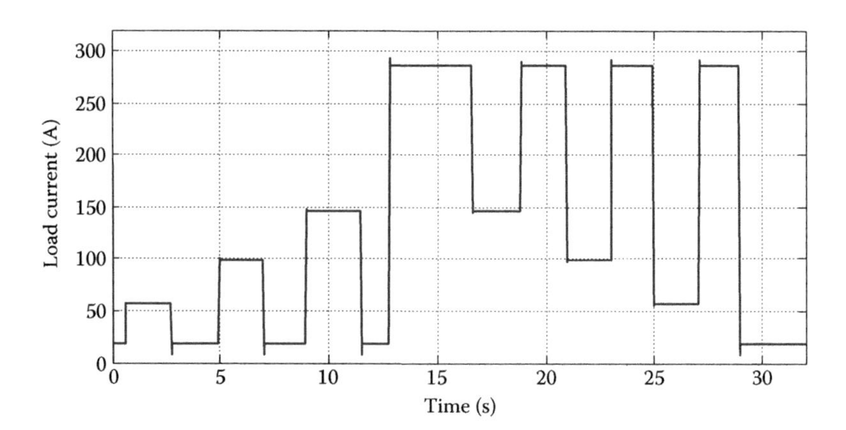

FIGURE 7.19
Load current (discharging mode).

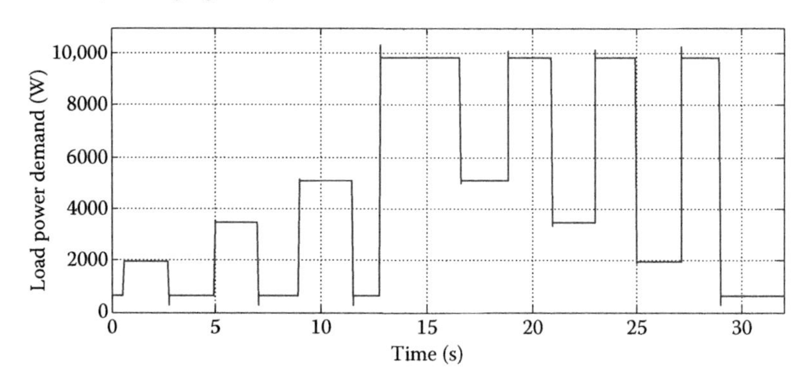

FIGURE 7.20
Load power demand (discharging mode).

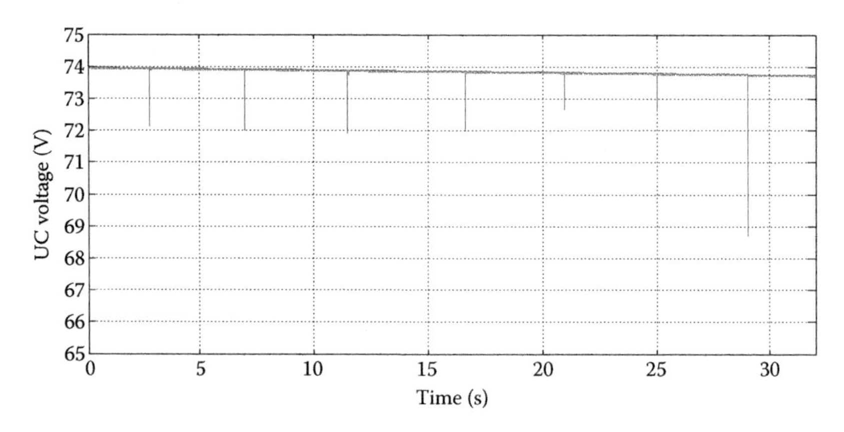

FIGURE 7.21
UC voltage (discharging mode).

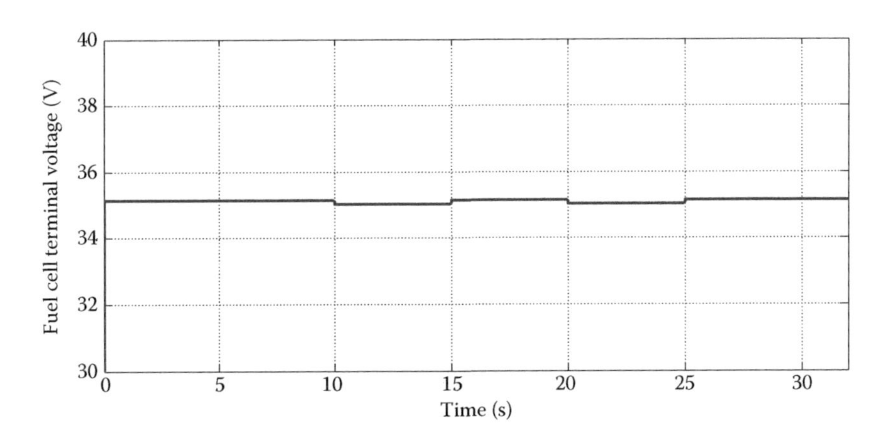

FIGURE 7.22
Fuel cell terminal voltage (charging mode).

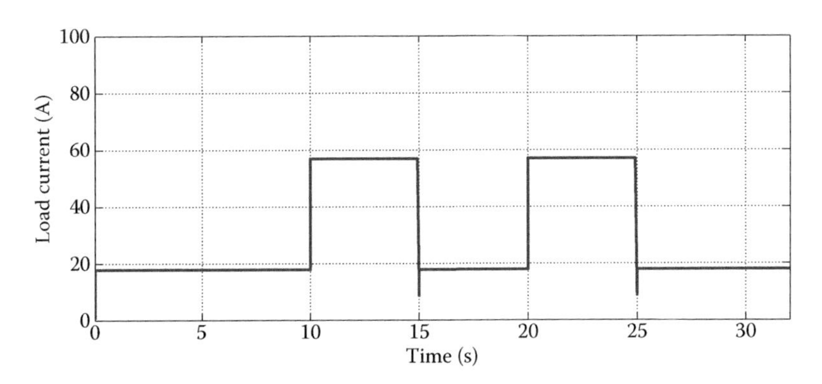

FIGURE 7.23
Load current (charging mode).

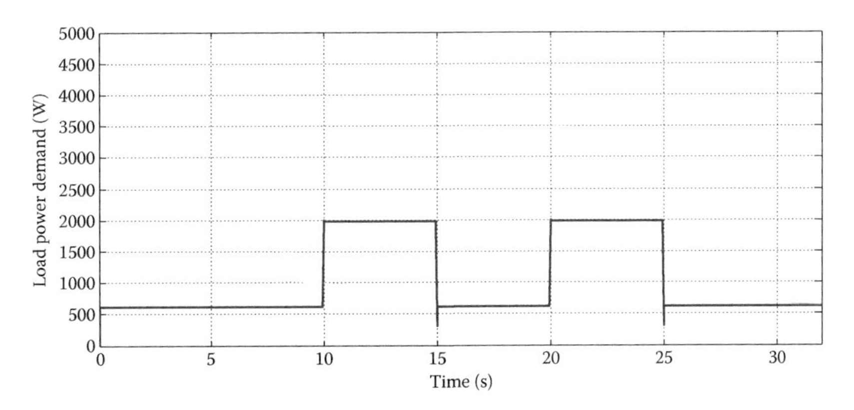

FIGURE 7.24
Load power demand (charging mode).

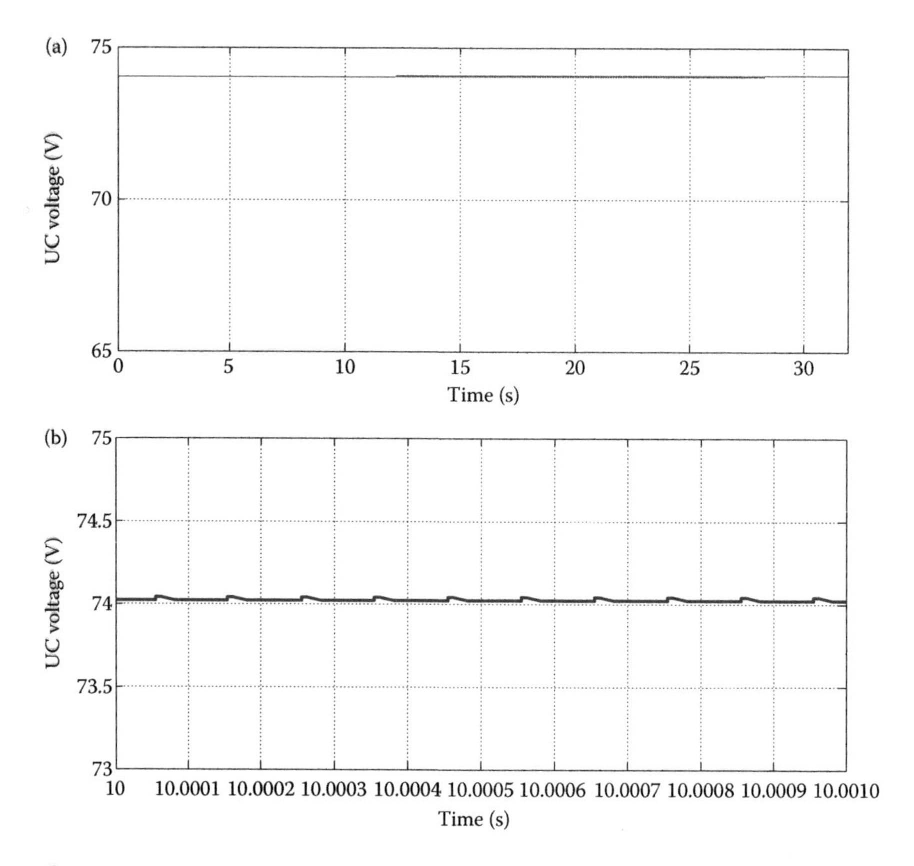

FIGURE 7.25
UC voltage (charging mode). (a) Total time duration. (b) Zoom in between 10 and 10.001 s.

To avoid a surge of current during the charging period, the initial condition of the UC is set to 70.3 V, 95% of the reference voltage. As seen in Figure 7.25b, a 10 kHz ripple can be found due to the switching frequency of the converter.

7.5 Conclusion

This chapter explained the applications of fuel cells to the utility power systems, stand-alone applications, and power electronics interface design for stand-alone fuel cell and UC-based power system. To validate the power interface design of the fuel cell-based hybrid system, the MATLAB–Simulink-based simulations have been carried out.

References

1. C. Wang, M.H. Nehrir, and H. Gao, Control of PEM fuel cell distributed generation systems, *IEEE Transactions on Energy Conversion*, 21, 586–595, 2006.
2. C. Wang, M.H. Nehrir, and S.R. Shaw, Dynamic models and model validation for PEM fuel cells using electrical circuits, *IEEE Transactions on Energy Conversion*, 20(2), 442–451, 2005.
3. A. Hirohisa, Y. Shigeo, I. Yoshiro, K. Junji, M. Tetsuhiko, Y. Hiroshi, M. Akinobu, and I. Itaru, Operational strategies of networked fuel cells in residential homes, *IEEE Transactions on Power Systems*, 21(3), pp. 1405–1414, 2006.
4. Fuel cell project team, senior vice-minister meeting: Report of Fuel Cell Project Team. Japanese Government (in Japanese), 2002.
5. DOE, Hydrogen posture plan, 2004, at https://www.hydrogen.energy.gov/pdfs/hydrogen_posture_plan_dec06.pdf.
6. M. Uzunoglu and M.S. Alam, Dynamic modeling, design, and simulation of a combined PEM fuel cell and ultracapacitor system for stand-alone residential applications, *IEEE Transactions on Energy Conversation*, 21(3), pp. 767–775, 2006.
7. M.Y. El-Sharkh, A. Rahman, M.S. Alam, A.A. Sakla, P.C. Byrne, and T. Thomas, Analysis of active and reactive power control of a stand-alone PEM fuel cell power plant, *IEEE Transactions on Power Systems*, 19(4), 2022–2028, 2004.
8. J. Larminie and A. Dicks, *Fuel Cell Systems Explained*, Wiley, New York, 2002.
9. Ballard Power System, Inc., Canada, at http://www.ballard.com.
10. A. Drolia, P. Jose, and N. Mohan, An approach to connect ultracapacitor to fuel cell powered electric vehicle and emulating fuel cell electrical characteristics using switched mode converter, *The 29th Annual Conference of the IEEE Industrial Electronics Society IECON '03*, Roanoke, VA, USA, Vol. 1, November 2–6, 2003, pp. 897–901.
11. P. Thounthing, S. Rael, and B. Davat, Control strategy of fuel cell/supercapacitor hybrid power sources for electric vehicle, *Journal of Power Source*, 158, 806–814, 2006.
12. M. Uzunoglu and M.S. Alam, Dynamic modeling, design and simulation of a PEM fuel cell/ultracapacitor hybrid system for vehicular applications, *Journal of Power Sources*, 48, 1544–1553, 2007.
13. J.C. Amphlett, R.M. Baumert, R.F. Mann, B.A. Peppy, P.R. Roberge, and A. Rodrigues, Parametric modeling of the performance of a 5-kW proton exchange membrane fuel cell stack, *Journal of Power Sources*, 49, 349–56, 1994.
14. R.W. Erickson and D. Maksimovic, *Fundamentals of Power Electronics*, Kluwer Academic Publishers, Dordrecht, 2000.
15. K.H. Hauer, *Analysis tool for fuel cell vehicle hardware software(control) with an application to fuel economy comparisons of alternative system designs*, PhD dissertation, Department Transport. Techno. Policy, University, Davis, California, 2001.
16. J. Larminie and A. Dicks, *Fuel Cell Systems Explained*, Wiley, New York, 2002.
17. N. Mohan, T.M. Underland, and W.P. Robibins, *Power Electronic: Converter, Applications, and Design* (2nd edition), John Wiley & Sons, Inc., New York.
18. L. Solero, A. Lidozzi, and J.A. Pomilio, Design of multiple-input power converter for hybrid vehicles, *IEEE Transactions on Power Electronics*, 20(5), 1007–1016, 2005.

19. S. Franco, *Design with Operational Amplifiers and Analog Integrated Circuits* (3rd edition), McGraw-Hill, New York, 2001.
20. Electric Double Layer Capacitor: BOOSTCAP Ultracapacitor, at http://www.maxwell.com/pdf/uc/datasheets/PC2500.pdf.
21. J. Hamelin, K. Abbossou, A. Laperriere, F. Laurencelle, and T.K. Bose, Dynamic behavior of a PEM fuel cell stack for stationary application, *International Journal of Hydrogen Energy*, 26, 625–29, 2001.

8

*Control and Analysis of Hybrid Renewable Energy Systems**

8.1 Introduction

Due to its capacity and operating characteristics, fuel cells normally work together with other alternative sources, for instance, the wind and solar. Several main issues including system configuration design, control methods, and power-conditioning system design need to be considered. Hybrid renewable energy power systems (HREPS) combine two or more energy conversion devices, or two or more fuels for the same device, that when integrated, overcome limitations inherent in either of them. Characteristics of distributed energy resources are (1) normally located at or near the point of use; (2) locational value; and (3) distribution voltage [1].

The advantages of hybrid renewable energy systems are [1]:

- Enhanced reliability

  Incorporating heat, power, and highly efficient devices (fuel cells, advanced materials, cooling systems, etc.) can increase the overall efficiency and conserve energy for a hybrid system when compared with individual technologies.
- Lower emissions

  HREPS can be designed to maximize the use of renewable resources, resulting in a system with lower emissions than traditional fossil-fueled technologies.
- Acceptable cost

  HREPS can be designed to achieve desired attributes at the lowest acceptable cost, which is the key to market acceptance.

Different alternative energy sources can complement each other to some extent, and multisource hybrid alternative energy systems (with proper control)

* This chapter was mainly prepared by Dr. Bei Gou, Smart Electric Grid, LLC and Dr. Woonki Na, California State University, Fresno.

have a great potential to provide higher quality and more reliable power to customers than a system based on a single resource. Because of this feature, hybrid energy systems have caught worldwide research attention [2–12].

8.1.1 Wind Power

Wind power has been available for centuries as a source of mechanical power, and since the 1890s as a source of electrical power. Producing power from the wind is now more important than ever. It is currently one of the fastest-growing sources of renewable energy in the United States with an approximately 30% annual increase over the last several years.

As a result of recent advances in technology, wind is becoming economically competitive with fossil fuels for power generation, especially in areas where electricity is expensive [13]. Lighter and stronger materials allow for greater energy capture and increase the lifespan of mechanical components. The newer generation of power and control electronics used in the conversion of wind into electricity includes the insulated gate bipolar transistor (IGBT) and the DSP. The latest IGBTs (insulated gate bipolar transistors) have both faster switching times and higher power ratings than previous generations, and the modern DSPs have more computing power and higher speeds than previous digital controllers. Faster switching and higher speed processing means higher-quality power is generated with less harmonic distortion. Higher power ratings and greater computing capability in a single chip allow for reduction in the total number of electronic components, which means smaller, less-expensive systems.

Technological advances are not the only reason for the recent growth in the wind industry. Today, we are much more aware of the environmental and political problems that are associated with using fossil fuels than we were a few decades ago. Although there is a wide spectrum of positions on environmental issues such as air pollution, global warming, and holes in the ozone layer, the debate now centers on the extent of the damage, and not whether the damage is actually occurring. No one seriously believes that humans are not doing at least some damage to our environment. We will inevitably have to address this problem. Wind is obviously a much more environmentally friendly source of energy than fossil fuels. No harmful emissions are produced and no water is consumed when generating power from the wind. One fact not widely known is that fossil fuel power generation consumes as much as 39% of the domestic freshwater usage in the United States [14,15]. In the drier areas of the country, for example, west Texas, optimizing water usage is critical for the sustainability and future growth of the populated areas, both urban and agricultural.

In addition to the environmental advantages of wind over fossil fuels, there are also political benefits to using wind power. Wind is both plentiful and renewable—the fuel is free and does not need to be conserved. There is more-than-enough wind available within our own borders to meet all of our power demands far into the future. With the continued development and expansion of electric and hydrogen-powered vehicles, we could reduce the amount of time, effort, and money spent on securing our foreign oil and gas supplies, which are also a part of the hidden costs of fossil fuels. By producing a larger portion of the country's required power from domestically available resources, we could greatly reduce the potential for problems related to international politics.

8.1.2 Hybrid Power

The energy content of the wind varies with the cube of the wind speed. From Newton's laws, it is known that the kinetic energy of a given mass is proportional to the square of the velocity. Added to this is the fact that increasing the speed of the wind also increases the mass of wind passing through a given plane proportionally. The result is that doubling the wind speed yields an eightfold increase in the available energy.

However, the main disadvantage of the wind as a source of energy is its inherent variability. Wind energy comes in a form that typically varies with time. The energy can be collected only when it is available, which makes it unsuitable as a primary source. The traditional solution to this problem, particularly for isolated rural systems, is to supplement the wind turbine with a diesel generator; however, the wind–diesel combination does not entirely eliminate the problems associated with fossil fuel use.

8.1.3 Fuel Cell Power

An alternative remedy is to supplement the wind with hydrogen fuel cells, which offer the availability lacking in the wind while retaining many of the advantages. Wind turbines are efficient for collection of energy, while fuel cells are efficient at producing electrical energy from hydrogen. Moreover, fuel cells continue to operate efficiently under part load conditions, while the efficiency of diesel generators drops substantially when operating away from rated capacity.

Pure hydrogen is a clean, nonpolluting fuel. But it should be noted that the source of the hydrogen is an important consideration for a truly renewable power source. Hydrogen is potentially renewable if an electrolyzer is used with renewable sources such as wind or solar power. Effectively, water (for the electrolyzer) is the only required fuel for such systems. Also, the hydrogen generated from electrolysis is generally of a much greater purity than hydrogen from steam reformation of hydrocarbons.

Fuel cells alone could supply clean energy reliably, but at a much higher cost than the wind. Fuel cells typically cost 3–5 times more per kilowatt of generating capacity than wind turbines currently, and they require fuel [16]. An integrated system would use wind and fuel cells in a complementary fashion, offering the chance to provide steady renewable power at a lower cost. This can be achieved when hydrogen is generated by an electrolyzer and stored whenever wind power exceeds the load demand. Such a hybrid system could be connected to the utility network, adding reliability and security to it through distributed generation. A stand-alone version of the system could also provide power on demand for locations that are remote from the utility grid.

In this chapter, we will introduce several typical designs of hybrid renewable energy systems.

8.2 Hybrid System Consisted of Wind and Fuel Cell Sources

8.2.1 Hybrid System Simulation Components and Equations

Figure 8.1 illustrates the hybrid wind turbine–fuel cell system model considered in a recent study [17], and Table 8.1 contains the associated system parameters and their values. The system is modeled using MATLAB and Simulink software and is similar to that previously described by Iqbal [18] with some modifications and corrections. In particular, the following changes were made to the simulation model presented in Reference 18:

- Fuel cell stack equations replaced with a model provided by the U.S. DoE [19,20] with the DoE fuel cell controller modified to regulate stack voltage instead of fuel utilization.
- DC load replaced with a DC–AC inverter and AC load model.
- A wind turbine and fuel cell system interconnection (not specified by Iqbal), consisting of a filter capacitor and inductor, was added.
- Mathematical errors in the wind turbine model derivation were corrected.

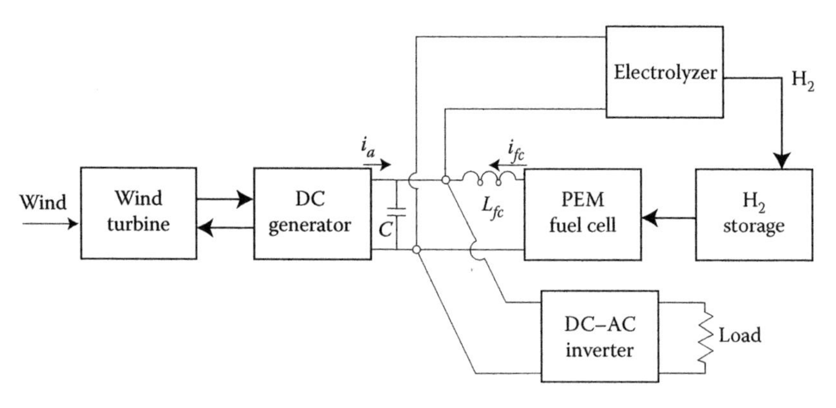

FIGURE 8.1
Complete hybrid wind turbine–fuel cell system.

TABLE 8.1

Parameter List for Wind Turbine–Fuel Cell System

Parameter	Value	Description
R	3.25 m	Wind turbine rotor radius
θ_P	2°	Fixed pitch of rotor blades
J_{eq}	85 kg m^2	Drivetrain moment of inertia
B_{eq}	0.5 N m s/rad	Drivetrain damping factor
N	14	Gearbox ratio
R_a	2.639 Ω	Armature resistance
L_a	34.6 mH	Armature inductance
R_f	335 Ω	Field-winding resistance
L_f	156 H	Field-winding inductance
K_g	0.08912	Generator constant
K_f	56	Field circuit voltage gain
n	220	Number of cells in the fuel cell stack
A_C	136.7 cm^2	Cell active area
V_a	6.495 cm^3	Volume of anode
V_c	12.96 cm^3	Volume of cathode
L_{fc}	25 mH	System interconnection inductance
C	25 mF	System interconnection capacitance

A description of each of the subsystems and the governing equations is presented in the subsequent sections.

8.2.1.1 Wind Turbine Subsystem

Wind energy is converted into electricity through the variable-speed turbine and DC generator. The point wind speed u_i, as measured by an anemometer, is first translated to an effective wind speed u_e for use in the wind turbine model to account for the variations in forces and torque applied to the entire swept area of the turbine rotor. The effective wind speed is found by using a spatial filter in state-space form [21]

$$
\begin{aligned}
du_1/dt &= -0.6467u_iu_1 - 0.084335u_i^2u_2 + u_i \\
du_2/dt &= u_1
\end{aligned}
\tag{8.1}
$$

where u_1 and u_2 are the state variables used to calculate the effective wind speed

$$
u_e = 0.2242u_iu_1 + 0.084335u_i^2u_2 \tag{8.2}
$$

The power coefficient C_P is the ratio of the mechanical power removed from the wind to the total power available in the wind passing through the turbine. The instantaneous C_P for a given wind turbine geometry is based on the tip speed ratio λ [22], which is the ratio of the speed of the rotor blade tip to the incoming wind speed: λ is one of the defining quantities used to determine the aerodynamic performance of a wind turbine. The tip speed ratio is thus given by

$$\lambda = \omega R / V_0 \tag{8.3}$$

where ω is the rotational speed of the turbine and R is the rotor radius [22]. By allowing the rotor to change its speed as the wind speed changes, λ can be held constant, which means the C_P also remains constant, an important factor for maximum energy capture. A typical maximum C_P for real wind turbines is 0.4 due to losses.

The corresponding torque coefficient C_q may be found by

$$C_q = C_P / \lambda \tag{8.4}$$

The aerodynamic torque T_w induced on the rotor is defined as

$$T_w = (1/2)\rho\pi R^3 u_e^{\,2} C_q(\lambda)\cos^2(\theta) \tag{8.5}$$

where C_q is a function of λ. The θ in Equation 8.5 refers to the yaw misalignment angle between the wind velocity and the turbine axis. In Reference 17, the yaw misalignment is assumed to be zero. The following equation is used to determine the torque coefficient C_q from the manufacturer's data for the considered wind turbine:

$$C_q(\lambda) = -0.0281 + 0.0385\lambda - 0.0046\lambda^2 + 0.000148\lambda^3 \tag{8.6}$$

The maximum value of C_q corresponds to a tip speed ratio of approximately 5.8. The torque calculation is subject to a modification due to unsteady aerodynamics. The expected torque T_{we} on the rotor shaft is calculated from the aerodynamic torque T_w using a lead-lag filter in state-space form given by [21]

$$\begin{aligned} dT_1/dt &= -0.0793 u_e T_1 + T_w \\ T_{we} &= -0.0293 u_e T_1 + 1.37 T_w \end{aligned} \tag{8.7}$$

Finally, the equation describing the complete drive train torque balance on the wind turbine side is given as

$$J_{eq}(d\omega/dt) + B_{eq}\omega = T_{we} - NT_e \tag{8.8}$$

where J_{eq} is the equivalent moment of inertia for the drive train and B_{eq} is the equivalent frictional damping coefficient. T_e is the electrical torque from the generator, which is multiplied by the gearbox ratio N: the gearbox is used to step up the shaft speed of the generator.

Wind speed input to the model is introduced as a constant speed, a step change in speed, or a variable-speed profile. The variable-speed wind is simulated by the SNwind v1.0 program, originally developed at Sandia National Laboratories and subsequently modified by the National Renewable Energy Laboratory (NREL). The SNwind program generates full-field turbulent wind data in three dimensions for specified meteorological conditions. The desired conditions for the SNwind simulation are given in an input parameter file. The parameters include turbulence intensity, turbulence spectral model, average wind speed at a reference height, and height of the turbine rotor hub.

8.2.1.2 DC Generator Subsystem

A 5 kW DC generator with a separately excited field circuit is assumed to be used for converting the mechanical energy from the wind turbine into electrical energy. The armature circuit is connected to an external load through a commutator and brushes for unidirectional current flow. The DC generator provides good speed and torque regulation over a wide operating range and is more easily implemented for variable-speed operation than an AC generator.

The field circuit is a series R–L circuit independent of the armature circuit. The differential equation describing the field circuit is

$$di_f/dt = (1/L_f)(K_f V_{fi} - R_f i_f) \tag{8.9}$$

where i_f is the field current, R_f the field-winding resistance, and L_f is the field inductance. V_{fi} is the input voltage to the field controller and is multiplied by the gain K_f to give the actual voltage V_f applied to the field circuit as

$$V_f = K_f V_{fi} \tag{8.10}$$

The armature circuit located on the rotor is also modeled as an inductance and a resistance in series with a voltage. The voltage E_a across the armature, also called the air gap voltage, is related to the current in the field windings by the equation

$$E_a = K_g \omega_m (206.7 i_f^3 - 760.75 i_f^2 + 982.3 i_f + 5.26) \tag{8.11}$$

Equation 8.11 is derived from the generator saturation characteristics provided by the manufacturer: K_g is a constant that depends on the construction of the generator—rotor size, number of rotor turns, and other details, and ω_m is the generator rotor speed, which is N times faster than the wind turbine rotor speed ω as a result of the gearbox. The differential equation describing the armature circuit is

$$di_a/dt = (1/L_a)(E_a - i_a(R_L + R_a)) \tag{8.12}$$

where i_a is the armature current, R_a is the armature resistance, and L_a is the armature inductance. Equation 8.12 is similar to Equation 8.9 for the field circuit; however, the equivalent load resistance R_L is added to the armature resistance R_a in the case of the armature circuit. The electrical torque generated is found by dividing the electrical power (voltage E_a times the armature current i_a) in the armature circuit by the generator rotor speed so that

$$T_e = E_a i_a / \omega_m \tag{8.13}$$

8.2.1.3 Wind Turbine and DC Generator Controller

The controllers used for the generator portion of the combined wind–fuel cell system model in Reference 17 are discrete-time PID controllers derived by the backward difference method. The backward difference method is based on Euler's method and is unconditionally stable [23]. The general transfer function for the controller is [24]

$$G(z) = (q_0 + q_1 z^{-1} + q_2 z^{-2})/(1 - z^{-1}) \tag{8.14}$$

The form of the controller used in the simulation is

$$y(k) = y(k-1) + q_0 e(k) + q_1 e(k-1) + q_2 e(k-2) \tag{8.15}$$

where $e(k)$ is the error (controller input) signal and $y(k)$ is the controller output signal. The q terms are obtained from the PID constants

$$\begin{aligned} q_0 &= K_p(1 + (T_d/T_s)) \\ q_1 &= K_p((T_s/T_i) - 2(T_d/T_s) - 1) \\ q_2 &= K_p(T_d/T_s) \end{aligned} \tag{8.16}$$

where K_p, T_i, and T_d are the proportional, integral, and derivative constants, respectively. A sampling time of T_s is used for the controller. The original form of the controller used by Iqbal was kept; only the gains were adjusted. All PID control parameters are listed in Table 8.2.

TABLE 8.2

Parameter List for Discrete- and Continuous-Time PID Controllers

Parameter	Value	Description
K_p	0.003	Omega-control proportional constant
T_i	5	Omega-control integral constant
T_d	1	Omega-control derivative constant
T_s	0.25	Omega-control sample period
K_{plm}	2.5×10^{-4}	Lambda-control proportional constant
T_{ilm}	0.08	Lambda-control integral constant
T_{dlm}	0	Lambda-control derivative constant
T_{slm}	0.25	Lambda-control sample period
K_{pfc}	4	Fuel flow-control proportional constant
K_{ifc}	1	Fuel flow-control integral constant
K_{dfc}	1	Fuel flow-control derivative constant

For the below-rated wind speed condition, the wind turbine is operated in a variable-speed mode. The lambda-controller attempts to maintain a constant tip speed ratio of $\lambda = 6.5$ by adjusting the field voltage applied to the generator. The value of $\lambda = 6.5$ rather than the optimal value of 5.8 was used for convenience, since proper regulation to this lambda value then results in a rotor speed (in rad/s) that is simply twice the wind speed (in m/s). Above the rated wind speed of $u_e = 8$ m/s, the wind turbine is regulated by the omega-controller to a constant rotor speed of $\omega = 16$ rad/s. Again, control is achieved by adjusting the generator field voltage.

8.2.1.4 PEMFC Subsystem

This is described here for the sake of completeness. The model of the PEMFC in Reference 17 begins with the ideal potential defined by the Nernst equation as described by Larminie and Dicks [25], which relates the concentrations of reactants and products to the potential difference produced by an electrochemical reaction under equilibrium conditions. The fuel cell model provided by the DoE for the 2001 Future Energy Challenge student design competition [19,20], was then used to construct the cell voltage equation as

$$V = N\left(OCV + \frac{R^*T}{2F}\ln\left\{ \frac{PH_2(PO_2/P_{std})}{PH_2O} \right\} - L \right) \tag{8.17}$$

which has also been previously described in Chapter 3.

The actual voltage produced by the PEMFC is derived by subtracting the irreversible losses L from the ideal potential defined above [25]. These irreversible losses fall into four primary categories: activation losses, ohmic losses, concentration losses, and internal current losses. The voltage losses L can be described by

$$L = (i + i_n)r + a\ln\left(\frac{i + i_n}{i_o}\right) - b\ln\left(1 - \frac{i + i_n}{i_l}\right) \tag{8.18}$$

as discussed in Chapter 3.

The partial pressures of hydrogen, oxygen, and water are defined as the three state variables of the system. Also, since water management affects the fuel cell's performance, humidifiers are used on both anode and cathode sides to control the humidity inside the cell. The consideration of water on the cathode side is more complicated than on the anode side because it includes not only the water supplied from the humidifiers, but also the by-product of the reaction.

Figure 8.2 is an illustration depicting how these gases flow in and out of the cell.

On the basis of the ideal gas law $P^*V = n^*R^*T$, the partial pressure of each gas is proportional to the amount of the gas in the cell, which is equal to the gas inlet flow rate minus gas consumption and gas outlet flow rate. Thus the state equations are

$$\begin{aligned}
\frac{dPH_2}{dt} &= \frac{R^*T}{V_a}(\mathrm{H}_{2\mathrm{in}} - \mathrm{H}_{2\mathrm{used}} - \mathrm{H}_{2\mathrm{out}}) \\
\frac{dPO_2}{dt} &= \frac{R^*T}{V_c}(\mathrm{O}_{2\mathrm{in}} - \mathrm{O}_{2\mathrm{used}} - \mathrm{O}_{2\mathrm{out}}) \\
\frac{dPH_2O_C}{dt} &= \frac{R^*T}{V_c}(\mathrm{H_2O}_{\mathrm{Cin}} + \mathrm{H_2O}_{\mathrm{Cproduced}} - \mathrm{H_2O}_{\mathrm{Cout}})
\end{aligned} \tag{8.19}$$

where H_{2in}, O_{2in}, and H_2O_{cin} are the inlet flow rates of hydrogen, oxygen, and water of cathode, respectively; H_{2out}, O_{2out}, and H_2O_{cout} are the outlet flow

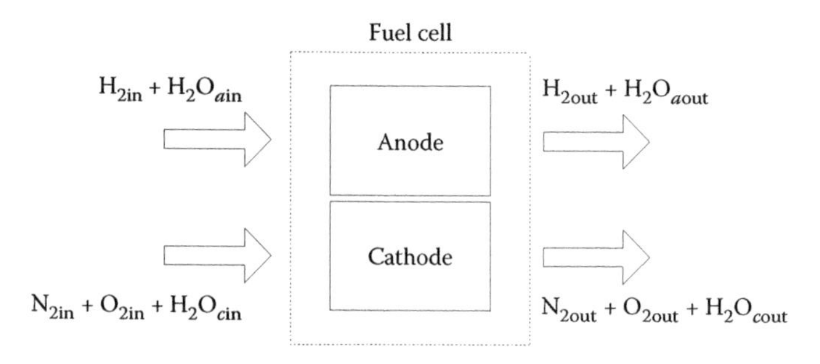

FIGURE 8.2
Illustration of gas flows of the PEMFC.

rates of each gas. Furthermore, H_{2used}, O_{2used}, and $H_2O_{cproduced}$ represent usage and production of the gases, which are related to output current I by

$$H_{2used} = 2O_{2used} = H_2O_{cproduced} = 2K_rI = 2K_rA_ci \tag{8.20}$$

where $K_r = N/4F$ in mol/C, A_c is the cell active area in cm^2, and i is the current density in A/cm^2.

Since the inlet flow rates and output current are measurable, the outlet flow rates can be defined by the equations

$$\begin{aligned} H_{2out} &= (Anode_{in} - 2K_rA_ci)\ FH_2 \\ O_{2out} &= (Cathode_{in} - K_rA_ci)\ FO_2 \\ H_2O_{cout} &= (Cathode_{in} + 2K_rA_ci)\ FH_2O_c \end{aligned} \tag{8.21}$$

where $Anode_{in}$ and $Cathode_{in}$ are the summations of anode inlet flow and cathode inlet flow, respectively, as defined in Figure 8.2; FH_2, FO_2, and FH_2O_c are the pressure fractions of each gas inside the fuel cell. For this DoE model

$$\begin{aligned} FH_2 &= \frac{PH_2}{P_{op}} \\ FO_2 &= \frac{PO_2}{P_{op}} \\ FH_2O_c &= \frac{PH_2O_c}{P_{op}} \end{aligned} \tag{8.22}$$

where it is assumed that the cell pressure remains constant at P_{op}, a steady-state operating pressure of 101 kPa.

8.2.1.5 Fuel Cell Controller

A continuous PID controller adjusts the fuel flow rate of the hydrogen at the fuel cell inlet to control the fuel cell stack voltage. The controller transfer function is described by

$$G(s) = K_{pfc} + (K_{ifc}/s) + sK_{dfc} \tag{8.23}$$

Again, the controller parameter values are listed in Table 8.2.

The flow rate for the air in the simulations performed in Reference 17 was fixed at a large-enough value that is more than sufficient to match the hydrogen flow needed to satisfy the given fuel cell current demands.

The fuel flow controller's objective is to regulate the inverter input voltage to a steady-state value of 200 V after the stack is activated to supplement the available wind power. If the voltage starts to fall below this value, the hydrogen flow rate is increased to compensate. If the voltage starts to rise above this value (when the wind power is sufficient to supply the load), the flow rate signal is reduced by the controller, and the fuel cell current drops to zero.

8.2.1.6 Electrolyzer Subsystem

When the available wind power is insufficient to supply the load, the fuel cell delivers additional electricity by converting the stored hydrogen. But if the wind is strong enough to completely supply the load, the fuel cell is inactive, and any excess wind energy can then be sent to the electrolyzer to recharge the hydrogen storage.

The unipolar Stuart electrolyzer's operation is governed by the equation

$$X_{H_2} = 5.18e^{-6} i_{elec} \tag{8.24}$$

where X_{H_2} is the flow rate of hydrogen in mol/s and i_{elec} is the current through the electrolyzer [26], which is equivalent to 0.0128 mol/s A.

8.2.1.7 Equivalent Load and System Interconnection

For the hybrid wind–fuel cell system described in Reference 17, the equivalent load is in parallel with a filter capacitor C. A series inductor L_{fc} is used to allow for a voltage difference at the interconnection of the DC generator and the fuel cell during transient events. The combined current from each subsystem flows to the load. The capacitor and inductor are ideal components and consume no real power. Figure 8.3 illustrates the details of this interconnection.

The next equation shows the calculation for the current i_{fc} drawn from the fuel cell through the inductor as

$$i_{fc} = (1/L_{fc})\int(V_{stack} - V_L)dt + i_{fc}(0) \tag{8.25}$$

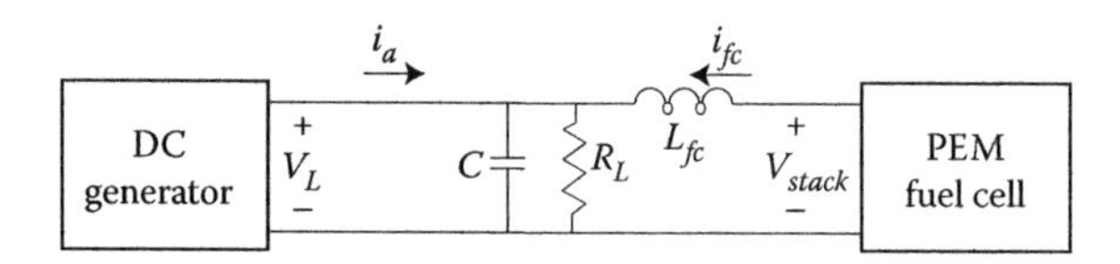

FIGURE 8.3
Equivalent load resistance with an interconnecting capacitor and inductor.

where V_{stack} and V_L are the stack voltage and load voltage, respectively, and $i_{fc}(0)$ is the inductor's initial current. V_L is equal to the voltage of the DC generator and is calculated by

$$V_L = (1/C)\int\{i_a + i_{fc} - (V_L/R_L)\}dt + V_L(0) \tag{8.26}$$

where i_a is the armature current from the generator, $V_L(0)$ is the initial voltage of the capacitor, and R_L is the equivalent load resistance representing the DC–AC inverter and AC load [27].

8.2.2 Simulation Results

Simulation runs were conducted with the wind–fuel cell system model in Reference 17 for above- and below-rated wind speed (8 m/s) conditions and for loads that were greater than and less than the available wind power. The different operating conditions simulated in these runs correspond to step changes in wind speed and equivalent load resistance and then to a more realistic wind speed profile.

8.2.2.1 Below-Rated Wind Speed Conditions (Wind Power > Load)

The fuel cell is not producing power when the available wind is greater than the load. For this case, it is assumed that the system is in operation at a wind speed of 6 m/s with an equivalent load resistance R_L of 30 Ω. Figure 8.4 shows the response of the rotor speed and inverter input voltage for a step change (at 10 s) in the input wind speed from 6 to 8 m/s. The lambda-controller regulates the rotor speed to 12 and 16 rad/s, respectively, for these two wind speeds.

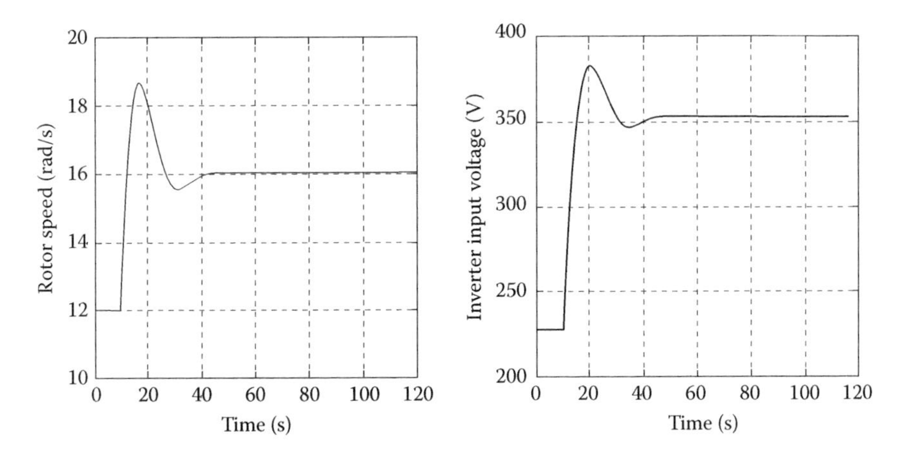

FIGURE 8.4
System response to step change in (below rated) wind speed (wind turbine generator only).

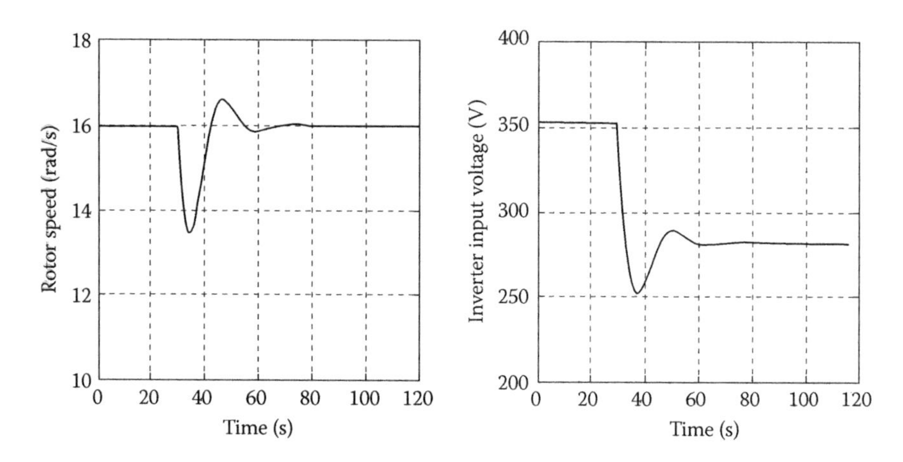

FIGURE 8.5
System response to step change in load resistance (wind turbine generator only).

Figure 8.5 shows the response of the rotor speed and inverter input voltage for a subsequent step change in R_L from 30 to 20 Ω with 8 m/s wind. The load resistance change emulates a change in the demand or the addition of the electrolyzer to take advantage of the "excess" available power. Again, note the effective regulation of the rotor speed provided by the lambda-controller to 16 rad/s.

8.2.2.2 Above-Rated Wind Speed Conditions (Wind Power > Load)

For this case, the fuel cell is again not producing power, and R_L is 20 Ω. It is assumed that the system is in operation at a wind speed of 8 m/s. Figure 8.6

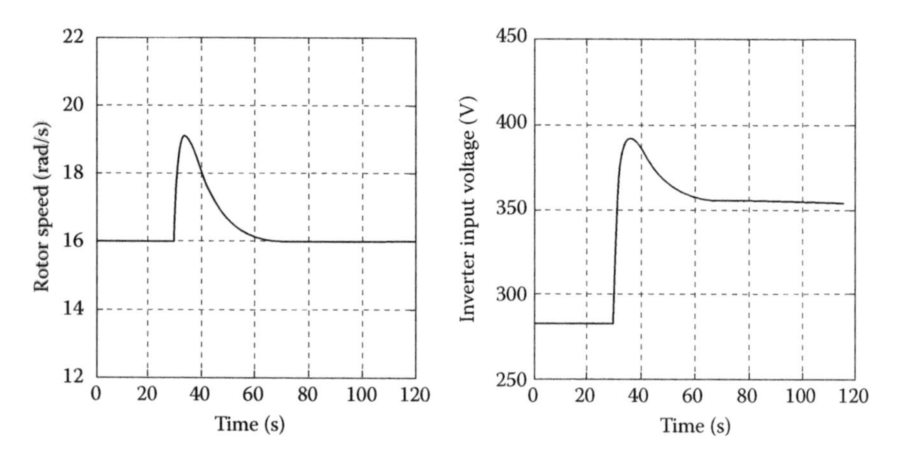

FIGURE 8.6
System response to step change in (above rated) wind speed (wind turbine generator only).

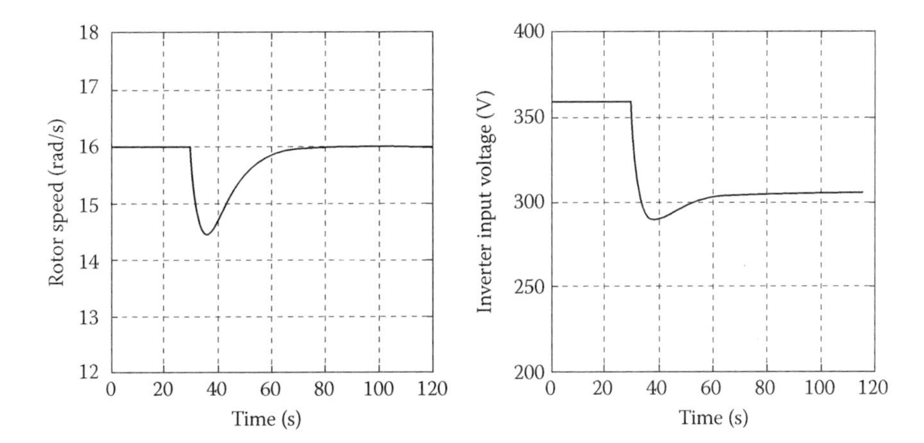

FIGURE 8.7
System response to step change in load resistance (wind turbine generator only).

shows the response of the rotor speed and inverter input voltage for a step increase (at 30 s) in the input wind speed from 8 to 10 m/s. Note that the rotor speed is now regulated by the omega-controller to 16 rad/s as the input wind speed exceeds the rated value.

Figure 8.7 shows the response of the rotor speed and inverter input voltage for a subsequent step change in the equivalent load resistance from 20 to 15 Ω. Again, note the effective regulation of the rotor speed provided by the omega-controller to 16 rad/s.

8.2.2.3 Below-Rated Wind Speed Conditions (Wind Power < Load)

For this case, the wind power is insufficient to meet the load so that the fuel cell produces the additional power needed to ensure a minimal inverter input voltage. It is assumed that the combined system is in operation at 6 m/s wind speed with an equivalent load resistance of 15 Ω. Figure 8.8 shows the response of the rotor speed, inverter input voltage, generator current, and fuel cell current for a step decrease (at 10 s) in input wind speed from 6 to 5 m/s. Note the regulation of the rotor speed provided by the lambda-controller to 12 and 10 rad/s, respectively, for these two wind speeds, as well as the regulation of the inverter input voltage with a maximum undershoot of 1.8%. It is preferred, but not essential, for the voltage deviation to be kept to ±3% of the nominal value; the inverter will further regulate the actual output delivered to the load. A total of 200 V is deemed to be the minimal voltage for effective operation of an inverter to produce utility-quality 115 Vrms output.

Figure 8.9 shows the response of the rotor speed, inverter input voltage, generator current, and fuel cell current for a subsequent step change in R_L from 15 to 20 Ω. The decrease in the demand results in a corresponding drop in the fuel cell current. Again, note the effective regulation of the inverter

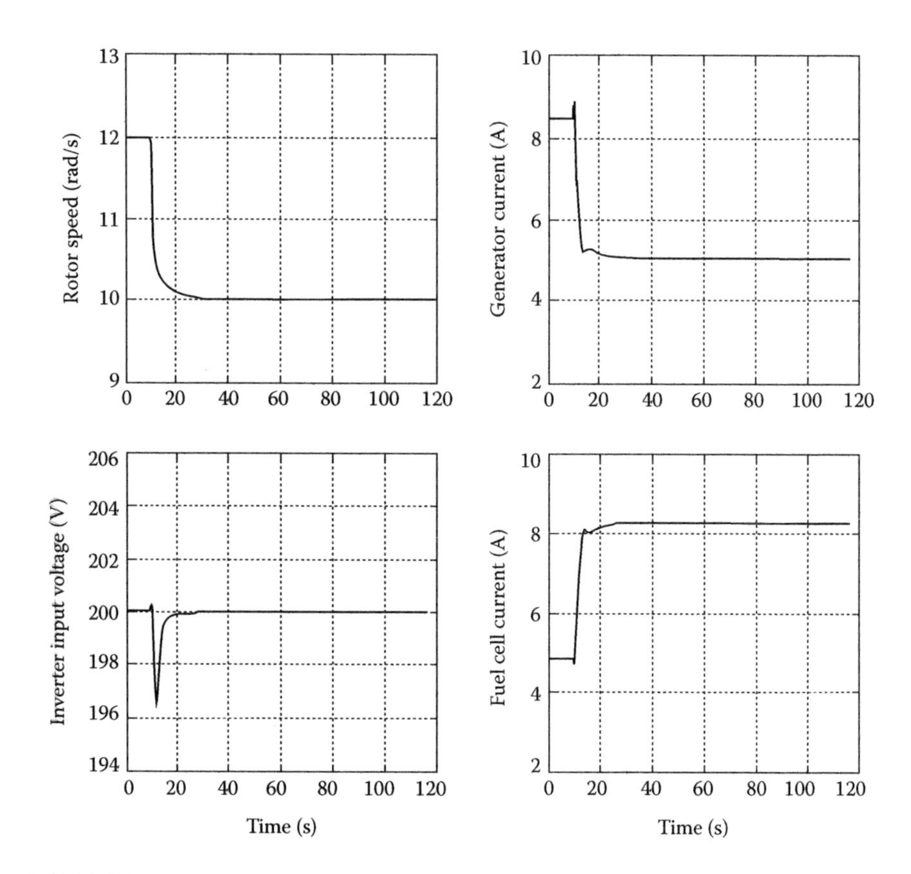

FIGURE 8.8
System response to step change in wind speed (combined system).

input voltage provided by the fuel cell controller to 200 V with a 3.8% maximum overshoot and a 3.6% maximum undershoot as the current drawn from the fuel cell is reduced.

8.2.2.4 Turbulent Wind, Below-Rated Wind Speed Conditions (Wind Power < Load)

For this case, the wind speed is established by the SNwind generated profile. The mean wind speed is set to 5 m/s, and the turbulence characteristics are determined by the SNwind input file parameters. The actual mean wind speed generated is 4.705 m/s. Figure 8.10 shows the response of the rotor speed, inverter input voltage, generator current, and fuel cell current for an initial equivalent load resistance of 15 Ω. Since the fuel cell is active, the fuel cell controller attempts to regulate the inverter input voltage to 200 V. The lambda-controller seeks to maintain an average rotor speed of 9.41 rad/s for the changing wind.

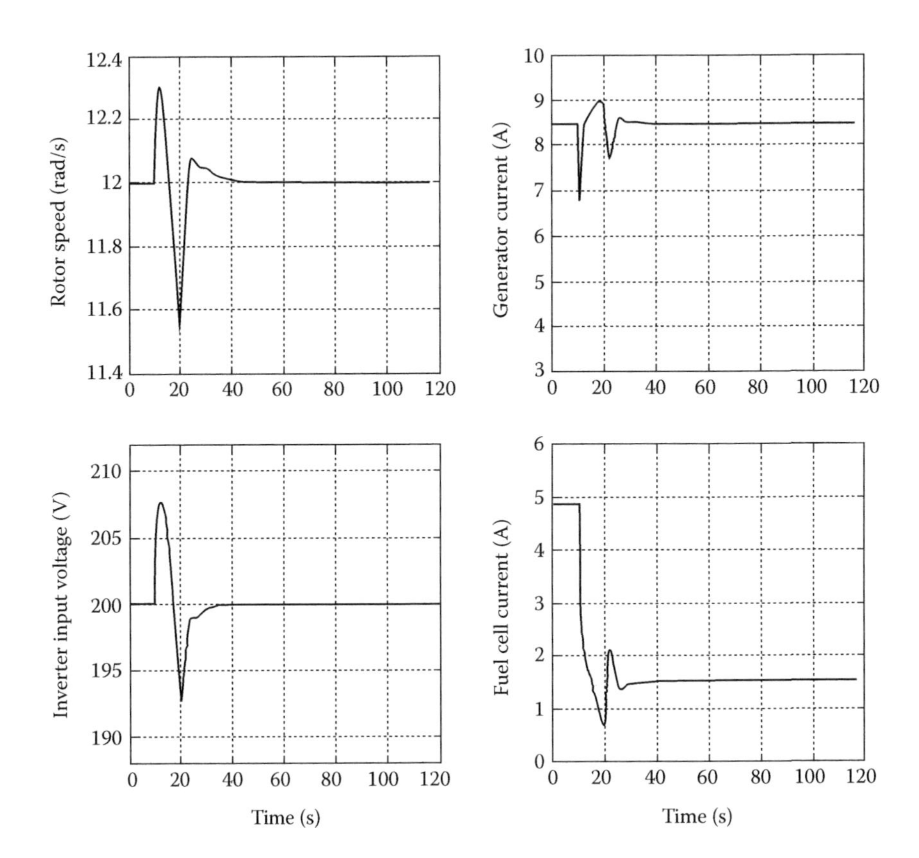

FIGURE 8.9
System response to step change in load resistance (combined system).

We can also see in Figure 8.10 that the generator and fuel cell currents are complementary. As the wind speed decreases, the generator current decreases and the fuel cell current increases accordingly. The combined current from the generator and the fuel cell is shown in Figure 8.11 along with the wind profile for the same case. With the voltage regulated to 200 V and an initial average combined current of 13.33 A, the system attempts to provide a constant power of 2.67 kW to the load.

Then at 30 s, a 33% step change in the value of R_L from 15 to 20 Ω is applied. The fuel cell controller again regulates the inverter input voltage to 200 V. The maximum overshoot is 4.5%, and the maximum undershoot is 5%. The lambda-controller seeks to maintain an average rotor speed of 9.41 rad/s for the changing wind conditions, corresponding to the desired lambda value of 6.5.

Figure 8.10 also shows that the generator current maintains an average value of approximately 4.25 A after the step change in equivalent load resistance while the fuel cell current's average value drops from approximately

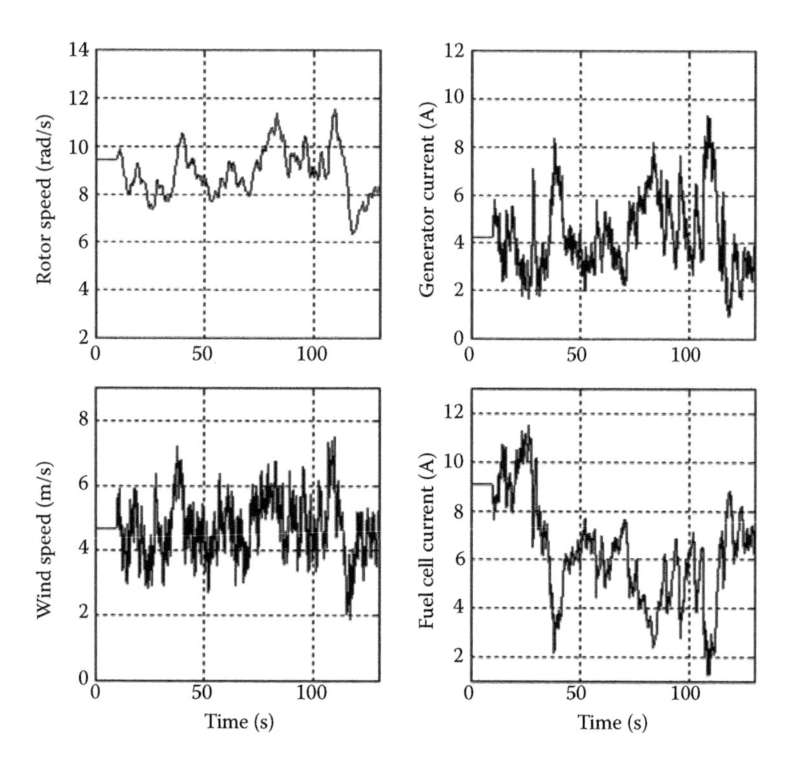

FIGURE 8.10
System response to the changing wind speed profile (combined system).

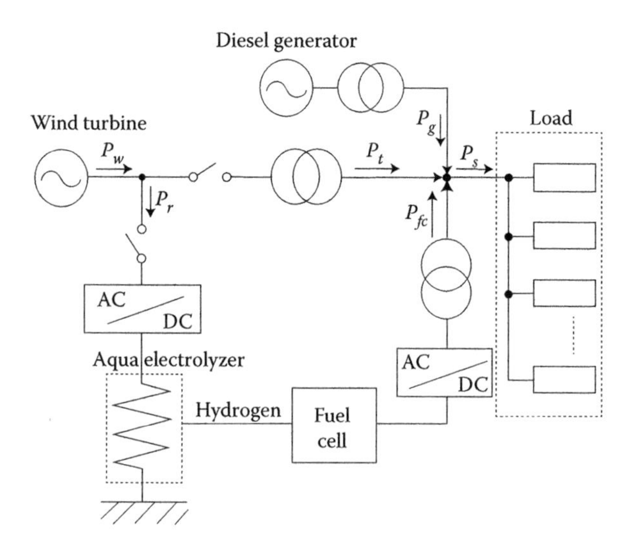

FIGURE 8.11
System configuration.

9.08 to 5.75 A. With the inverter input voltage regulated to 200 V and a step change in the average combined current (fuel cell plus generator current) from 13.33 to 10 A as shown in Figure 8.11, the power output to the load changes from 2.67 to 2 kW.

Inverter input voltage and fuel cell plus generator combined current response (combined system).

8.2.3 Conclusions

Wind is a renewable energy resource that is growing in importance as a means to address the national and global issues of air pollution, grid reliability, dependence on foreign oil, climate change, etc. However, the wind varies over time and there is less of it (on average) at sites around major load centers, which are two of the technical and economic challenges to the establishment of power generation facilities solely relying on the wind. But hybrid systems, such as wind turbine–fuel cell systems, have the potential to rectify these shortcomings and also make the greater decentralization of electric power generation in the United States possible, thus easing the transmission and distribution system's bottlenecks while increasing its overall reliability and security without sacrificing the quality of power delivered to the customer. Furthermore, wind turbine–fuel cell systems can achieve these results in an environment-friendly manner without producing harmful emissions.

The study by Carter and Diong [17] described the development of a simulation model and the associated control schemes for a small-scale, prototype wind turbine–fuel cell system that will enable the further study and optimization of the performance, sizing, cost, etc., of such hybrid systems. For this study, models of a variable-speed wind turbine and DC generator, a PEM fuel cell, and an electrolyzer unit were combined to simulate a regenerative hybrid wind–fuel cell AC power-generating system. The system model with appropriately designed controllers was then exercised and found to perform effectively, both below and above the rated wind speed of 8 m/s, for delivering power to an AC load and/or electrolyzer while regulating to the necessary minimal inverter input DC voltage of 200 V. Simulations of this model have also shown that the system adequately responds to significant changes in wind speed (both step and turbulent) and load demand for various representative cases.

The simulations indicated that such a hybrid system can be made to deliver utility-grade electricity reliably and effectively from the combined energy sources. These results are encouraging from the standpoint of utilizing regenerative hybrid wind turbine–fuel cell systems as a means to address some of the current challenges facing the electric power industry, the nation, and the world, such as air pollution, grid reliability, dependence on foreign oil, and climate change.

8.3 Hybrid Renewable Energy System for Isolated Islands

The output power of wind turbine generators is mostly fluctuating and has an effect on system frequency. Senjyu et al. [28] tried to solve this problem by using hybrid renewable energy systems. Its configuration is given in Figure 8.11.

Where P_s is the power supply to the load; P_g is the generating power of diesels; P_{fc} is the actual generating power of a fuel cell; P_w is the output power of a wind turbine; and P_r is the power flowing to an aqua electrolyzer.

This system consists of wind generators, diesel generators, fuel-cell generators, and aqua electrolyzers. The aqua electrolyzers are used to absorb the rapidly fluctuating output power from wind turbine generators and generate hydrogen as fuels for fuel cells. Power supplied to the load is P_s.

8.3.1 Simulation Models

To simulate the proposed system configuration in Figure 8.11, the simulation model in Figure 8.12 is used to test it.

It should be noted that the wind power is modeled as an input power to the load. A detailed model and control method for the wind turbine are not presented in Reference 28.

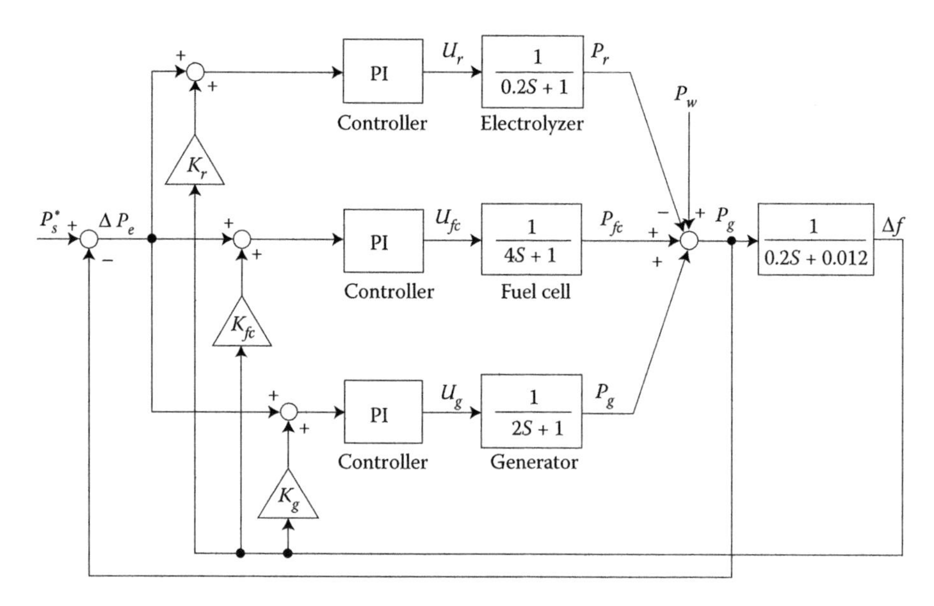

FIGURE 8.12
Simulation model.

8.3.2 Control Methods

From Figure 8.12, we can see that a proportional–integral (PI) controller is applied to the fuel cell, the electrolyzer, and the diesel generator, which are shown in Figures 8.13 through 8.15.

8.3.3 Simulation Results

In this simulation study, four cases, whose conditions are shown in Table 8.3, have been simulated. The gains of PI controllers for each case are

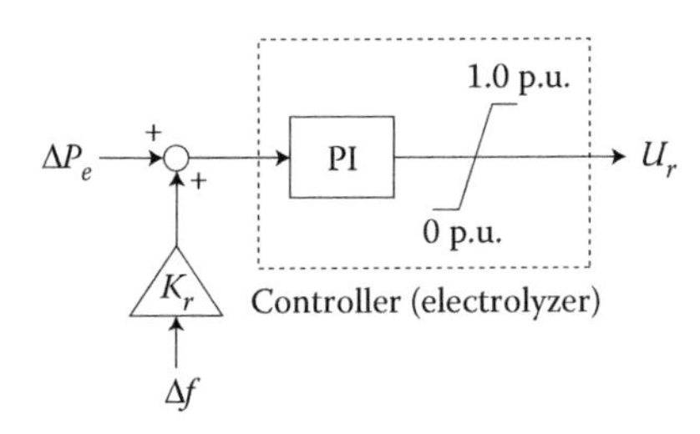

FIGURE 8.13
PI controller for an electrolyzer.

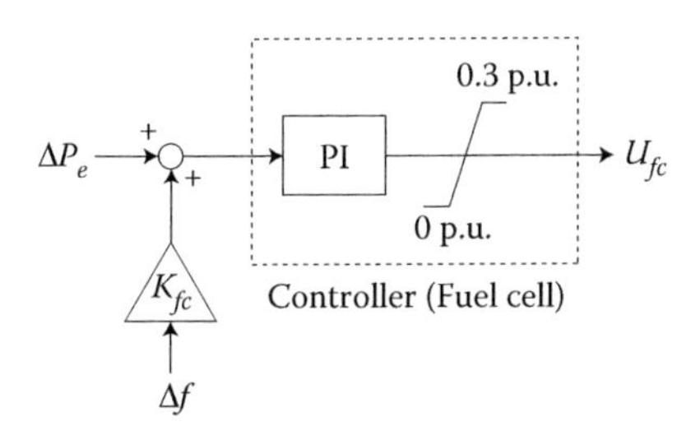

FIGURE 8.14
PI controller for a fuel cell.

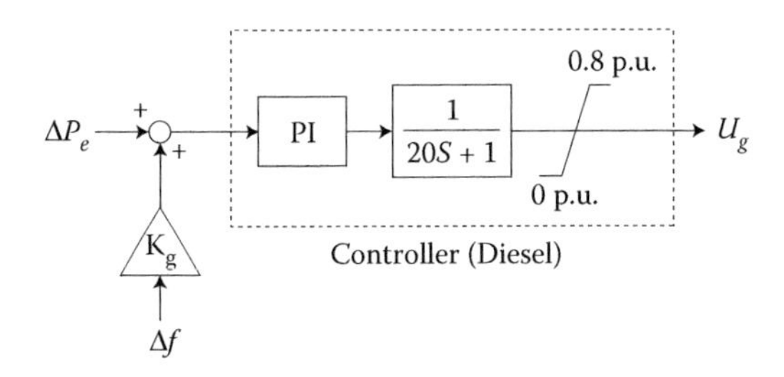

FIGURE 8.15
PI controller for a diesel generator.

TABLE 8.3

Four Cases of Simulations

	Case 1	Case 2	Case 3	Case 4
Diesel generators	○	○	○	○
Fuel cells	×	○	○	×
Aqua electrolyzers	×	○	○	×
Batteries	×	×	×	○
Wind turbines	○	○	△	○

Note: ○, in use; ×, no use; and △, limited use such as wind turbines use for aqua electrolyzers only.

TABLE 8.4

Gain Values in PI Controllers

Case	Equipment	Propositional Gain	Integral Gain
Case 1	Diesel	4.7	1.3
Case 2	Diesel	4.7	2.5
	Fuel cell	11.5	3.7
	Electrolyzer	16.8	5.8
Case 3	Diesel	8.7	5.7
	Fuel cell	5.5	2.3
Case 4	Diesel	4.7	2.3
	Battery	1.4	0.4

shown in Table 8.4. These parameters are determined by the trial-and-error method such that supply error and the frequency deviation of the power system is small. In Case 1, the system consists of only diesel generators and wind turbines; the system of Case 2 consists of diesel generators, fuel-cell generators, wind turbines, and aqua electrolyzers, which is the system proposed in Reference 28. The system of Case 3 uses the same equipment as the system of Case 2. The system of Case 4 consists of diesel generators, wind turbines, and batteries. In Case 4, the supply power is represented as follows:

$$P_s = P_w + P_g + P_b \tag{8.27}$$

where P_b is the charge or discharge power of batteries.

Simulation results for Case 1, Case 2, Case 3, Case 4–1, and Case 4–2 have been given in Figures 8.16 through 8.20, respectively.

8.3.4 Remarks and Discussion

From the above simulations, Table 8.5 shows the feature of each power-supply system, from Case 1 to Case 4. Among these four cases, Case 1 simulation

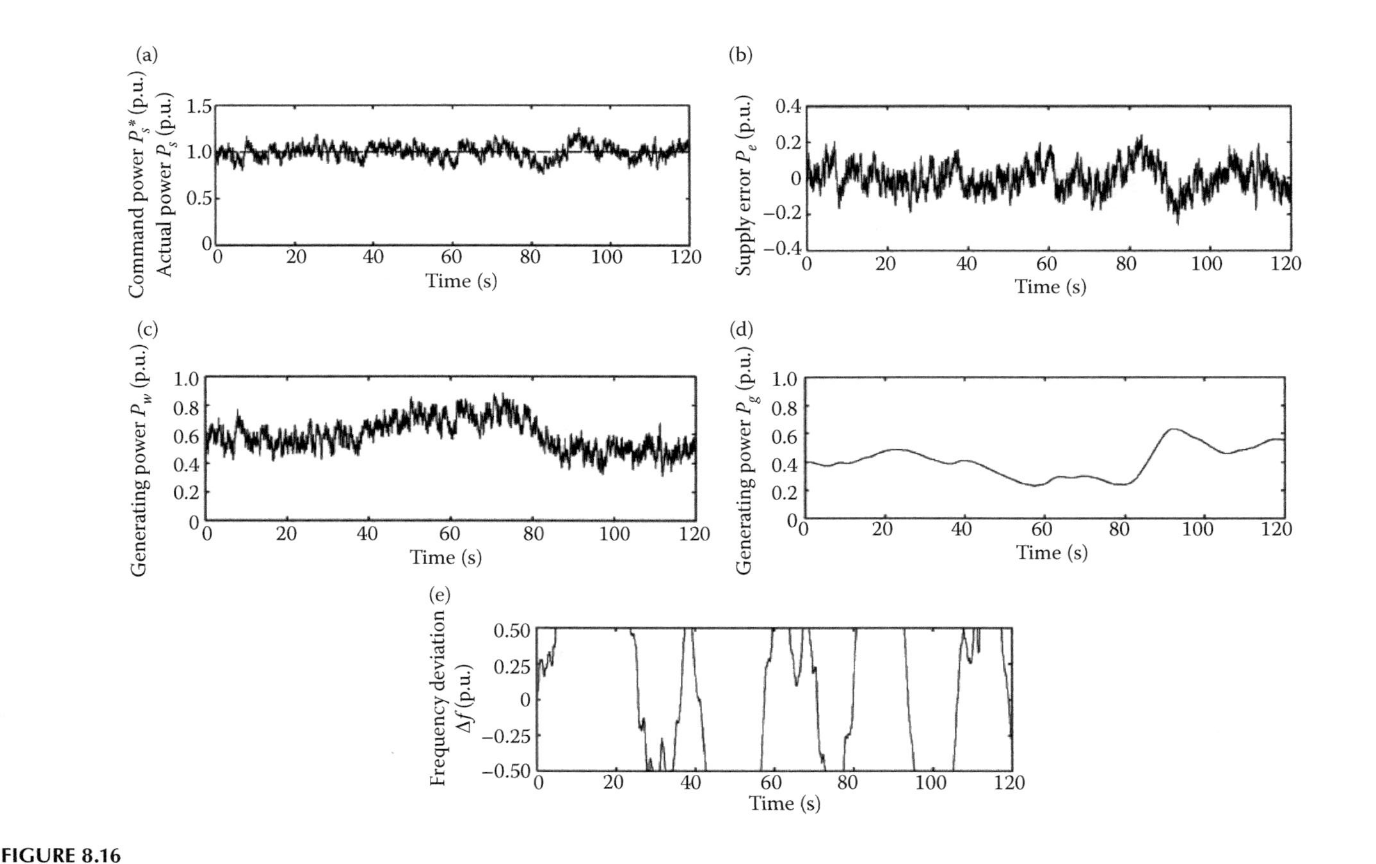

FIGURE 8.16
Simulation results with a diesel generator only (Case 1). (a) Demand power P_S^* and supply power P_S. (b) Error in supply demand ΔP_e. (c) Generating power of a wind turbine P_w. (d) Generating power of a diesel generator P_g. (e) Frequency deviation of a power system Δf.

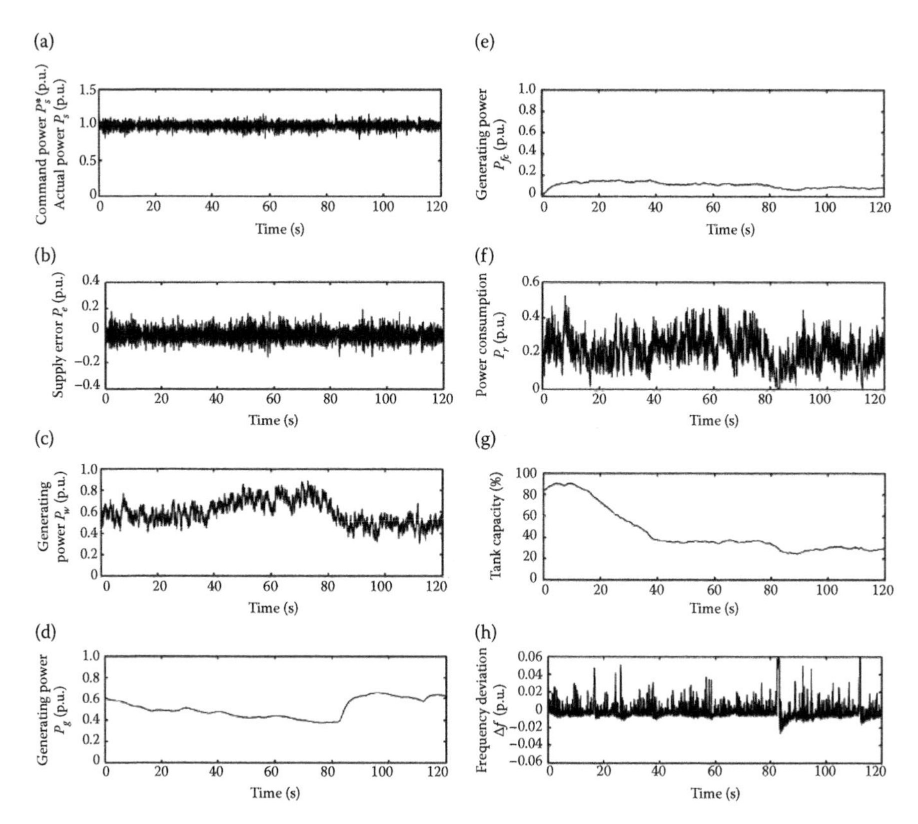

FIGURE 8.17
Simulation results of the proposed system (Case 2). (a) Demand power P_S^* and supply power P_s. (b) Error in supply demand ΔP_e. (c) Generating power of a wind turbine P_w. (d) Generating power of a diesel generator P_g. (e) Generating power of a fuel cell P_{fc}. (f) Input power to aqua electrolyzers P_r. (g) Fuel tank capacity. (h) Frequency deviation of power systems Δf.

system is the most inexpensive system, but it cannot supply high-quality power to load demand when the output power of wind turbines changes suddenly. Case 3 can supply very high-quality power to load demand when the output power of wind turbines changes suddenly, but this system is not an effective system since all of the total output power generated from wind turbine generators is used only in electrolysis.

In addition, this system needs a large capacity of diesel generators, fuel-cell generators, aqua electrolyzers, and a fuel tank. The Case 4 simulation system can supply high-quality power to load demand. However, this system is very costly since this system needs a large battery capacity. In addition, the battery charges and discharges many times, which makes the lifetime of the battery very short. The Case 2 simulation system is a more effective and inexpensive system compared with the system in Cases 3 and 4, and maintains the load frequency of the system in Case 1. However, the configuration of Case 2 is expensive compared with the system in Case 1.

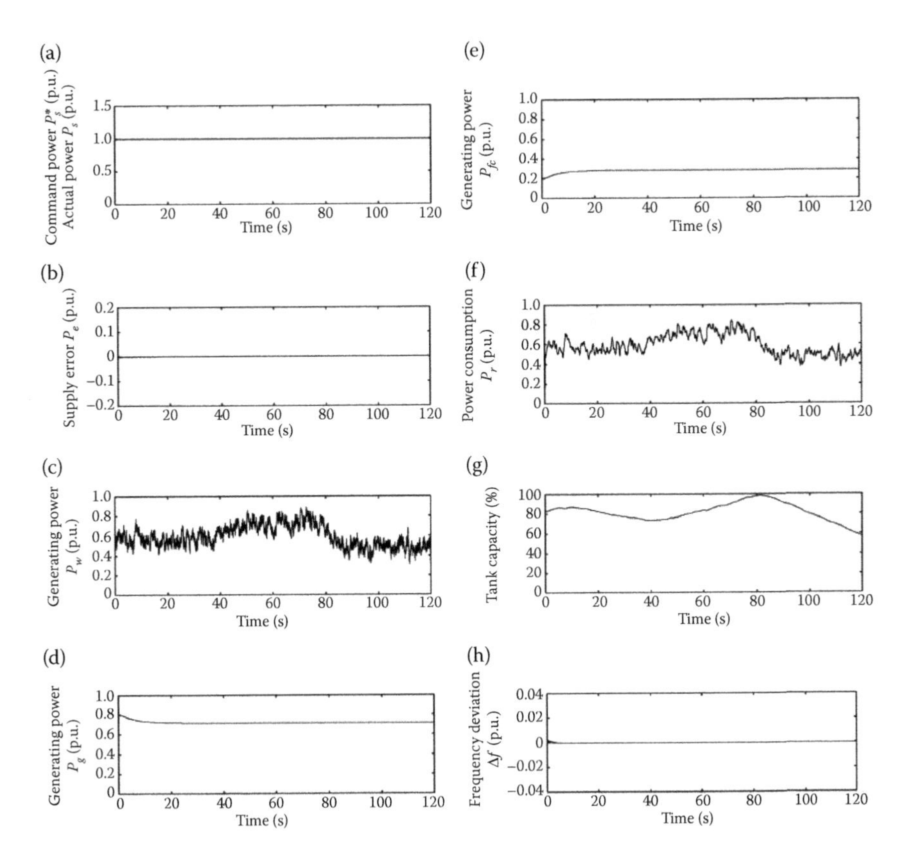

FIGURE 8.18
Simulation results for Case 3. (a) Demand power P_s^* and supply power P_s. (b) Error in supply demand ΔP_e. (c) Generating power of a wind turbine P_w. (d) Generating power of a diesel generator P_g. (e) Generating power of a fuel cell P_{fc}. (f) Input power to aqua electrolyzers P_r. (g) Fuel tank capacity. (h) Frequency deviation of power systems Δf.

8.4 Power Management of a Stand-Alone Wind–Photovoltaic–Fuel Cell Energy System

Due to the ever-increasing energy consumption, the soaring cost and the exhaustible nature of fossil fuel, and the worsening global environment, power industry becomes more and more interested in green (renewable and/or fuel cell-based energy sources) power generation systems [29]. As we know, wind and solar power generation are two of the most promising renewable power generation technologies. The growth of wind and photovoltaic (PV) power generation systems has exceeded the most optimistic estimation [30–32]. Fuel cells (FCs) also show a great potential to be green power sources of the future because they have many merits (such as high efficiency,

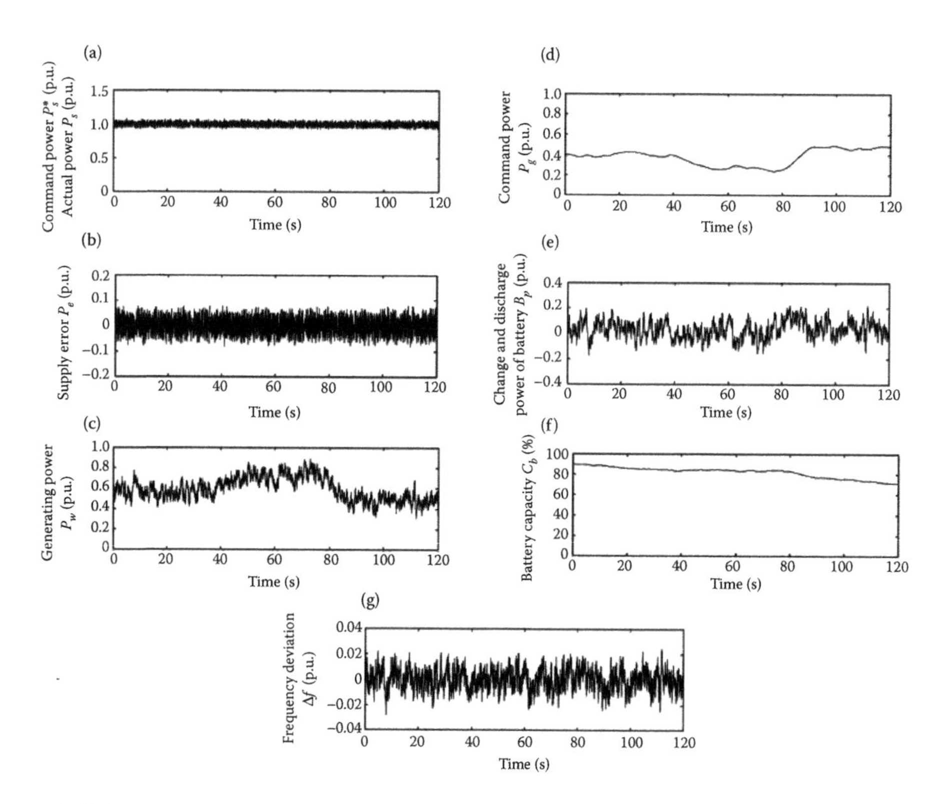

FIGURE 8.19
Simulation results for Case 4–1. (a) Demand power P_s^* and supply power P_s. (b) Error in supply demand ΔP_e. (c) Generating power of a wind turbine P_w. (d) Generating power of a diesel generator P_g. (e) Generating power of a fuel cell P_{fc}. (f) Battery capacity. (g) Frequency deviation of power systems Δf.

zero or low emission of pollutant gases, and a flexible modular structure) and the rapid progress in FC technologies. However, each of the aforementioned technologies has its own drawbacks. For instance, wind and solar power are highly dependent on climate while FCs need hydrogen-rich fuel. Nevertheless, because different alternative energy sources can complement each other to some extent, multisource hybrid alternative energy systems (with proper control) have a great potential to provide higher quality and more reliable power to customers than a system based on a single resource. Because of this feature, hybrid energy systems have caught worldwide research attention [2–12].

Many alternative energy sources including wind, PV, FC, diesel system, gas turbine, and microturbine can be used to build a hybrid energy system [2–12]. Nevertheless, the major renewable energy sources used and reported are wind and PV power [2–12]. Due to the intermittent nature of wind and solar energy, stand-alone wind and PV energy systems normally require energy-storage devices or some other generation sources to form a hybrid system.

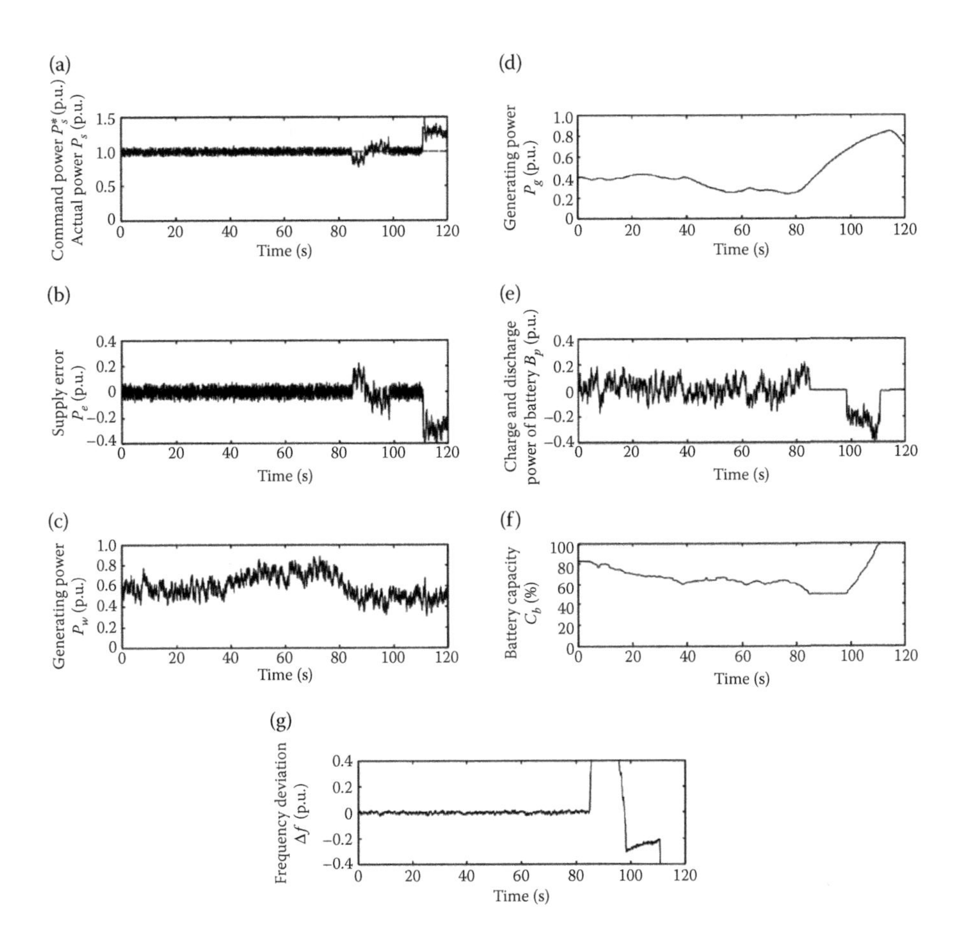

FIGURE 8.20
Simulation results for Case 4–2. (a) Demand power P_s^* and supply power P_s. (b) Error in supply demand ΔP_e. (c) Generating power of a wind turbine P_w. (d) Generating power of a diesel generator P_g. (e) Generating power of a fuel cell P_{fc}. (f) Battery capacity. (g) Frequency deviation of power systems Δf.

TABLE 8.5
Simulation Results for Each Case

	Effective Utilization of Wind Energy	Quality of Electronic Power	Cost
Case 1	○	×	◎
Case 2	○	○	○
Case 3	×	◎	△
Case 4	○	○	×

Note: ○, effective; ×, ineffective; ◎, very effective; and △, moderate.

The storage device can be a battery bank, supercapacitor bank, superconducting magnetic energy storage (SMES), or an FC–electrolyzer system.

Wang and Nehrir [29] proposed a stand-alone hybrid alternative energy system consisting of wind, PV, FC, electrolyzer, and battery. Wind and PV are the primary power sources of the system to take full advantage of renewable energy, and the FC–electrolyzer combination is used as a backup and a long-term storage system. A battery bank is also used in the system for short-time backup to supply transient power. The different energy-storage sources in the proposed system are integrated through an AC link bus. The details of the system configuration, system unit sizing, and the characteristics of the major system components are also discussed in this chapter. An overall power management strategy is designed for the system to coordinate the power flows among the different energy sources. Simulation studies have been carried out to verify the system performance under different scenarios using a practical load profile and real weather data.

8.4.1 System Configuration

The system configuration for the proposed hybrid alternative energy system is shown in Figure 8.21. The wind and PV power play the primary roles in this design while the FC–electrolyzer combination serves as a backup and storage system. If excess wind and/or solar generation becomes available, the electrolyzer starts to produce hydrogen, which is then delivered to the hydrogen-storage tanks. On the other hand, when there is a deficit in power generation, the FC stack will begin to produce energy using hydrogen from the reservoir tanks, or in case they are empty, from the backup hydrogen tanks. The design also uses a battery bank to supply transient power to load transients, ripples, and spikes. Different energy sources are connected to the AC bus through appropriate power electronic-interfacing circuits. A detailed system design can be found in Reference 29.

In the design, DC/AC converters are used for the wind energy, solar energy, and FC stack because this hybrid renewable energy system is applied in utility power systems. The details of the components in the system can be found in Reference 29.

8.4.2 Power Management Strategies

In the overall control scheme shown in Figure 8.22, the wind energy conversion system is controlled by a pitch angle controller, while a PV electricity generation unit is controlled by a maximum power point tracking (MPPT) controller. The wind energy conversion system and PV electricity generation unit are the main energy sources to feed the load. The power difference between the generation sources and the load demand is calculated as

$$P_{net} = P_{wind} + P_{PV} - P_{load} - P_{sc} \tag{8.28}$$

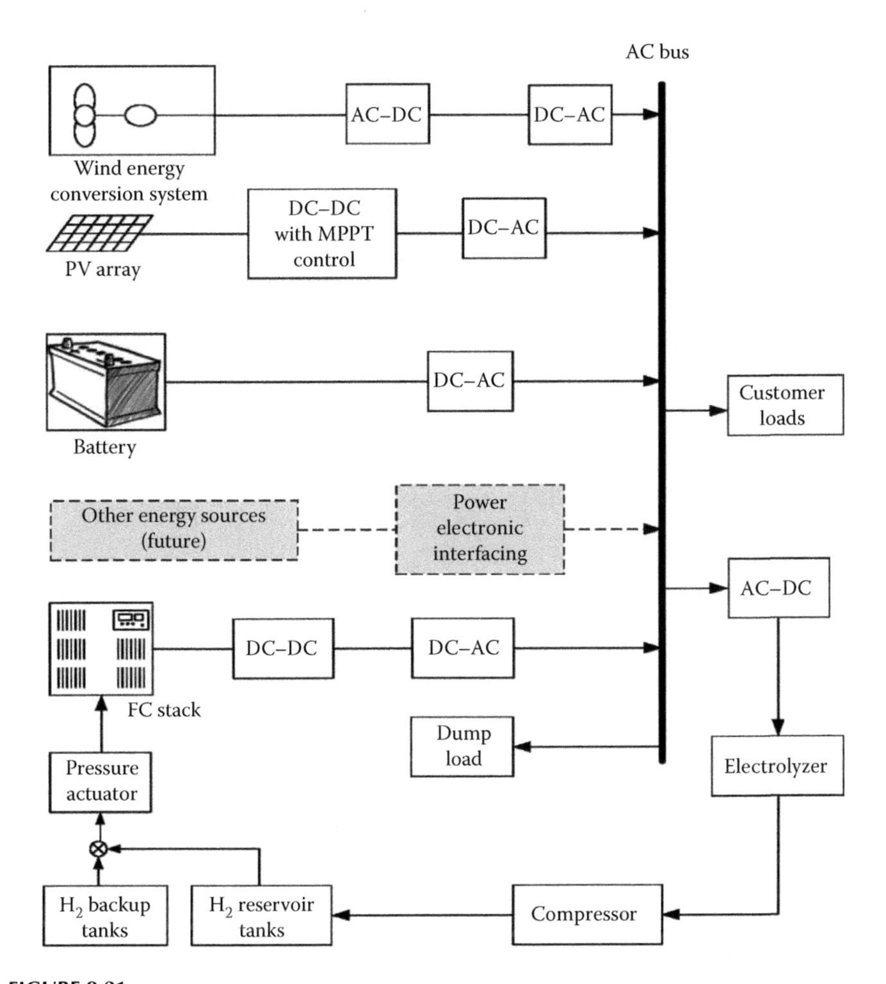

FIGURE 8.21
System configuration of a hybrid renewable energy system.

where P_{sc} is the self-consumed power for the operating system.

Two modes are needed to be considered in the operations: excess mode and deficit mode.

1. Excess mode

   When the power difference between the generation and the demand P_{net} is larger than zero, the excess wind and PV-generated power P_{net} is supplied to the electrolyzer to generate hydrogen that is delivered to the hydrogen-storage tanks through a gas compressor. The power balance equation becomes

$$P_{wind} + P_{PV} = P_{load} + P_{elec} + P_{comp}, \quad P_{net} > 0 \tag{8.29}$$

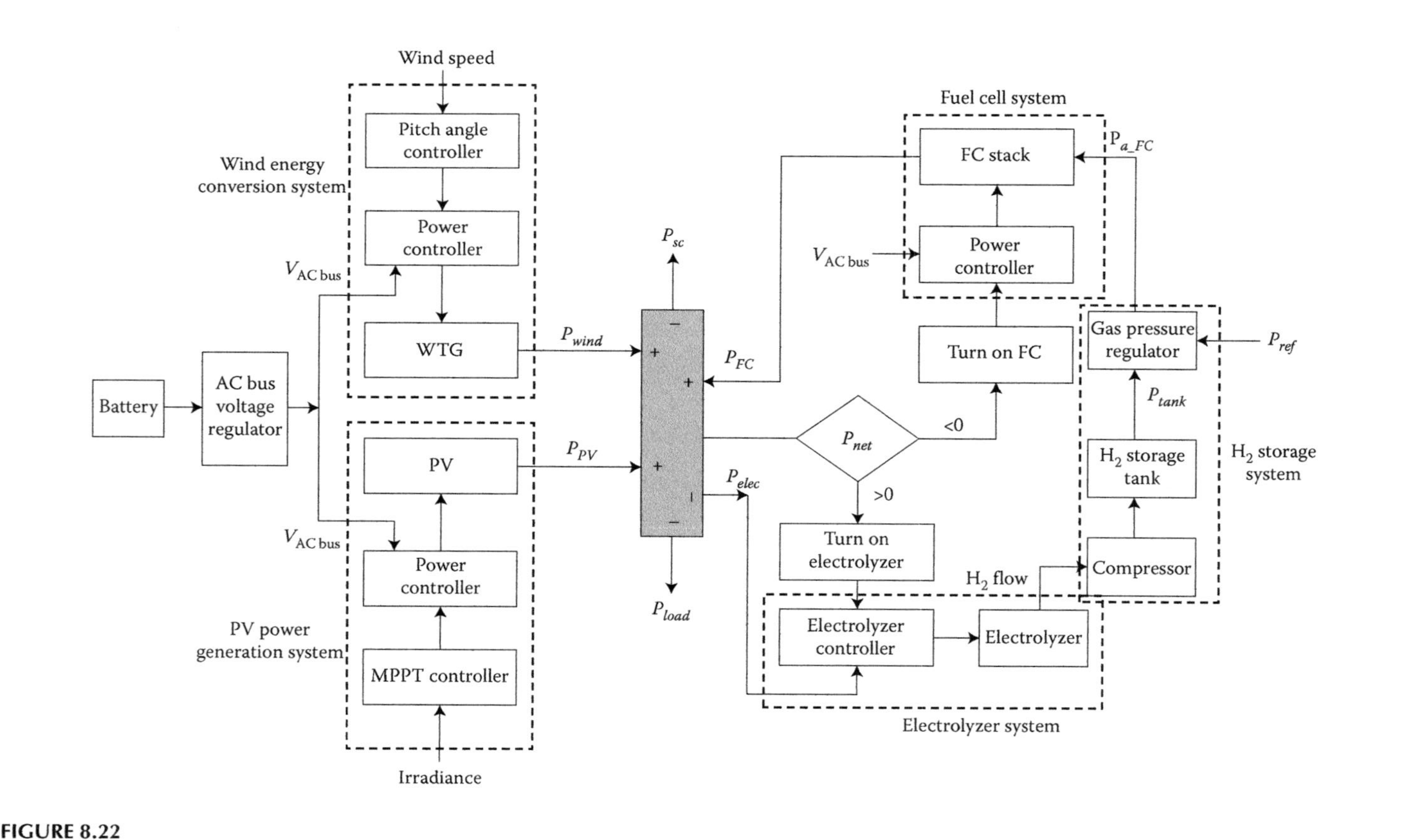

FIGURE 8.22
Block diagram of the overall control scheme for the proposed hybrid alternative energy system. (Adapted from C. Wang and M.H. Nehrir, *IEEE Transactions on Energy Conversion*, 23(3), 957–967, 2008.)

where P_{elec} is the power consumed by the electrolyzer to generate hydrogen and P_{comp} is the power consumed by the gas compressor.

2. Deficit mode

When there is a deficit in power generation ($P_{net} < 0$), the fuel cell stack begins to produce energy for the load using hydrogen stored in the tanks. The power balance equation is

$$P_{wind} + P_{PV} + P_{FC} = P_{load}, \quad P_{net} < 0 \tag{8.30}$$

where P_{FC} is the power generated by the fuel cell stack.

8.4.3 Simulation Results

The proposed wind–PV–FC–electrolyzer energy system has been developed using MATLAB–Simulink. In order to verify the system performance under different situations, simulation studies have been carried out using practical load-demand data and real weather data (wind speed, solar irradiance, and air temperature). The system is designed to supply electric power demand of five houses in the southwestern part of Montana. A typical hourly average load demand for a house in the Pacific Northwest regions provided in Reference 33, is used in this simulation study. The total hourly average load-demand profile of five houses over 24 h is shown in Figure 8.23. The weather data are obtained from the online records of the weather station at Deer Lodge, Montana, affiliated with the Pacific Northwest Cooperative Agricultural Weather Network (AgriMet) [34].

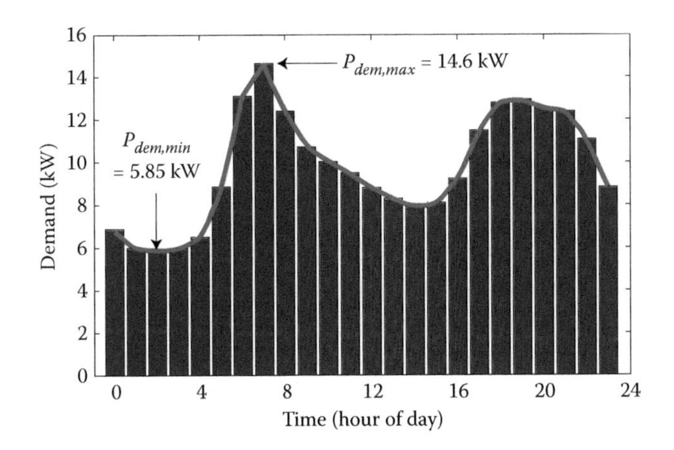

FIGURE 8.23
Hourly average demand of five typical homes in the Pacific Northwest area.

Simulation studies are conducted for the proposed power management scheme during a typical winter day and a summer day. The load demand is kept the same for the two cases.

A. Winter scenario: The weather data for the winter scenario simulation were collected on February 1, 2006. Figure 8.24 shows the output power from the wind energy conversion unit in the hybrid energy system over the 24-h simulation period. Figure 8.25 shows the effect of temperature on the PV performance. Used by the electrolyzer to generate H_2, the available power profile over the 24-h simulation period is given in Figure 8.26. The corresponding DC voltage applied to the electrolyzer and the electrolyzer current is shown in Figure 8.27. Figure 8.28 shows the actual power delivered by the FC stack.
B. Summer scenario: The output power from the wind energy conversion system and the PV array in the hybrid energy system over the 24-h simulation period are shown in Figures 8.29 and 8.30, respectively. When $P_{net} > 0$, there is excess power available for H_2 generation. Figure 8.31 shows the H_2 generation rate over the simulation period. When $P_{net} < 0$, the sum of the wind- and the PV-generated power is not sufficient to supply the load demand. Under this scenario, the FC stack turns on to supply the power shortage by using the H_2 stored in the storage tank. Figure 8.32 shows the corresponding H_2 consumption rate. Figure 8.33 shows the tank pressure variations over the 24-h simulation period for the summer scenario study.

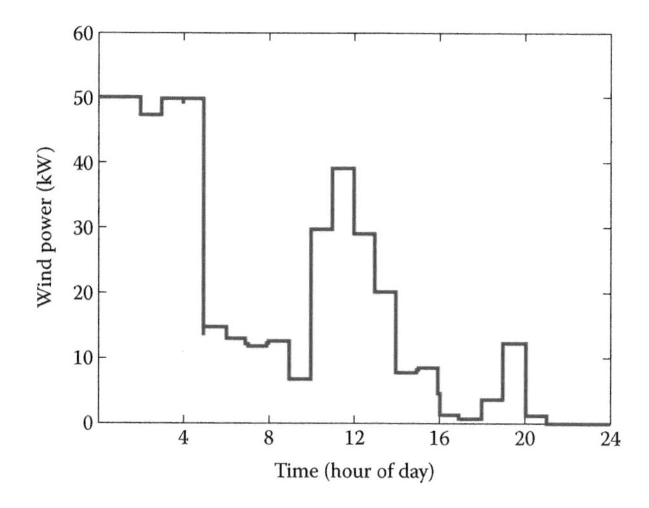

FIGURE 8.24
Wind power for the winter scenario study.

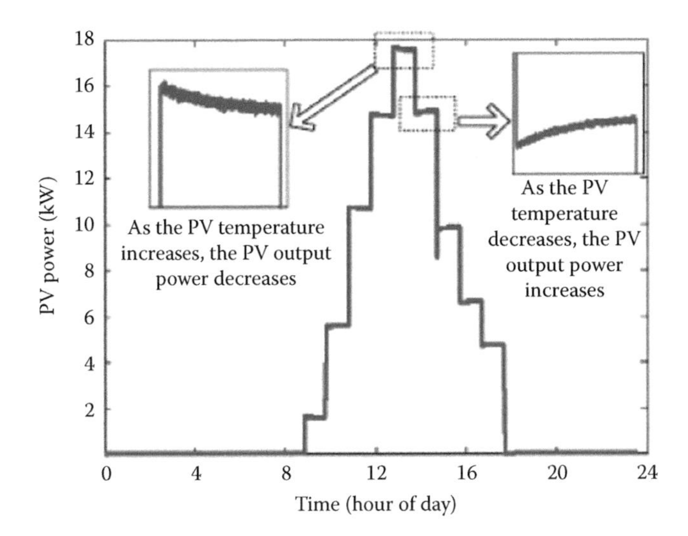

FIGURE 8.25
PV power for the winter scenario.

8.4.4 Summary

This section proposes an AC-linked stand-alone wind–PV–FC alternative energy system. The wind and PV generation systems serve as the main power generation sources, the electrolyzer acts as a dump load using any excess power available to produce hydrogen, and the FC system serves as the

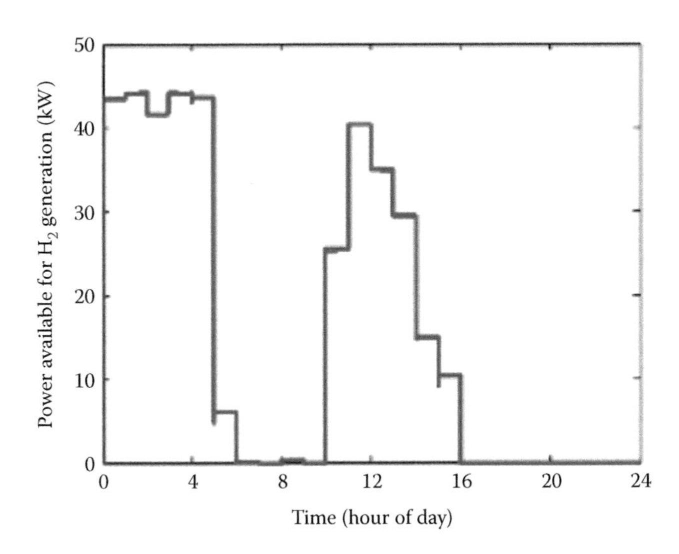

FIGURE 8.26
Power available for H_2 generation scenario study.

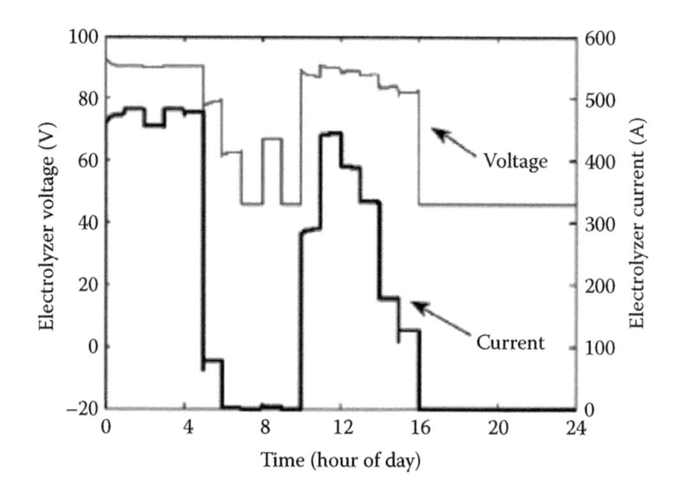

FIGURE 8.27
Electrolyzer voltage and current for the winter scenario.

backup and supplies power to the system when there is power deficit. The characteristics of the main components in the system including the WECS, PV, FC, and electrolyzer are provided, and the overall control and power management strategy for the proposed hybrid energy system is described. The simulation model of the hybrid system has been developed using MATLAB/Simulink. Simulation studies have been carried out to verify the system performance under different scenarios (winter and summer) using the practical load profile in the Pacific Northwest regions and the real weather data collected at Deer Lodge, MT. The simulation results show the effectiveness of

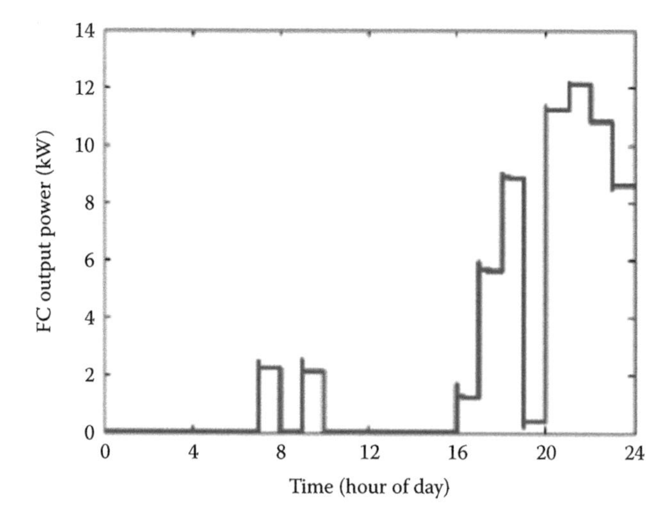

FIGURE 8.28
Power supplied by the FC stack for the winter scenario study.

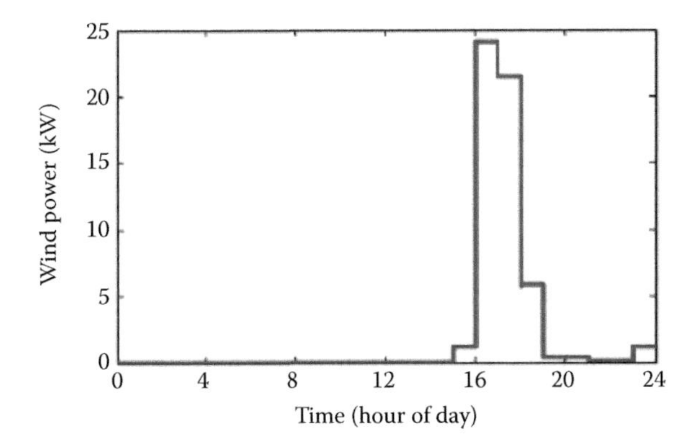

FIGURE 8.29
Wind power generated for the summer scenario study.

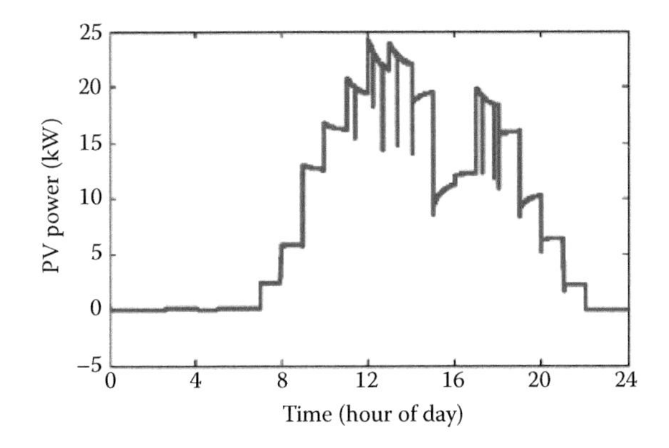

FIGURE 8.30
PV power generated for the summer scenario study.

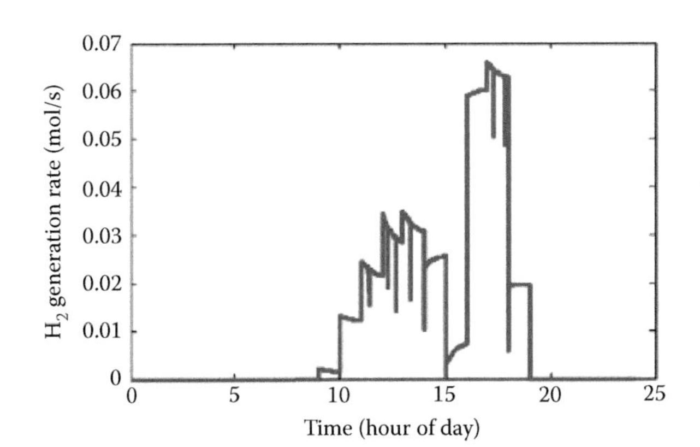

FIGURE 8.31
H_2 generation rate for the summer scenario study.

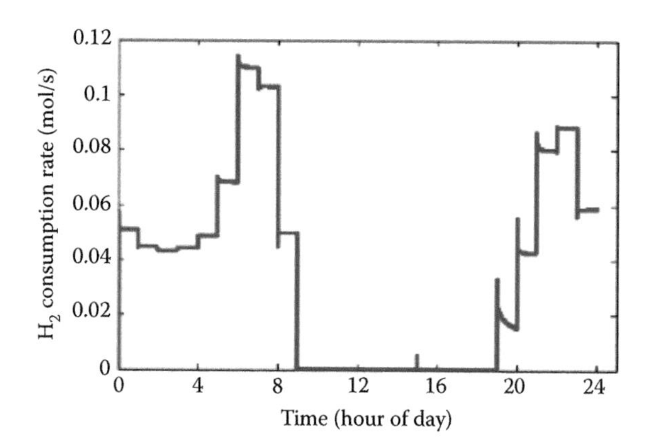

FIGURE 8.32
H_2 consumption rate for the summer scenario study.

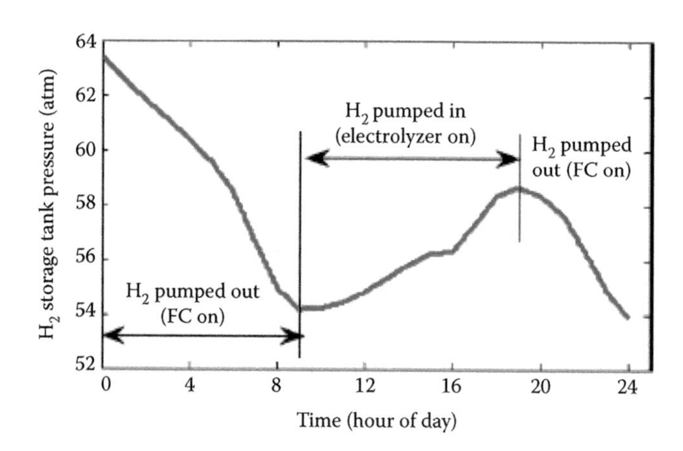

FIGURE 8.33
Tank pressure over 24 h for the summer scenario study.

the overall power management strategy and the feasibility of the proposed hybrid alternative energy system.

8.5 Hybrid Renewable Energy Systems in Load Flow Analysis

Power distribution system is one of the three parts in power grids. Traditional power distribution systems normally consist of a power source (substation) at the root of the feeder and loads along the feeder. Load flow analysis is a powerful tool for distribution automation (DA) and demand-side management

(DSM). Several typical load flow algorithms have been proposed for traditional power distribution systems.

With the advent of distributed generation (DG), the advantages of distributed generators make them to be a promising source to be installed in power distribution systems. The installation of distributed generators in distribution systems may cause the reverse power flow and power injections to the feeder. The problems involving the integration of DA, DGs, and DSM are not easy to deal with. Furthermore, the installation of DGs in distribution feeders and the participation in system operations, makes it necessary to restudy the problems including power flow, power quality, ferroresonance, voltage control, loss reduction, protection device coordination, voltage flicker, etc. Since voltage profiles are directly related to power quality and customer satisfaction, an efficient and robust three-phase load flow method with the considerations of DGs should be developed first [35].

This chapter serves as a brief introduction to load flow analysis with the considerations of distributed generators. In order to achieve that goal, a brief introduction to load flow analysis and its typical algorithms is provided first.

8.5.1 Load Flow Analysis for Power Distribution Systems

Load flow analysis has been well developed for power transmission systems. Typical approaches include Gauss–Seidel approach, Newton–Raphson, and fast decoupled approaches. All these approaches can be directly applied to power distribution systems. However, due to its own characteristics of radial structure, new approaches have also been developed for power distribution systems.

Four typical load flow analysis approaches have been developed for power distribution systems: Gauss–Seidel approach, Newton–Raphson approach, DistFlow approach, and a rigid approach. In this section, we will provide a brief introduction to these three approaches.

1. Gauss–Seidel approach: For transmission networks, the Gauss–Seidel algorithm is a basic iterative numerical procedure with a merit of a simple procedure. It can be directly used for power distribution systems. The Gauss–Seidel algorithm is an iterative numerical procedure that attempts to find a solution to the system of linear equations by repeatedly solving the linear system until the iteration solution is within a predetermined acceptable bound of error. It is a robust and reliable load flow method that provides convergence to extremely complex power systems. The iterative procedure of Gauss–Seidel algorithm is given as follows:

$$V_k^{(i+1)} = \frac{1}{Y_{kk}}\left(\frac{P_k - iQ_k}{(V_k^{\,i})^*} - \sum_{n=1}^{N} Y_{kn}V_n^{(i)}\right) \tag{8.31}$$

where $V_k^{(i+1)}$ and $V_k^{(i)}$ are the voltage values at bus k for $(i + 1)$th and ith iterations; P_k and Q_k are the real power and reactive power injection at bus k; and Y_{kk} and Y_{kn} are the elements in the admittance matrix.

2. Newton–Raphson approach: One of the disadvantages of Gauss–Seidel approach is that each bus is handled independently, that is, each correction to a bus requires subsequent correction to all the buses to which it is connected. However, Newton–Raphson approach is based on the idea of calculating the corrections while considering all the interactions.

To solve a nonlinear equation $f(x) = b$, Newton's method is to drive the error in the function $f(x)$ to zero by making the adjustment Δx to the independent variable associated with the function.

The Tylor expansion of the function $f(x)$ about a specific x^0 is

$$f(x^0)+\frac{df(x)}{dx}\bigg|_{x=x^0}\Delta x+\varepsilon=b \tag{8.32}$$

By setting the error ε to zero, we get

$$\Delta x=\left(\frac{df(x)}{dx}\bigg|_{x=x^0}\right)^{-1}(b-f(x^0)) \tag{8.33}$$

The above equation is the calculation of the adjustment in each iteration.

By applying the above idea to power systems, we could obtain the Newton–Raphson method derivation for power flow analysis.

For each bus i, we have the following power balance equation:

$$P_i+jQ_i=V_iI_i^* \tag{8.34}$$

where $P_i + jQ_i$ is the complex power of bus i, and V_i and I_i are the voltage and injection current phasors of bus i.

The variables are the voltage magnitudes $|V_i|$ and voltage angles θ_i. Thus, we need to write the power balance equation as follows:

$$P_i+jQ_i=V_i\left(\sum_{k=1}^{N}Y_{ik}V_k\right)^*=|V_i|^2Y_{ii}^*+\sum_{\substack{k=1\\k\neq i}}^{N}Y_{ik}^*V_iV_k^* \tag{8.35}$$

Obtaining the derivative of the above equation and considering the real and imaginary parts of this equation, we obtain the following equation:

$$\Delta P_i = \sum_{k=1}^{N} \frac{\partial P_i}{\partial \theta_k} \Delta\theta_k + \sum_{k=1}^{N} \frac{\partial P_i}{\partial |V_k|} \Delta |V_k| \qquad i = 1,2,\ldots,N \tag{8.36}$$
$$\Delta Q_i = \sum_{k=1}^{N} \frac{\partial Q_i}{\partial \theta_k} \Delta\theta_k + \sum_{k=1}^{N} \frac{\partial Q_i}{\partial |V_k|} \Delta |V_k|$$

According to Equation 8.36, we can obtain the adjustment calculation for iterations

$$\Delta X = [J]^{-1} \Delta Z \tag{8.37}$$

where

$$\Delta X = \begin{bmatrix} \Delta\theta_1 \\ \Delta|V_1| \\ \Delta\theta_2 \\ \Delta|V_2| \\ \vdots \end{bmatrix}, \quad \Delta Z = \begin{bmatrix} \Delta P_1 \\ \Delta Q_1 \\ \Delta P_2 \\ \Delta Q_2 \\ \vdots \end{bmatrix}, \quad [J] = \begin{bmatrix} \frac{\partial P_1}{\partial \theta_1} & \frac{\partial P_1}{\partial |V_1|} & \cdots \\ \frac{\partial Q_1}{\partial \theta_1} & \frac{\partial P_1}{\partial |V_1|} & \cdots \\ \vdots & \vdots & \vdots \end{bmatrix}$$

For details of Newton–Raphson method of power flow analysis, the readers can refer to any books of power system analysis or calculations.

3. DistFlow approach: This approach has been developed by Mesut E. Baran and Felix F. Wu, for the purpose of solving the optimal capacitor placement.

According to the structure nature of radial distribution systems, the power flow equations, called the DistFlow equations, of a radial distribution system, comprise branch flow equations and the associated terminal conditions for each lateral including the main feeder that is treated as the 0th lateral. They are of the following form:

$$x_{ki+1} = f_{ki+1}(X_{ki}, u_{ki+1}) \quad k = 0,1,\ldots,l \tag{8.38}$$

$$x_{k0} = x_{0k}, x_{kn} = x_{kn_2} = 0 \quad i = 0,1,\ldots,nk-1 \tag{8.39}$$

where

$$X = \left[x_i^T \cdots x_l^T x0^T\right]^T, \quad x_k = \left[x_{k0}^T \cdots x_{kn}^T\right]^T$$

For a radial distribution system consisting of a main feeder and m primary laterals, we need to solve the following $2(m+1)$ equations:

$$P_{0n} = \hat{P}_{on}(Z_{00}, Z_{10}, \ldots, Z_{k0}, |V_{00}|) = 0$$

$$P_{kn_k} = \hat{P}_{kn_k}(Z_{00}, Z_{10}, \ldots, Z_{j0}, |V_{00}|) = 0 \quad j \le k, \quad \text{and} \quad k = 1, 2, \ldots, m \qquad (8.40)$$

$$Q_{kn_k} = \hat{Q}_{kn_k}(Z_{00}, Z_{10}, \ldots, Z_{j0}, |V_{00}|) = 0$$

or in vector form

$$H(Z) = 0 \qquad (8.41)$$

where $Z = [Z_{10}{}^T \cdots Z_{l0}{}^T Z_{00}{}^T]$ is the state variable vector.

Using Newton–Raphson method, we can solve this load flow problem:

Step 1: For a given estimate Z^0, calculate the mismatch of $H(Z^0)$.
Step 2: Build the Jacobian matrix $J(Z^0) = (\partial H/\partial Z)|_{Z=Z^0}$
Step 3: The solution of the following system of equations to update the state Z:

$$J(Z^0)\Delta Z^0 = -H(Z^0).$$

Details of DistFlow method can be referred to in References 36 and 37.

8.5.2 Modeling of Distributed Generators in Load Flow Analysis

Modeling distributed generators in load flow analysis is an important problem for the distribution system operations and control. Some efforts have been made to solve parts of the problems [38–40]. Feijoo and Cidras [40] proposed some useful DG models for distribution load flow. Some works have been proposed to satisfy some of those requirements [41–43]. Chen et al. [41] used the Gauss implicit Z-matrix method to solve the three-phase load flow problem. The transformer and cogenerator were also modeled for rigorous system analysis. The Gauss method is traditionally used for the load flow solution of general meshed networks; however, distribution systems typically have a radial or weakly meshed structure. Therefore, research reveals some new ideas to deal with the special network characteristics of distribution feeders [42,43]. Cheng and Shirmohammadi [42] proposed a compensation-based technique, where the forward/backward-sweep algorithm was adopted in the solution scheme. A sensitivity matrix, used to calculate the incremental relation between the voltage magnitude and current injections, was derived and used for a voltage-specified generator bus. A direct-approach technique for solving the three-phase distribution power flow that fully exploits the

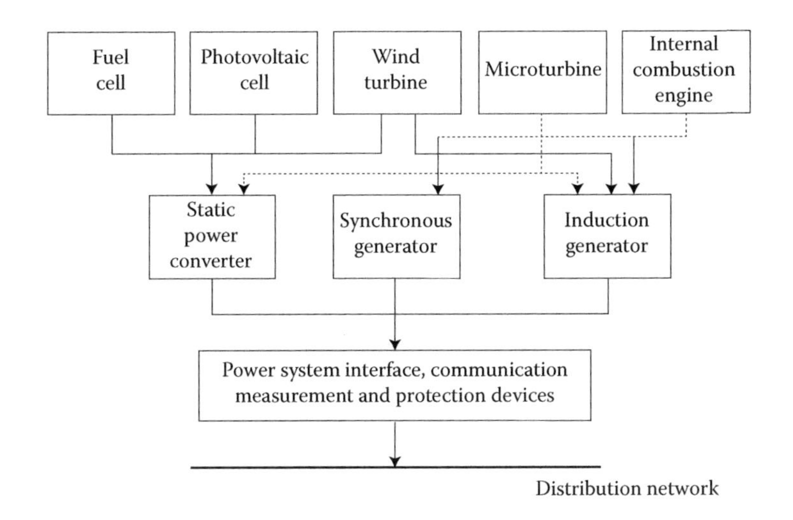

FIGURE 8.34
An example of the combination of DGs.

network characteristic of a distribution feeder was proposed in Reference 43. Two developed matrices, the bus-injection to branch-current (BIBC) matrix and the branch-current to bus-voltage (BCBV) matrix, and matrix multiplications are utilized to obtain the power flow solution. However, the integration of different types of DGs was not proposed in Reference 43.

Figure 8.34 shows an example of the combination of distributed generators for distribution systems. A fuel cell, PV cell, and wind turbine forming a hybrid renewable energy system, connect to the distribution network through a static power converter and a power system interface.

Teng [35] proposed three models of distributed generators for load flow analysis.

8.5.2.1 Several Models for Distributed Generators

8.5.2.1.1 A Constant Power Factor Model

A constant power factor model is a commonly used model in power systems. It is applicable for the controllable DGs, such as synchronous generator-based DGs and power electronic-based DGs. In this model, the values of real power output $P_{i,g}$ and the power factor $pf_{i,g}$ of the DG need to be specified. Once these two values are given, the reactive power of the distributed generator and its injected current can be calculated accordingly

$$Q_{i,g} = P_{i,g}\tan(\cos^{-1}(pf_{i,g})) \tag{8.42}$$

$$I_{i,g} = \left(\frac{P_{i,g} + jQ_{i,g}}{V_{i,g}^{k}}\right) \tag{8.43}$$

where $Q_{i,g}$, $I_{i,g}$, and $V_{i,g}^{k}$ are the reactive power, injected current, and the output voltage at the kth iteration of the distributed generator, respectively.

8.5.2.1.2 Variable Reactive Power Model

Distributed generators using induction generators as power conversion devices act nearly like variable reactive power generators. A typical example is the induction generator-based wind turbine, whose real power output can be calculated by the wind turbine power curve. The reactive power consumed by a wind turbine can be represented as a function of its real power [40]

$$Q^{1}{}_{i,g} = -Q_0 - Q_1 P_{i,g} - Q_2 P_{i,g}^{2} \tag{8.44}$$

where Q_0, Q_1, and Q_2 are experimentally obtained.

According to the equivalent circuit given in Figure 8.35, the reactive power output can be calculated as

$$Q_{i,g} = Q_{i,g}^{1} + Q_{i,g}^{c} \tag{8.45}$$

8.5.2.1.3 Constant Voltage Model

This model is normally used for large-scale controlled distributed generators. The real power output and bus voltage magnitude are specifically valued. For the three methods introduced in Section 8.4.1, the reactive power output value needs to be calculated for iterations

$$\Delta Q_{i,g}^{k,m} = V_{i,g}^{mis}(2\,|\,X_g\,|\,)^{-1} \tag{8.46}$$

$$V_{i,g}^{mis} = \left(\left|V_{i,g}^{spec}\right|\right)^{2} - \left(\left|V_{i,g}^{k,m}\right|\right)^{2} \tag{8.47}$$

$$[X_g] = img([BCBV_i][BIBC_i]) \tag{8.48}$$

where $\Delta Q_{i,g}^{k,m}$ is the required reactive power variation for the mth inner iteration and the kth outer iteration, $V_{i,g}^{mis}$ is the square voltage mismatch between the specified voltage and the calculated voltage for the mth inner iteration

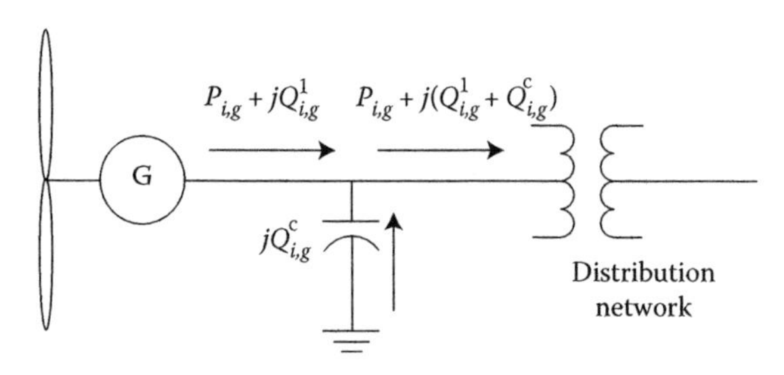

FIGURE 8.35
Equivalent circuit of an induction generator-based wind turbine.

and the *k*th outer iteration, $[BIBC_i]$ the column vector of $[BIBC]$ corresponding to bus *i*, and $[BCBV_i]$ is the row vector of $[BCBV]$ corresponding to bus *i*.

Two matrices $[BIBC]$ and $[BCBV]$ are defined as follows:

1. $[BIBC]$ describes the relationship between current injections $[I]$ and branch currents $[B]$.

$$[B] = [BIBC][I] \tag{8.49}$$

2. $[BCBV]$ describes the relationship between branch currents $[B]$ and bus voltages $[V]$:

$$[V_0] - [V] = [BCBV][B] \tag{8.50}$$

By applying the three load flow analysis models to the load flow analysis, one can calculate the state of distribution systems.

8.5.2.2 Test Results

The proposed models have been implemented by Borland C++ language on a Windows-XP-based Pentium-III PC. The following feeder is used for the test (Figure 8.36).

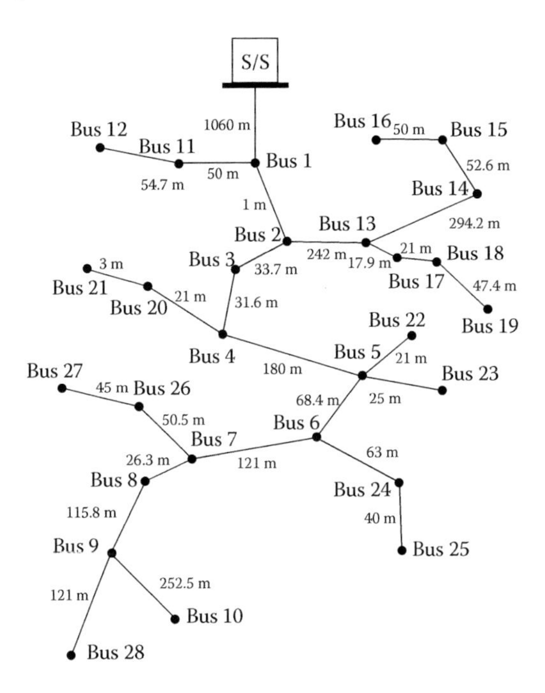

FIGURE 8.36
Test feeder.

Tables 8.6 and 8.7 are the load data and line length for this feeder, respectively. Figure 8.37 shows the bus voltage profiles for phases A, B, and C without the distributed generator installation. Figure 8.37 also shows that this feeder is unbalanced, and therefore, a three-phase load flow analysis can give an exact solution.

A test that two wind turbines (WTs) are replaced by power converter-based fuel cells and operated at constant voltage, the voltage profiles from the

TABLE 8.6

Load Data of Test Feeder

	Phase A		Phase B		Phase C	
Bus Number	**P (kW)**	**Q (kVAR)**	**P (kW)**	**Q (kVAR)**	**P (kW)**	**Q (kVAR)**
1	139.9	90.4	31.9	20.6	31.9	20.6
2	0	0	0	0	0	0
3	28.4	48.3	56.0	–0.4		
4	154.1	73.8	74.6	24.5	74.6	24.5
5	101.5	76.2	101.5	76.2	183.4	124.7
6			29.9	69.9	75.5	9.1
7	26.1	12.6	26.1	12.6	84.0	40.7
8			37.7	67.1	77.0	0.9
9	105.0	20.2	34.6	56.2		
10	84.0	63.0	84.0	63.0	141.5	98.7
11	113.2	61.1	113.2	61.1	212.4	109.1
12	165.1	91.3	81.4	46.1	81.4	46.1
13	367.3	133.3	367.3	133.3	367.3	133.3
14	30.5	13.9	30.5	13.9	108.9	58.3
15			75.3	85.9	88.8	–6.5
16	20.5	33.2	104.6	39.6		
17	74.4	64.6	34.0	1.1		
18	67	44.8	14.2	0.2		
19			67.5	109.5	128.5	– 3.7
20	98.4	33.6	36.2	13.1	36.2	13.1
21	7.8	8.8	11.5	–2.4		
22	29.9	16.1	73.9	49.1	29.9	16.1
23	173.7	68.3	72.1	34.9	72.1	34.9
24			47.9	–3.5	75.4	69.7
25	88.2	23.2	26.0	48.5		
26	75.7	19.0	75.7	19.0	207.0	51.9
27	23.5	1.2			87.2	66.3
28	79.5	49.3	79.5	49.3	187.7	130.4
Total demand	2053.7	1046.2	1787.1	1088.4	2350.7	1038.2

simulation as shown in Figure 8.38 can be obtained. In this case, the specified voltage magnitude is 0.97 p.u. Therefore, from Figure 8.38a, it can be seen that the voltages at Bus 29 are 0.97 p.u. and the unbalanced condition of this feeder is improved. Figure 8.38b shows the voltage profiles for Buses 1–10. The proposed distributed generator constant voltage model was used in the load flow analysis, and the results are displayed in Figure 8.38.

8.5.2.3 Summary

This section proposed the mathematical models of distributed generators for the application in the three-phase distribution load flow program. According to the characteristics of power output, distributed generators can be specified as a constant power factor model, constant voltage model, or variable

TABLE 8.7

Line Length of Test Feeder

Line Number	From Bus	To Bus	Length (m)
1	S/S	1	1060
2	1	2	1
3	2	3	33.7
4	3	4	31.6
5	4	5	180
6	5	6	68.4
7	6	7	121
8	7	8	26.3
9	8	9	115.8
10	9	10	252.5
11	1	11	50
12	11	12	54.7
13	2	13	242
14	13	14	294.2
15	14	15	52.6
16	15	16	50
17	13	17	17.9
18	17	19	21
19	18	19	47.4
20	4	20	21
21	20	21	3
22	5	22	21
23	5	23	25
24	6	24	63
25	24	25	40
26	7	26	50.5
27	26	27	45
28	9	28	121

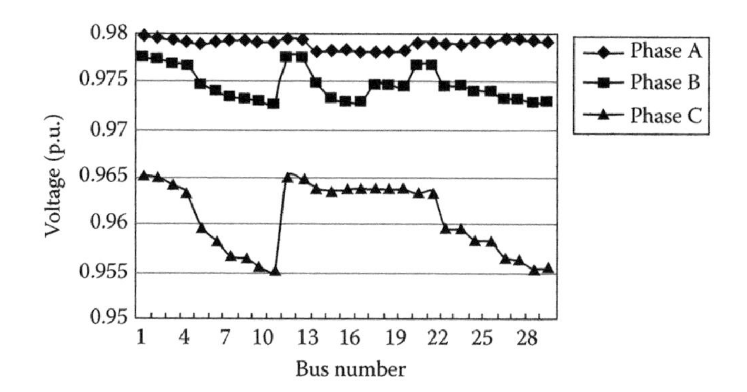

FIGURE 8.37
Voltage profiles of a test feeder without DG installation.

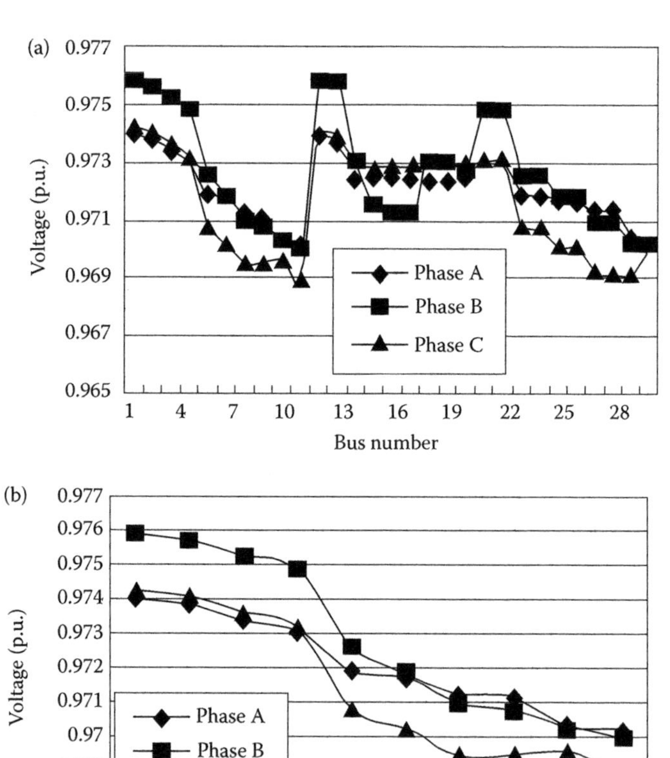

FIGURE 8.38
Voltage profiles of DGs operated in a constant voltage model. (a) Bus voltage profiles of a test feeder. (b) Bus voltage profiles for buses 1–10.

reactive power model in the load flow analysis. Test results show that the proposed method can be used to analyze the penetration of distributed generators to distribution feeders effectively and efficiently. Besides, the structures of the proposed matrices and the performance of the original load flow method are maintained. The proposed BIBC and BCBV matrices have the potential to be applied in other applications such as an asymmetrical short-circuit current calculation, optimal capacitor placement, feeder reconfiguration problem, and so on.

References

1. D.B. Gary, Hybrid Renewable Energy Systems. http://www.netl.doe.gov/publications/proceedings/01/hybrids/Gary%20Burch%208.21.01.pdf.
2. K. Agbossou, M. Kolhe, J. Hamelin, and T.K. Bose, Performance of a stand-alone renewable energy system based on energy storage as hydrogen, *IEEE Transactions on Energy Conversion*, 19(3), 633–640, 2004.
3. D.B. Nelson, M.H. Nehrir, and C. Wang, Unit sizing and cost analysis of stand-alone hybrid Wind/PV/fuel cell systems, *Renewable Energy*, 31(10), 1641–1656, 2006.
4. S.H. Chan, H.K. Ho, and Y. Tian, Multi-level modeling of SOFC-gas turbine hybrid system, *International Journal of Hydrogen Energy*, 28(8), 889–900, 2003.
5. L.A. Torres, F.J. Rodriguez, and P.J. Sebastian, Simulation of a solar hydrogen-fuel cell system: Results for different locations in Mexico, *International Journal of Hydrogen Energy*, 23(11), 1005–1010, 1998.
6. F. Bonanno, A. Consoli, A. Raciti, B. Morgana, and U. Nocera, Transient analysis of integrated diesel–wind–photovoltaic generation systems, *IEEE Transactions on Energy Conversion*, 14(2), 232–238, 1999.
7. K. Strunz and E.K. Brock, Stochastic energy source access management: Infrastructure-integrative modular plant for sustainable hydrogen electric co-generation, *International Journal of Hydrogen Energy*, 31, 1129–1141, 2006.
8. E.S. Abdin, A.M. Osheiba, and M.M. Khater, Modeling and optimal controllers design for a stand-alone photovoltaic-diesel generating unit, *IEEE Transactions on Energy Conversion*, 14(3), 560–565, 1999.
9. F. Giraud and Z.M. Salameh, Steady-state performance of a gridconnected rooftop hybrid wind–photovoltaic power system with battery storage, *IEEE Transactions on Energy Conversion*, 16(1), 1–7, 2001.
10. W.D. Kellogg, M.H. Nehrir, G. Venkataramanan, and V. Gerez, Generation unit sizing and cost analysis for stand-alone wind, photovoltaic, and hybrid wind/PV systems, *IEEE Transactions on Energy Conversion*, 13(1), 70–75, 1998.
11. H. Sharma, S. Islam, and T. Pryor, Dynamic modeling and simulation of a hybrid wind diesel remote area power system, *International Journal of Renewable Energy Engineering*, 2(1), 19–25, 2000.
12. R. Chedid, H. Akiki, and S. Rahman, A decision support technique for the design of hybrid solar-wind power systems, *IEEE Transactions on Energy Conversion*, 13(1), 76–83, 1998.

13. D.L. Elliott and M.N. Schwartz. Wind energy potential in the United States, National Renewable Energy Laboratory (NREL). 1993. http://www.nrel.gov/wind/wind_potential.html (current May 2004).
14. P. Gipe, *Wind Energy Comes of Age*, John Wiley & Sons, New York, 1995.
15. S.S. Hutson, N.L. Barber, J.F. Kenny, K.S. Linsey, D.S. Lumia, and M.A. Maupin, 2004, Estimated use of water in the United States in 2000 Reston Virginia, U.S. Geological Survey, Circular 1268, 46pp.
16. American Wind Energy Association (AWEA). Wind energy fast facts, AWEA. http://www.awea.org/pubs/factsheets/FastFacts2003.pdf (current May 2004).
17. W. Carter and B. Diong, Model of regenerative fuel cell-supported wind turbine AC power generating system, *Proceedings of the IEEE Industry Applications Society Annual Meeting*, Seattle, Washington, 2004.
18. M.T. Iqbal, Modeling and control of a wind fuel cell hybrid energy system, *Renewable Energy*, 28(2), 223–237, 2003.
19. Grainger Center for Electric Machinery and Electromechanics, University of Illinois at Urbana-Champaign. The 2001 International Future Energy Challenge. http://www.energychallenge.org/.
20. L.-Y. Chiu, B. Diong, and R. Gemmen, An improved small-signal model of the dynamic behavior of PEM fuel cells, To appear in *IEEE Transactions on Industry Applications*, 40(4), 970–977, 2004.
21. W.E. Leithead, S.A. De La Selle, and D. Reardon. Classical control of active pitch regulation of constant speed horizontal axis wind turbine, *International Journal of Control*, 55(4), 845–876, 1992.
22. D.M. Eggleston and F.S. Stoddard, *Wind Turbine Engineering Design*, Van Nostrand Reinhold, New York, 1987.
23. E.W. Weisstein, Euler backward method. From *MathWorld—A Wolfram Web Resource*. http://mathworld.wolfram.com/EulerBackwardMethod.html (current May 2004).
24. R. Isermann, *Digital Control Systems*, Vol. 2 (2nd edition), Springer-Verlag, Berlin, Germany, 1989.
25. J. Larminie and A. Dicks, *Fuel Cell Systems Explained*, John Wiley & Sons, Inc., Chichester, United Kingdom, 2000.
26. K. Sapru, N.T. Stetson, and S.R. Ovshinsky, Development of a small scale hydrogen production storage system for hydrogen applications, *Proceedings of Intersociety Energy Conversion Engineering Conference*, 3–4, 1947–1952, 1997.
27. N. Mohan, T.M. Undeland, and W.P. Robbins, *Power Electronics: Converters, Applications, and Design*, John Wiley & Sons, Inc., New York, 1995.
28. T. Senjyu, T. Nakaji, K. Uezato, and T. Funabashi, A hybrid power system using alternative energy facilities in isolated Island, *IEEE Transactions on Energy Conversion*, 20(2), 406–414, 2005.
29. C. Wang and M.H. Nehrir, Power management of a stand-alone wind/photovoltaic/fuel cell energy system, *IEEE Transactions on Energy Conversion*, 23(3), 957–967, 2008.
30. Global Wind 2007 report, Global Wind Energy Council [Online]. Available: http://www.gwec.net/index.php?id=90.
31. *Wind Power Today—Federal Wind Program Highlights*, NREL, DOE/GO-102005-2115, 2005.

32. *Trends in Photovoltaic Applications: Survey Report of Selected IEA Countries between 1992 and 2004*, International Energy Agency Photovoltaics Power Systems Programme (IEA PVPS), 2005.
33. J. Cahill, K. Ritland, and W. Kelly, *Description of Electric Energy Use in Single Family Residences in the Pacific Northwest 1986–1992*, Office Energy Resour., Bonneville Power Admin., Portland, Oregon 1992.
34. [Online]. Available: http://www.usbr.gov/pn/agrimet/webaghrread.html, 2006.
35. J.H. Teng, Modeling distributed generations in three-phase distribution load flow, *IET Generation, Transmission & Distribution*, 2(3), 330–340, 2008.
36. M.E. Baran and F.F. Wu, Optimal sizing of capacitors placed on a radial distribution system, *IEEE Transactions on Power Delivery*, 4(1), 735–743, 1989.
37. H. Chiang and M.E. Baran, On the existence and uniqueness of load flow solution for radial distribution power networks, *IEEE Transactions on Circuits and Systems*, 37(3), 410–416, 1990.
38. M. Rabinowitz, Power systems of the future, *IEEE Power Engineering Review*, 20(1), 5–16, 2000.
39. S. Persaud, B. Fox, and D. Flynn, Impact of remotely connected wind turbines on steady state operation of radial distribution networks, *IEE Proceedings, Generation, Transmission and Distribution*, 147(3), 157–163, 2000.
40. A.E. Feijoo and J. Cidras, Modeling of wind farms in the load flow analysis, *IEEE Transactions on Power System*, 15(1), 110–115, 2000.
41. T.H. Chen, M.S. Chen, T. Inoue et al., Three-phase cogenerator and transformer models for distribution system analysis, *IEEE Transactions on Power Delivery*, 6(4), 1671–1681, 1991.
42. C.S. Cheng and D. Shirmohammadi, A three-phase power flow method for real-time distribution system analysis, *IEEE Transactions on Power System*, 10(2), 671–679, 1995.
43. J.H. Teng, A direct approach for distribution system load flow solutions, *IEEE Transactions on Power Delivery*, 18(3), 882–887, 2003.
44. M. Korpas and A.T. Holen, Operation planning of hydrogen storage connected to wind power operating in a power market. *IEEE Transactions on Energy Conversion*, 21(3), 742–749, 2006.

9

Optimization of PEMFCs

9.1 Introduction

Owing to these multiple advantages of PEMFCs mentioned in Chapter 1, the PEMFCs become popular for an alternative power source in transportation and stationary power systems. First of all, in order to commercialize the PEMFC, the cost and the efficiency need to be taken into account simultaneously. So, an optimal PEMFCs system design considering cost and efficiency of PEMFCs together has become a hot topic in recent years. For stationary and transportation applications, the efficiency of fuel cells is required to achieve higher or equal to 40% comparing the internal combustion engine [1].

Since the efficiency decreases as the power output increases [1], more cells are to be integrated to achieve a high efficiency for the maximum power output by considering its economic aspect. In this chapter, using a multiobjective optimization technique, the SQP (sequential quadratic programming) method, the efficiency, and the cost of a fuel cell system have been optimized under various operating conditions. The system pressure, hydrogen and air stoichiometric ratios, the fuel cell current density, and the fuel cell temperature are defined as design variables.

Section 9.2 gives a concept of PEMFCs' efficiency. Section 9.3 addresses the design of PEMFCs cost model for the optimization. In Section 9.4, the multiobjective optimization for the PEMFC is presented. Section 9.5 provides the results and discussion in regard to the optimization.

9.2 PEMFCs Efficiency Model

In practice, especially, the Ballard mark V [2], its fuel cell voltage per cell is described in Equation 9.1, which has specific coefficients given in Table 9.1.

$$V_c = E_{oc} - r \cdot i - C\ln(i) - m\exp(ni) \tag{9.1}$$

TABLE 9.1

Ballard Mark V PEMFC Coefficient

Coefficients	Values (T in °C)
E_{oc} (*V*)	1.05
C (*V*)	$4.01 \times 10^{-2} - 1.4 \times 10^{-4} T$
r (*k*Ω/cm²)	$4.77 \times 10^{-4} - 3.32 \times 10^{-6} T$
m (*V*)T ≥ 39°C	$1.1 \times 10^{-4} - 1.2 \times 10^{-6} T$
m (*V*)T ≤ 39°C	$3.3 \times 10^{-3} - 8.2 \times 10^{-5} T$
n (cm²/mA)	8.0×10^{-3}

Source: Adapted from J. Hamelin et al., *International Journal of Hydrogen Energy*, 26, 625–29, 2001.

The detailed explanation of each voltage loss can be found in Reference 2. In the Nernst equation, the ideal standard potential E_o for a PEMFC is 1.229 V with liquid water product, or 1.18 V with gaseous water product [4]. Under the assumption that pressure on both the cathode and the anode is approximately the same, the Nernst equation is transferred into a function of the system pressure P_{sys} [2,5] given as follows:

$$E = N\left[E_o + \frac{R^*T}{2F}\ln\left(\frac{\alpha \cdot \beta^{1/2}}{\delta}\right) + \frac{R^*T}{4F}\ln(P_{sys})\right] \tag{9.2}$$

where α, β, and δ are constants depending on the molar masses and concentrations of H_2, O_2, and H_2O. Each partial pressure can be expressed by these constants and the system pressures.

$$\begin{aligned} P_{H_2} &= \alpha P_{sys} \\ P_{O_2} &= \beta P_{sys} \\ P_{H_2O} &= \delta P_{sys} \end{aligned} \tag{9.3}$$

With assuming that α, β, and δ are constants, Equation 9.3 shows that the electromotive force (EMF) of a fuel cell is increased by the system pressure P_{sys}. Although P_{sys} is able to use optimization design variables, the predefined α, β, and δ are required for the optimization, so that the system pressure can be considered in the compressor model of Equation 9.7. For the multiobjective optimization, the specification of the fuel cell stacks has to be identified in advance and then each optimization model will be delivered (Table 9.2).

First, the fuel cell efficiency optimization model is derived based on Reference 6, and the output power of the fuel cell system is described by the following equations:

TABLE 9.2

Specification of the Fuel Cell System

Items	Specification
Nominal power output	50 kW
Stack temperature	353 K (80°C)
Inlet H_2/air humidity	100%
Cell open voltage, E_o	1.05 V
Entry air temperature, T_e	288 K (15°C)
Specific heat constant, c_p	1004 J/K kg
Compressor efficiency, η_c	0.75
Compressor connecting efficiency, η_m	0.85
Inlet pressure, P_{in}	10^5 Pa

Source: Based on J. Larminie and A. Dicks, *Fuel Cell Systems Explained*, Wiley, New York, 2002.

$$P_{fcs} = P_{stack} - P_{prs} \tag{9.4}$$

$$P_{stack} = N \cdot i \cdot V_c \cdot A = 50\,\text{kW} \tag{9.5}$$

$$P_{prs} = P_{comp} + P_{oth} \tag{9.6}$$

$$P_{comp} = c_p \frac{T_e}{\eta_m \cdot \eta_{mt}} \left(\left(\frac{P_{sys}}{P_{in}} \right)^{0.286} - 1 \right) \cdot \dot{m} \tag{9.7}$$

$$\dot{m} = 3.57 \times 10^{-7} \times \lambda_{air} \times i \times A \times N\,\text{kg/s} \tag{9.8}$$

where P_{fcs} is the net power of the fuel cell system and P_{stack} is the stack output power. The parasitic power consumed by the compressor is P_{comp} and the power consumed by others is P_{oth}. Even though P_{oth} was assumed to be a constant of 2 kW in Reference 6 based on 62.5-kW-rated stack power, here, P_{oth} is assumed to be 2.5 kW that is 5% of the nominal power out of 50 kW by including the unexpected power consumption. The flow rate of air $\dot{m}$ is related to the air stoichiometry, the cell current density, and active cell area. Before proceeding to build the efficiency optimization model, let us consider how to decide the optimal cell number and cell area. Once an optimal current density and a cell voltage have been achieved, the total active cell area ($N \times A$) is able to be calculated by using optimal power density, the product of V_c and i, namely, we can decide a number of cells as long as a single active cell area is given. The system pressure of the fuel cell is always higher than the atmospheric pressure in a certain range because the compressor cannot provide a pressure under the atmospheric pressure.

According to Barbir [7], P_{sys} must be 0.02 MPa higher than the inlet air pressure P_{in} and therefore, this requirement is included as one of the constraints in the optimization study. Thus, if using the lower heating value (LHV), the fuel cell efficiency optimization model is obtained to achieve a maximum efficiency of the fuel cell as follows:

$$\max \eta_{fc}(P_{sys}, \lambda_{H_2}, \lambda_{air}, V_c, i) = \frac{V_c \cdot u_f \cdot (P_{stack} - P_{prs})}{1.25 \cdot P_{stack}}$$

s.t.

$$\begin{aligned} & P_{sys} \geq 0.12\ \text{MPa} \\ & \lambda_{H_2} \geq 1 \\ & \lambda_{air} \geq 1 \\ & V_c \geq 0\ \text{V} \\ & i \geq 0\ \text{mA/cm}^2 \end{aligned} \tag{9.9}$$

where u_f is the fuel utilization rate, which is the reverse of hydrogen stoichiometric ratio [8]. The air stoichiometric ratio must be over the minimum limit in order to prevent the depletion of oxygen at this minimum limit. The hydrogen stoichiometric ratio is also greater than 1 unless it runs in the hydrogen dead-ended mode [8]. Normally, higher air and hydrogen stoichiometric ratio are preferred in low-power ranges. The ranges of cell voltage and current density will be based on the *V–I* polarization curve. Since the cell voltage is a function of the cell current density as seen in Equation 9.1, we can reduce the four optimization parameters such as system pressure, air and hydrogen stoichiometric ratios, and cell current density. By using four optimization parameters, the fuel cell efficiency optimization model has been built. In the following section, the fuel cell cost optimization model will be described.

9.3 PEMFCs Cost Model

For this analysis, we are particularly interested in the small- and middle-size fuel cell systems. We use a 50 kW PEM fuel cell system for transportation applications as the example of our study. For the fuel cell cost model for optimization, the cost of fuel cell stack and BOP components for water, thermal, and fuel management have been assessed. Owing to a lack of latest data about hydrogen storage, power electronics, electric drive motor, and hybrid batteries for PEM fuel cell system, the fuel storage and fuel generation components were excluded from the scope of this study.

The target cost is the cost of the fuel cell stack and BOP system given as follows:

$$C = C_{st} + C_{BOP} \tag{9.10}$$

where C_{st} is the cost of the fuel cell stack and C_{BOP} is the cost of balance of plants (BOPs). For assessing of fuel cell stack cost C_{st}, currently, two types of fuel cell stack cost models are available in References 9 and 10. One is represented by the following equation [10]:

$$C_{st1} = M \cdot \left[\left(\frac{A - 105.4}{10} + \frac{17.56 \cdot L_p \cdot C_p}{380} \right) \cdot \frac{P_G \cdot (1+d)^N}{P_d} + B \right] \tag{9.11}$$

where
- M = Fixed cost markup (1.1 default)
- A, B = Cost parameter that depends on production volume (see Table 9.3)
- L_p = Fuel cell platinum loading for both electrodes (mg/cm²)
- C_p = Cost of platinum ($/troy ounce)
- P_G = Fuel cell gross DC peak power (kW)
- P_d = Fuel cell power density (W/cm²)
- d = Annual fuel cell degradation (%/year)
- N = Planned fuel cell lifetime (years)

The parameter A is the power-dependent term in terms of dollar per square meter of the membrane area, and the parameter B is the fixed cost for the fuel cell stack.

In Reference 9, the annual fuel cell degradation is assumed to be a 6% drop per year and the planned fuel cell lifetime is assumed to be 87,600 h, 10 years. The platinum loading L_p is defined as 0.4 mg/cm² and the cost of platinum C_p is 1160 $/troy ounce. Although this C_{st1} ($) model is used in Reference 12, it is not logically understandable because suddenly many constants and parameters M, A, and B are involved in Equation 9.11 without justification.

TABLE 9.3
Fuel Cell Stack Cost Parameters

Production Volume	Cost Parameter, A ($/m²)	Cost Parameter, B ($)
100	811.77	1311.3
1000	722.54	363.33
10,000	454.45	428.51
30,000	329.24	405.79
60,000	312.26	160.98

Source: Adapted from L. You and H. Liu, *International Journal of Hydrogen Energy*, 26, 991–999, 1991.

Thus, a more reasonable stack cost model C_{st2} [10,13] is chosen for our study. The C_{st2} (\$/kW) is described as follows:

$$C_{st2} = \frac{(C_m + C_e + C_b + C_{pt} + C_o)}{P} + C_a \tag{9.12}$$

$$C_{pt} = C_{wpt} \times Y_{pt} \tag{9.13}$$

$$P = 10 \times V_c \times i \tag{9.14}$$

where C_{st2} is the fuel cell stack cost per kW (\$/kW), C_m is the membrane cost (\$/m^2), C_e is the electrode cost (\$/m^2), C_b is the bipolar plates cost (\$/m^2), C_{pt} is the cost of platinum catalyst loading (\$/m^2), C_{wpt} is the weight of platinum catalyst loading (g/m^2), Y_{pt} is the unit cost of platinum (\$/g), C_o is the cost of peripheral materials (\$/m^2) that include end plates, a plastic frame, and thrust volts, C_a the assembly cost (\$/kW), V_c the cell voltage, and i is the cell current density (A/cm^2). In References 10 and 13, the cost of each component of a 50 kW PEM fuel cell stack was estimated based on an automatic production line with an annual production capacity of 18,000 vehicles. Table 9.4 shows the specific cost for components in a PEM fuel cell stack.

With respect to the cost of BOPs C_{BOP}, the cost model can be found in Reference 9. C_{BOP}, including the air blower, humidification, radiator, a stainless-steel pump, iron pump, control electronics, actuation, piping, and valves, is approximated by a quadratic equation in the fuel cell output power and varies with the production volume as the following equations [9]:

For 100 production units:

$$C_a = 3343.5 + 39.942 \cdot P_G - 0.0454 \cdot P_G^2 \tag{9.15}$$

TABLE 9.4

Specific Cost for Components in PEM Fuel Cell Stack

Components	Cost
Nafion membrane	500 \$/m^2
Platinum (2–4 g/m^2)	32–64 \$/m^2
Electrode (max. 0.8 mm for single cell)	177 \$/m^2
Bipolar plate (max. 4 mm)	1650 \$/m^2
Peripheral parts	15.6 \$/m^2
Assembly	7.7 \$/kW

Source: Adapted from E.J. Carlson et al., *Cost Analysis of PEM Fuel Cell Systems for Transportation*, TIAX LLC, Cambridge, Massachusetts, 2005; S. Kamarudin et al., *Journal of Power Source*, 157, 641–649, 2006.

For 10,000 production units:

$$C_a = 2980.2 + 35.654 \cdot P_G - 0.0422 \cdot P_G^2 \tag{9.16}$$

Unfortunately, since Equations 9.15 and 9.16 are not only with respect to a stationary PEM fuel cell, but also out of date, which was published in 1999, they cannot be applied to this analysis. Therefore, using most updated data [13], C_{BOP} is estimated to 34% of the fuel cell system cost as C_{st} including assembly, that is assumed to be contributing to approximately 66% of the fuel cell system cost. Even though the breakdown of the fuel cell system is for an 80 kW direct hydrogen system, the same breakdown is applied to a 50 kW fuel cell system because the cost analysis is only integrated in the stack and BOP costs (Figure 9.1).

In order to build the cost model sharing the same optimization parameters, the cost model is likely to be a function of the efficiency as in Equation 9.17, which is able to investigate the impact of the cell voltage, V_c and current density, i as well as other optimization parameters on the fuel cell cost. In our study, the costs of the fuel cell stack and BOP are considered. And the maximum fuel cells system cost is obtained as the maximum efficiency is achieved. Thus, this cost optimization model can share the same optimization parameters with the system efficiency optimization model and each feasible range of the parameters will be used as a constraint of this optimization

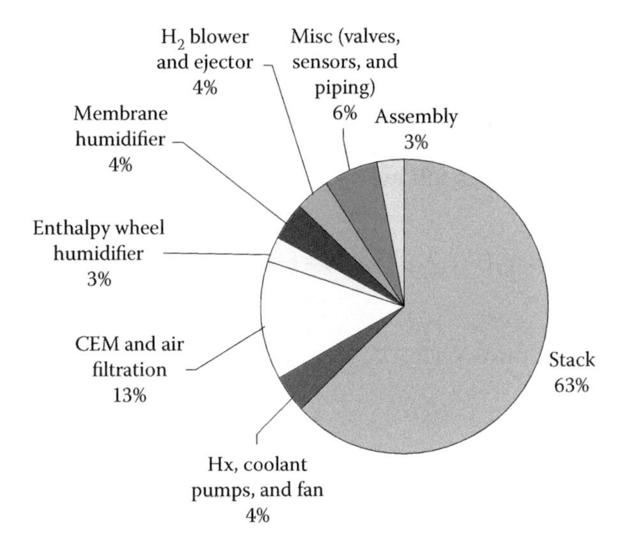

FIGURE 9.1
Breakdown in stack and BOP component cost contribution for an 80 kW direct hydrogen fuel cell system. (Adapted from C.E. Tomas et al., Cost analysis of stationary fuel cell systems including hydrogen cogeneration ACG-8-18012-02, National Renewable Energy Laboratory, Colorado, 1999.)

problem. The base production volume is chosen as 18,000 units, which can be a mass production for the mobile 50 kW PEM fuel cell system.

$$\min C_{FC}(P_{sys}, \lambda_{H_2}, \lambda_{air}, V_c, i) = (C_{st} + C_{BOP}) \times \eta_{fc}$$

s.t.

$$\begin{aligned} &P_{sys} \geq 0.12 \text{ MPa} \\ &\lambda_{H_2} \geq 1 \\ &\lambda_{air} \geq 1 \\ &V_c \geq 0 \text{ V} \\ &i \geq 0 \text{ mA/cm}^2 \end{aligned} \tag{9.17}$$

As explained in the section on the cell voltage and current density, two of the optimization parameters of the cost model could be reduced to just the cell current density.

In the following section, the multiobjective optimization will be presented with the consideration of both the efficiency and the cost optimization.

9.4 Multiobjective Optimization of PEMFCs

According to the above considerations, the multiobjective optimization problem is formulated as follows:

$$\min \eta_{fc}(P_{sys}, \lambda_{H_2}, \lambda_{air}, i) = (-1) \cdot \frac{V_c \cdot u_f \cdot (P_{stack} - P_{prs})}{1.25 \cdot P_{stack}}$$

$$\min C_{FC}(P_{sys}, \lambda_{H_2}, \lambda_{air}, i) = (C_{st} + C_{BOP}) \times \eta_{fc}$$

s.t.

$$\begin{aligned} &P_{sys} \geq 0.12 \text{ MPa} \\ &\lambda_{H_2} \geq 1 \\ &\lambda_{air} \geq 1 \\ &i \geq 0 \text{ mA/cm}^2 \end{aligned} \tag{9.18}$$

In the multiobjective optimization problem, both objective functions have to be minimized simultaneously. Normally, the objectives in such a problem often have the issue of the conflict between the objectives. From Equation 9.18, when the efficiency is increased, the cost is increased as well. Thus, during the optimization process for such a problem, there is no single optimum solution to improve both the objectives, which means a number of optimal solutions exit. As the solution to the multiobjective problem is a set of points that represent the best tradeoffs between the objective functions, for each solution, there is no way to further improve an objective function without worsening at least another one. Such points are called Pareto optimal points or noninferior points. The set of all the Pareto optimal points is called the Pareto optimal set or the Pareto frontier.

In this study, the MATLAB Optimization Toolbox [14] for a multiobjective optimization problem is used to find the Pareto frontier solution set. MATLAB has two functions to solve a multiobjective problem: fminimax and fgoalattain. Even though both the methods use the popular nonlinear-programming algorithm, a SQP, the fminimax method is more appropriate to be applied to our optimization study than the fgoalattain method because the fgoalattain method is more complicated than the fminimax due to the defining weighting coefficients. The general form of fminimax method is

$$\min_{x} \max_{f} \{f_1, f_2, \ldots, f_m\}$$

such that

$$\begin{aligned} &Ax \leq b \\ &A_{eq}x = b_{eq} \\ &C(x) \leq 0 \\ &C_{eq}(x) = 0 \\ &L_b \leq x \leq U_b \end{aligned} \tag{9.19}$$

where x is the design variable vector; $f_1, f_2, \ldots, f_m$ are the objective functions; matrix A and vector b are the coefficients of the linear inequality constraints; matrix A_{eq} and vector b_{eq} are the coefficients of the linear equality constraints; C contains the nonlinear inequality constraints; C_{eq} contains the nonlinear equality constraints; and L_b and U_b are the lower and upper bounds, respectively.

In order to search for an optimal design value x, the fminimax method iteratively minimizes the worst-case value (or maximum) of the objective functions subject to the constraints. The advantage of this method is to easily find the optimum design point from an arbitrary initial design point. Furthermore, less function and gradient evaluation are required compared to other constrained nonlinear optimization. However, the main disadvantage of both methods, fminimax and fgoalattain, is that the objective functions must be continuous and each method has a limitation to search for global solutions.

9.5 Results and Discussion

In this section, the fminimax method is executed to solve the multiobjective optimization. To avoid the unrealistic design criterion, the upper bounds of the system pressure, air and hydrogen stoichiometric ratios are specified at 10 MPa, 10 and 10, respectively. According to the polarization *V–I* curve in Figure 1.2, the cell current density will lie within the range of 0–1 A/cm^2, which is used as one of the bound limits for the optimization. The bound limits of the design variables are given as follows:

$$\begin{aligned} &0.12\,\text{MPa} \le P_{sys} \le 10\,\text{MPa} \\ &1 \le \lambda_{H_2} \le 10 \\ &1 \le \lambda_{air} \le 10 \\ &0 \le i \le 1\,\text{A/cm}^2 \end{aligned} \tag{9.20}$$

With various initial conditions of the design parameters, P_{sys}, λ_{H_2}, λ_{air}, and i, the corresponding tradeoff (Pareto) solutions are obtained. So to speak, these solutions are in the Pareto set, that is, as one objective is improved in the set, the other is worsened.

For simplicity, the initial conditions are able to be described in the column vector, such as $[P_{sys}, \lambda_{H_2}, \lambda_{air}, i]$. If an arbitrary initial condition is defined as the vector, In 1 = [0.12 MPa, 2, 2, 800 mA/cm^2], Figure 9.2 shows the tradeoff solution based on the vector In 1. In Figure 9.2, as the efficiency is improved in the set, the cost is increased as well. In changing the initial condition, this Pareto frontier will be changed because the fminimax method will find a local solution in the changed initial condition.

First, when In 1 is changed to In 2 = [0.24 MPa, 2, 2, 800 mA/cm^2], that is, the system pressure becomes 2 times bigger than that of the In 1, Figure 9.2 shows how this change affects the optimization of the fuel cell system. As seen in Figure 9.2, the higher system pressure is able to achieve the cost-effective and high-performance model compared with the model given by the In 1. For instance, in Figure 9.2, at the efficiency of 0.45, the In 2, Pareto frontier is corresponding to about 420 \$/kW, whereas, In 1 Pareto frontier is almost over 700 \$/kW, that means In 1 is a less economic condition than the In 2.

However, in the case of a much higher pressure, the change from the In 1 to In 3 = [0.36 MPa, 2, 2, 800 mA/cm^2] as shown in Figure 9.3, although the better optimum model than In 2 is obtained, the efficiency range of the In 3 becomes unrealistic because the fuel cell system efficiency is not normally greater than 0.6 [1]. Hence, the initial condition In 2 is more recommendable

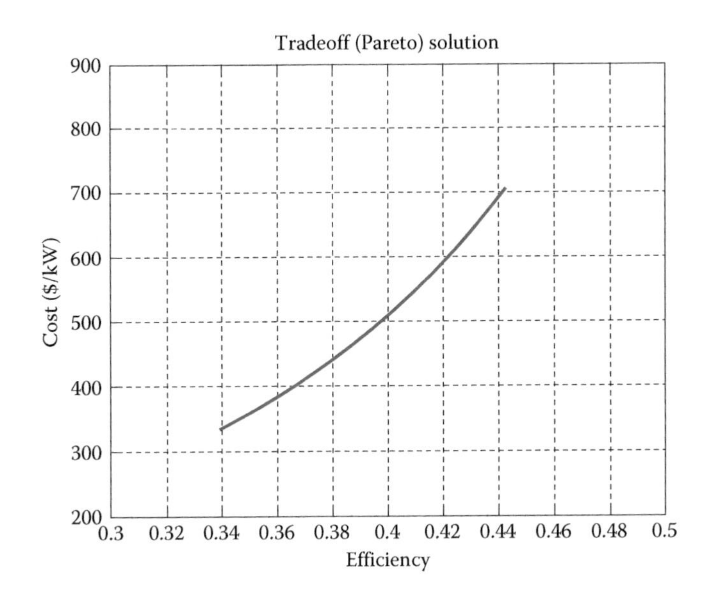

FIGURE 9.2
Pareto frontier based on the In 1(In 1 = [0.12 MPa, 2, 2, 800 mA/cm²]). (Adapted from W. Na and B. Gou, *Journal of Power Source*, 166(2), 411–418, 2007.)

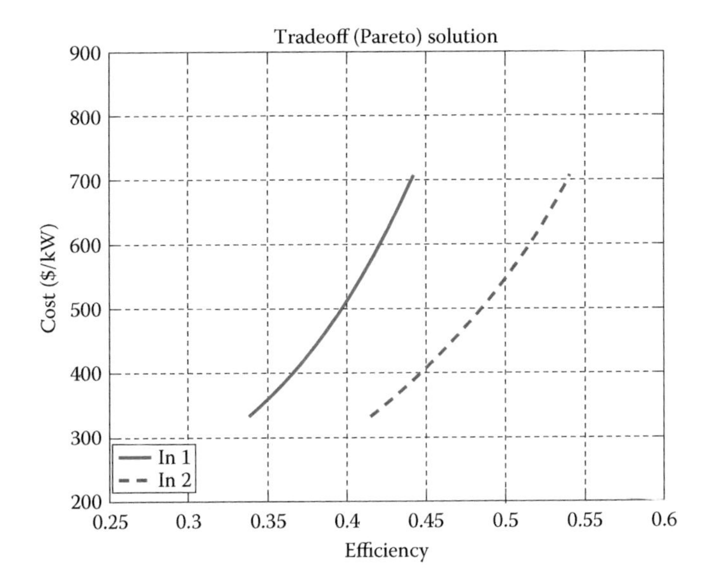

FIGURE 9.3
Pareto frontier change from In 1 to In 2 (In 2 = [0.24 MPa, 2, 2, 800 mA/cm²]). (Adapted from W. Na and B. Gou, *Journal of Power Source*, 166(2), 411–418, 2007.)

than the In 3. Second, if the initial hydrogen stoichiometric ratio is changed from 2 to 1.5 such as In 4 = [0.12 MPa, 1.5, 1, 800 mA/cm^2], the less cost-effective and bad performance model is found in Figure 9.5. Hence, around the hydrogen stoichiometric ratio, 2 is more preferable than 1.5.

In the case that the air stoichiometric ratio is changed from 2 to 1.5 as the In 5 = [0.12 MPa, 1.5, 1, 800 mA/cm^2], a more cost and efficiency effective model is achieved in Figure 9.6. However, if the air stoichiometric ratio keeps decreasing to 1, the efficiency is not applicable to the real system as seen in Figure 9.6. Therefore, the ratio around 1.5 is more preferable than 2 and 1. For the current density, as it decreases, the cost and efficiency effective model is achieved. Through the trial and error, the optimal current density is approximately estimated to 450 mA/cm^2. By comparing with In 1, the In 7 has a better efficiency and cost-effective model as shown in Figure 9.8. With the recommended current density of 450 mA/cm^2 and *V–I* polarization curve in Figure 3.2, the optimal cell voltage can be calculated to 0.72 V and the power density will be 3.2 kW/m^2. If 50-kW rate power output is selected, then the total active cell area is 15.625 m^2, which means the stack will need to contain 174 layers of a single cell with 30 cm × 30 cm active cell area. Thus, the optimal current density allows us to determine the total active cell area and even provide the information about the cell number to be required for the target stack if a single cell area is given.

Figures 9.3–9.10 indicate that the change of each design variable gives a severe impact on the cost and efficiency of the fuel cell. Specially, since the current density is closely associated with the fuel cell area, it affects more directly on the fuel cell cost and efficiency rather than any other variables.

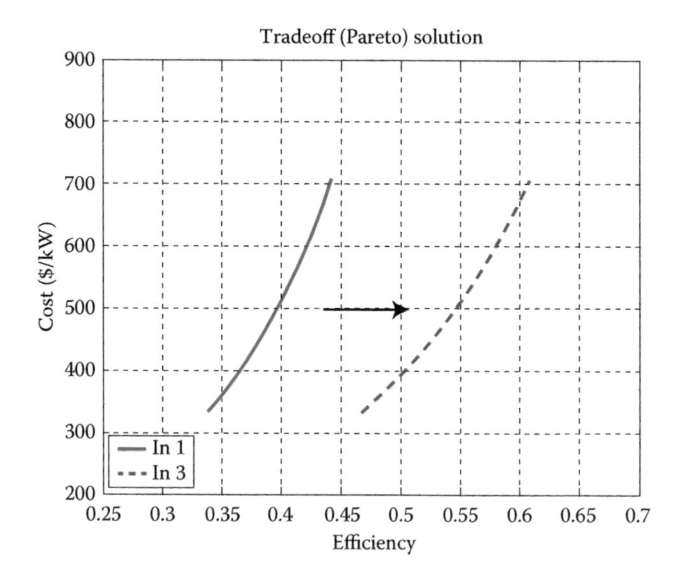

FIGURE 9.4
Pareto frontier change from In 1 to In 3 (In 3 = [0.36 MPa, 2, 2, 800 mA/cm^2]). (Adapted from W. Na and B. Gou, *Journal of Power Source*, 166(2), 411–418, 2007.)

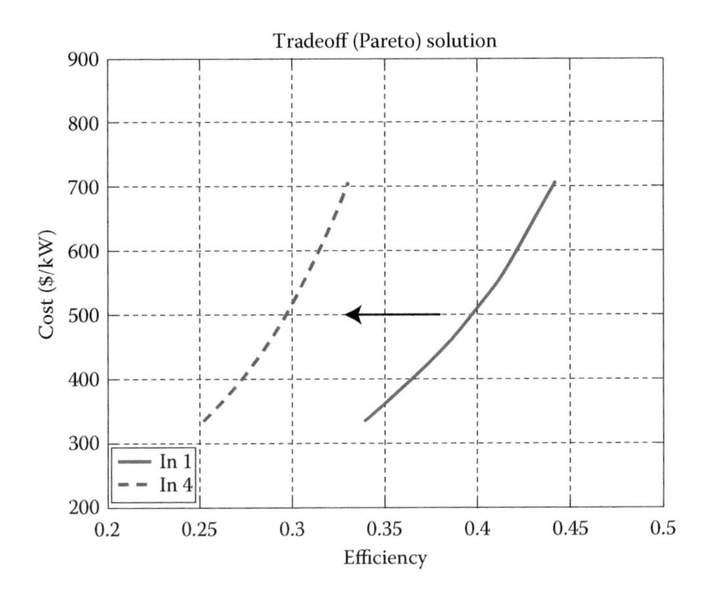

FIGURE 9.5
Pareto frontier change from In 1 to In 4 (In 4 = [0.12 MPa, 1.5, 2, 800 mA/cm^2]). (Adapted from W. Na and B. Gou, *Journal of Power Source*, 166(2), 411–418, 2007.)

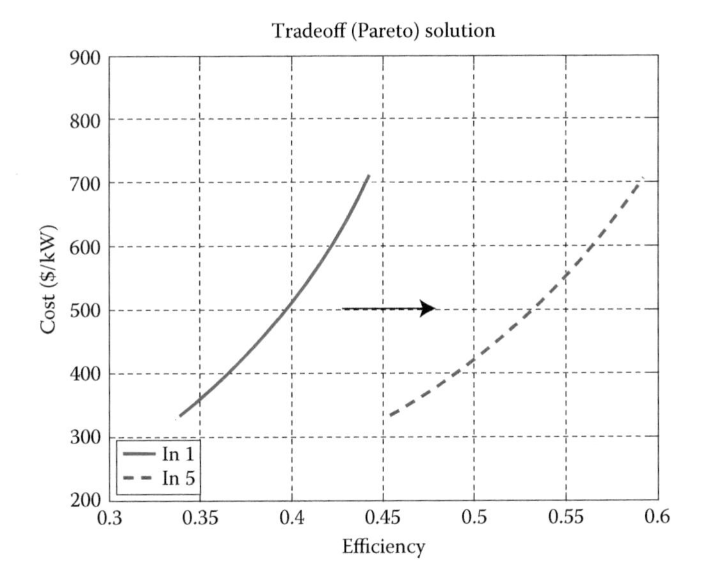

FIGURE 9.6
Pareto frontier change from In 1 to In 5 (In 5 = [0.12 MPa, 2, 1.5, 800 mA/cm^2]). (Adapted from W. Na and B. Gou, *Journal of Power Source*, 166(2), 411–418, 2007.)

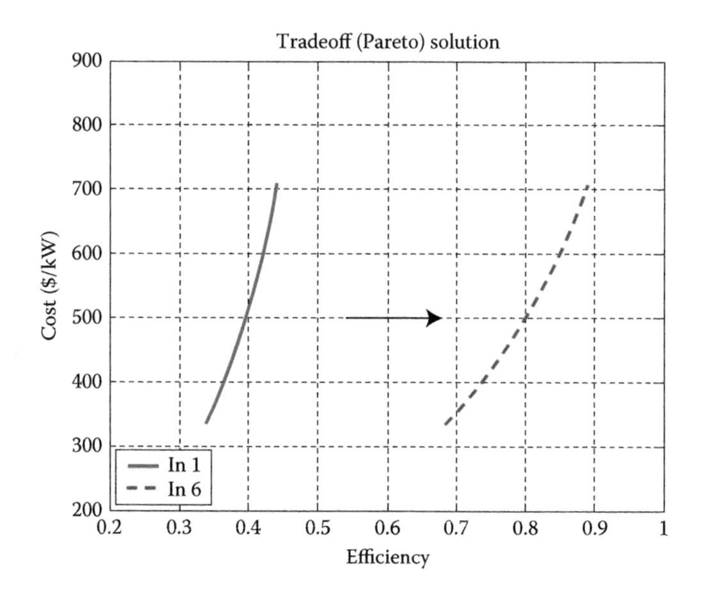

FIGURE 9.7
Pareto frontier change from In 1 to In 6 (In 6 = [0.12 MPa, 2, 1, 800 mA/cm^2]). (Adapted from W. Na and B. Gou, *Journal of Power Source*, 166(2), 411–418, 2007.)

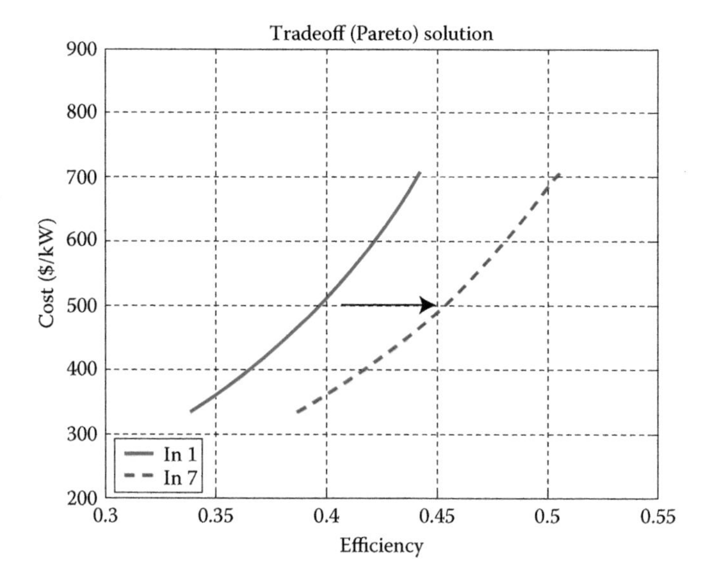

FIGURE 9.8
Pareto frontier change from In 1 to In 7 (In 7 = [0.12 MPa, 2, 2, 450 mA/cm^2]). (Adapted from W. Na and B. Gou, *Journal of Power Source*, 166(2), 411–418, 2007.)

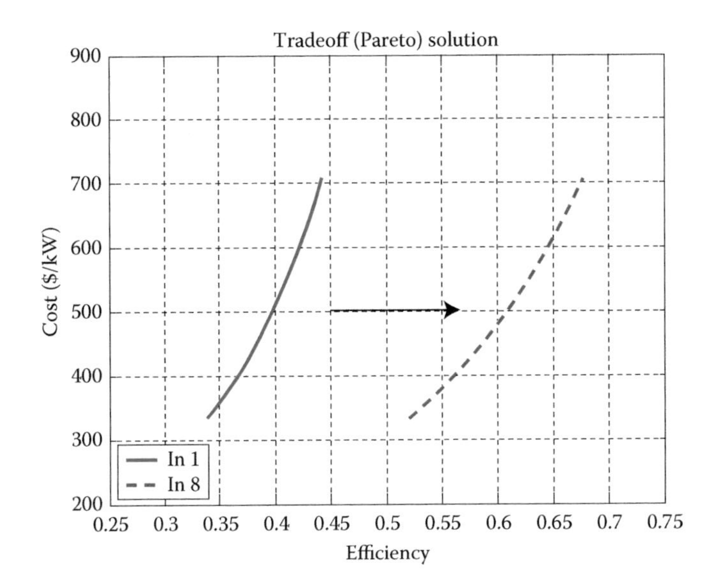

FIGURE 9.9
Pareto frontier change from In 1 to In 8 (In 8 = [0.24 MPa, 2, 1.5, 450 mA/cm^2]). (Adapted from W. Na and B. Gou, *Journal of Power Source*, 166(2), 411–418, 2007.)

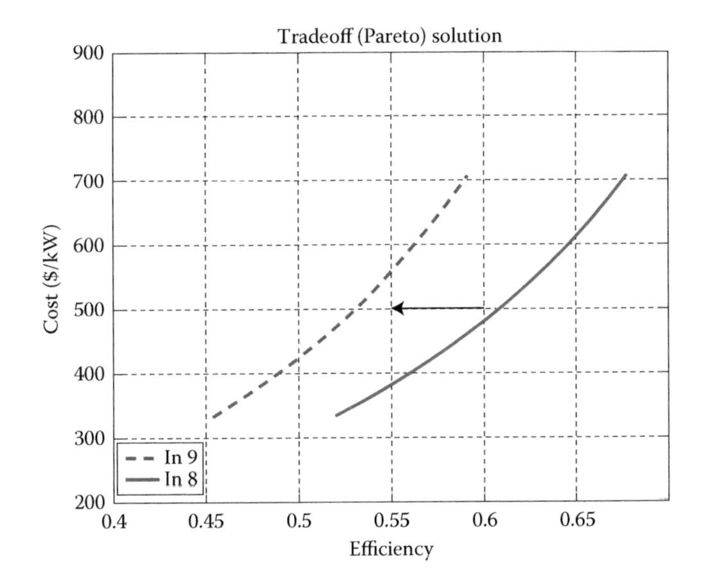

FIGURE 9.10
Pareto frontier change from In 8 to In 9 (In 9 = [0.24 MPa, 1.75, 1.5, 450 mA/cm^2]). (Adapted from W. Na and B. Gou, *Journal of Power Source*, 166(2), 411–418, 2007.)

When selecting the preferable initial condition, In 8 = [0.24 MPa, 2, 1.5, 450 mA/cm^2] based on the above discussion, the Pareto solution is achieved in Figure 9.9, but it is definitely not applicable to the real system due to an impractical high efficiency. Therefore, In 8 must be adjusted such that the Pareto solution lies in the realistic ranges. As In 8 = [0.24 MPa, 2, 1.5, 450 mA/cm^2] is adjusted to In 9 = [0.24 MPa, 1.75, 1.5, 450 mA/cm^2], this condition can be used for the design of an economic and high-performance fuel cell. Any initial condition can be chosen as long as it is within the bound limits and the corresponding Pareto solutions are applicable in practice.

References

1. F. Barbir and T. Gomez, Efficiency and economics of PEM fuel cells, *International Journal of Hydrogen Energy*, 22(10/11), 1027–1037, 1997.
2. J. Larminie and A. Dicks, *Fuel Cell Systems Explained*, Wiley, New York, 2002.
3. J. Hamelin, K. Abbossou, A. Laperriere, F. Laurencelle, and T.K. Bose, Dynamic behavior of a PEM fuel cell stack for stationary application, *International Journal of Hydrogen Energy*, 26, 625–29, 2001.
4. U.S. Department of Energy, *Fuel Cell Handbook* (6th edition), EG&G Technical Service Inc., Morgantown, WV, USA, 2002.
5. B. Blunier and A. Miraoui, Optimization and air supply management of a polymer electrolyte fuel cell, *Vehicle Power and Propulsion, 2005 IEEE Conference*, Chicago, IL, USA.
6. P. Pei, W. Yang, and P. Li, Numerical prediction on an automotive fuel cell driving system, *International Journal of Hydrogen Energy*, 31, 361–369, 2006.
7. F. Barbir, *PEM Fuel Cells: Theory and Practice*, Elsevier Academic Press, Waltham, MA, USA, 2005.
8. C.C. Boyer, R.G. Anthony, and A.J. Appleby, Design equations for optimized PEM fuel cell electrodes, *Journal of Applied Electrochemistry*, 30, 777–786, 2000.
9. C.E. Tomas, J.P. Barbour, B.D. James, and F.D. Lomax, Cost analysis of stationary fuel cell systems including hydrogen cogeneration ACG-8-18012-02, National Renewable Energy Laboratory, Colorado, 1999.
10. E.J. Carlson, P. Kopf, S.S. Sinha, and Y. Yang, *Cost Analysis of PEM Fuel Cell Systems for Transportation*, TIAX LLC, Cambridge, Massachusetts, 2005.
11. L. You and H. Liu A parametric study of the cathode catalyst layer of PEM fuel cells using a pseudo homogeneous model, *International Journal of Hydrogen Energy*, 26(9), 991–999, 1991.
12. H. Tsuchiya and O. Kobayashi, Mass production cost of PEM fuel cell by learning curve, *International Journal of Hydrogen Energy*, 29, 985–990, 2004.
13. S. Kamarudin, W. Daud, A. Som, M. Takriff, and A. Mohammad, Technical design and economic evaluation of a PEM fuel cell system, *Journal of Power Source*, 157, 641–649, 2006.
14. *Optimization toolbox for use with Matlab ver.2.* Mathwork. http://www.mathworks.com/
15. W. Na and B. Gou, The efficient and economic design of PEM fuel cell systems by multi-objective optimization, *Journal of Power Source*, 166(2), 411–418, 2007.

10

Power Electronics Applications for Fuel Cells[*]

10.1 Introduction

Lately, the plug-in-hybrid electrical vehicle (PHEV) and fuel cell vehicles (FCVs) have emerged in the market by incorporating hybrid technology, and even adding the function of plug-in-charging from the utility grid. The main driving force of PHEVs and FCVs in a market is that PHEVs and FCVs can not only improve the fuel economy, but also produce less pollutants compared to conventional ICE vehicles. Also, renewable energy systems such as solar and wind systems have drawn attention to the increasing demand for renewable energy sources. Therefore, wind and solar power generation, which are both environmentally friendly and economically competitive, have become a popular option for electricity power generation. Both electric vehicles (EVs) including FCVs and renewable energy systems have a similar power converter structure consisting of a DC–DC power converter, an inverter, and a power factor corrected (PFC) converter for a PFC specially in PHEV applications [1–6]. In this chapter, designs of sliding mode controllers (SMCs), and linear controllers for a PFC and bidirectional (buck/boost) converter are described. Also, controls of predictive controller for a FCVs applications are explained.

10.2 Linear Controllers

10.2.1 PFC Converter

In terms of PHEVs, the high-energy battery pack is to be charged with power factor correction from an AC outlet [3,7]. The most common topology for the PHEV battery charger is a two-stage approach with cascaded PFC AC–DC

[*] This chapter was mainly prepared by Dr. Woonki Na, California State University, Fresno, USA.

and DC–DC converters [4]. In the PFC stage, AC single phase is rectified and boosted with power factor correction. The output of the boost converter is connected to the DC bus, and a bidirectional converter for the battery charger is also connected to this DC bus.

The power circuit and its control configuration for the PFC boost converter are realized with a two-cascade control structure as shown in Figure 10.1.

The main objective of PFC is to make the input sinusoidal current, i_s, in phase with the utility voltage, V_s, [7] as shown in Figure 10.2. The voltage controller in Figure 10.1 generates a reference inductor current by multiplying the voltage controller output with the absolute value of the input voltage $|V_s(t)|$ for PFC. The current error between the reference and the measured inductor current becomes the input for the current controller, and then the output of the current controller will determine the control voltage to generate a PWM signal for the switching device. Since the motion rate of the inductor current is much faster than that of the output voltage in practice, and the current controller is normally very sensitive to system parameters and disturbances [8], this chapter simply focusses on developing current controllers rather than voltage controllers. The error between the reference current i_L^* and measured inductor current i_L is amplified by a current controller,

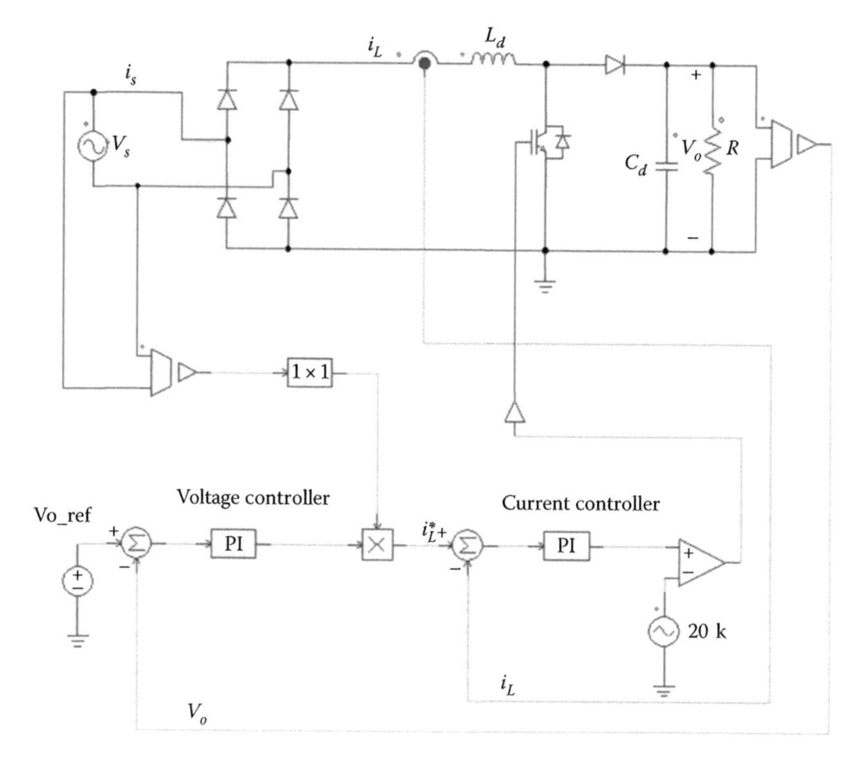

FIGURE 10.1
PFC boost converter with controllers.

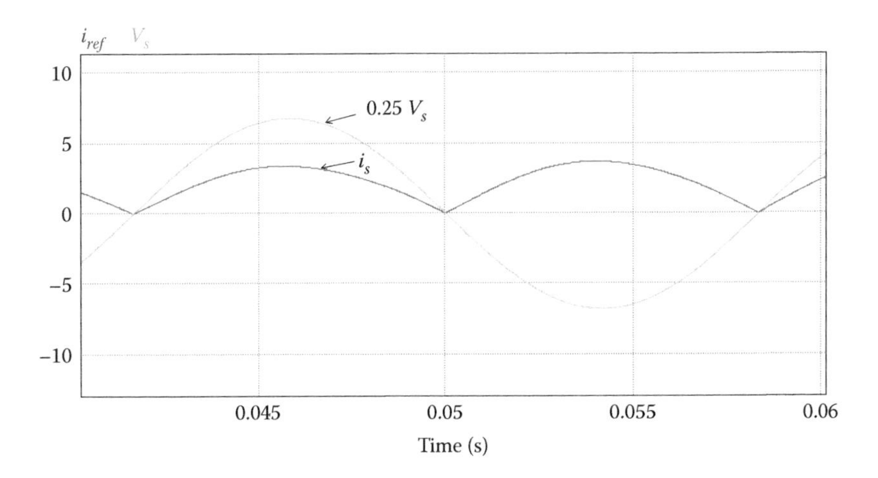

FIGURE 10.2
PFC V_s and i_s waveforms.

and then the current controller produces the control voltage between 0 and 10 V, that is compared with a triangular waveform 10 V peak to peak with the switching frequency 30 kHz to produce the PWM switching signal. The implementation of generating PWM as well as the current controllers can be easily realized by any DSP such as TMS320F2812 or 28335 DSP from Texas Instruments and so on. In order to design a current controller, a simplified small signal transfer function, G_{ps_pfc} [7] for the inductor current to the duty ratio in the PFC boost converter is used in the following equation:

$$G_{ps\_pfc} = \frac{\tilde{i}_L(s)}{\tilde{d}(s)} \cong \frac{V_o}{sL_d} \tag{10.1}$$

where $\tilde{i}_L$ is the inductor current perturbation, $\tilde{d}$ the duty ration perturbation, L_d the inductor, and V_o the DC-bus voltage are defined.

The current controller, $G_{c\_pfc}(s)$, with a PI controller with gain k_i and k_p for this PFC current control has to be designed such that it has a desired phase margin as $\phi_{PM1} = 60°$ of the loop transfer function, $G_{L1}(s)$, at the crossover frequency, and also the actual current value tracks the reference current as quickly as possible with minimum oscillations [7]. The loop transfer function is defined as

$$G_{L1}(s) = G_{ps\_pfc}(s) \cdot G_{c\_pfc}(s) \tag{10.2}$$

However, in this case, since the PI current controller has one zero pole, the phase margin, ϕ_{PM1} $\angle G_{L1}(s) + 180°$ at the crossover frequency, 90° is acceptable based upon the Bode plot of $G_{L1}(s)$ (Figure 10.3).

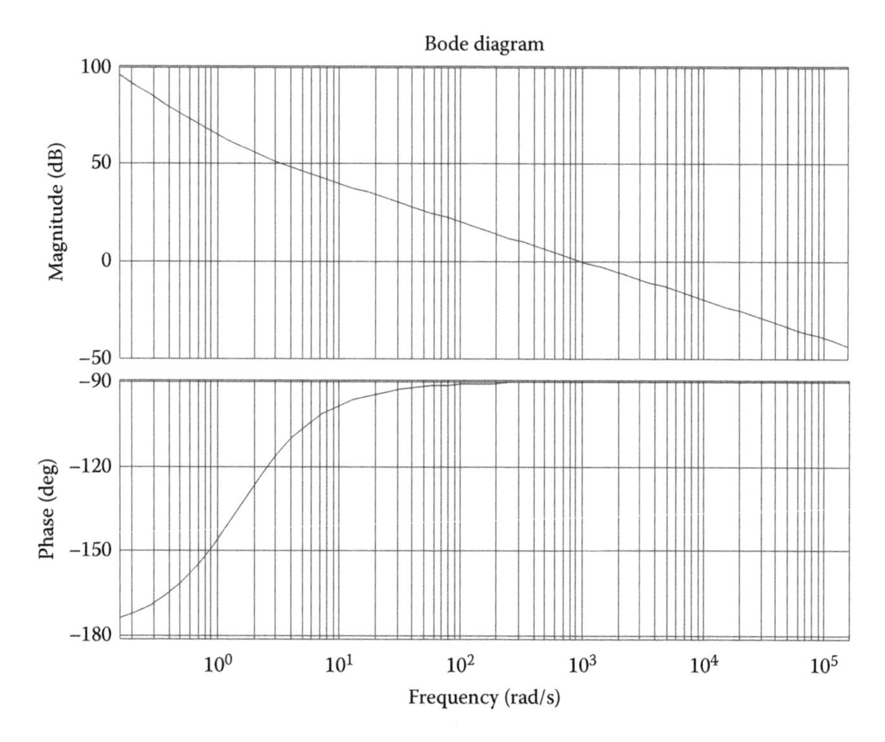

FIGURE 10.3
Bode plot of $G_{L1}(s)$.

The PFC current loop crossover frequency, f_{ci_pfc}, is set to 1 kHz, ($\omega_{ci\_pfc} = 2\pi \times 1 \times 10^3$ rad/s) because it is recommended to select 1–2 orders of magnitude smaller than the switching frequency of the boost converter in order to prevent interference in the control loop from the switching frequency noise [7]. The required phase angle of $G_{c\_pfc}(s)$, the PFC current controller at the crossover frequency, 1 kHz is calculated as follows [7]:

$$\angle G_{L1}(s)\big|_{f_{ci\_pfc}} = \angle G_{ps\_pfc}(s)\big|_{f_{ci\_pfc}} + \angle G_{c\_pfc}(s)\big|_{f_{ci\_pfc}} \tag{10.3}$$

$$\angle G_{L1}(s)\big|_{f_{ci\_pfc}} = -180^\circ + \phi_{PM1} \tag{10.4}$$

and the current controller gain for the PFC control can be calculated using the following equation:

$$\left|G_{L1}(s)\right|_{f_{c\_pfc}} = \left|G_{c\_pfc}(s)\right|_{f_{c\_pfc}} \times \left|G_{PWM1}(s)\right|_{f_{c\_pfc}} \times \left|G_{ps\_pfc}(s)\right|_{f_{c\_pfc}} \times k_{FB} = 1 \tag{10.5}$$

where the feedback gain k_{FB} is set to 1 and $|G_{PWM1}(s)| = 1/V_r$ (V_r is the triangular waveform peak to peak voltage, 10 V). According to Equation 10.5 and the Bode plot analysis of Figure 10.3, we can determine the gains of the PI controller for the PFC current controller.

10.2.2 Bidirectional Converter (Boost + Buck Converter)

By using a Buck and Boost converter, a bidirectional converter can be designed as shown in Figure 10.4.

The output V_o of the PFC boost converter in Figure 10.4 is connected to the DC bus across the capacitor C_d. From the DC bus, the 48 V, Li-ion battery can be charged through the buck converter. As mentioned earlier, designing of current controllers are focussed for the PFC boost converter and bidirectional converter, specially buck (charging) mode, not considering battery discharging mode because the design procedure of this boost converter is almost identical to the PFC boost converter. The buck (charging) mode can be achieved by utilizing the switching S1 and the antiparallel diode D2 in Figure 10.4.

As shown in Figure 10.4, like the PFC boost converter, a two-cascade control structure is required to regulate the battery voltage to 48 V, precisely with a good control performance. The error between the measured battery voltage V_{bat} and its reference V_{bat}^* can be amplified to produce the battery, or the inductor current reference, i_{bat}^* which is compared with the measured inductor current. To design the current controller G_{c_bi}, the small signal transfer function G_{ps_bi} [7] for the output voltage to the inductor current in the bidirectional converter is used as follows:

$$G_{ps\_bi}(s) = \frac{\tilde{v}_o}{\tilde{i}_L}(s) = \frac{R(1+srC)}{1+sRC} \tag{10.6}$$

where R is the small signal resistance connected to the battery in parallel, given as 100 kΩ, and the value of the battery capacitor value to build a small signal model is set to 10 F, r is equivalent series resistance (ESR) of the battery as 0.1 Ω.

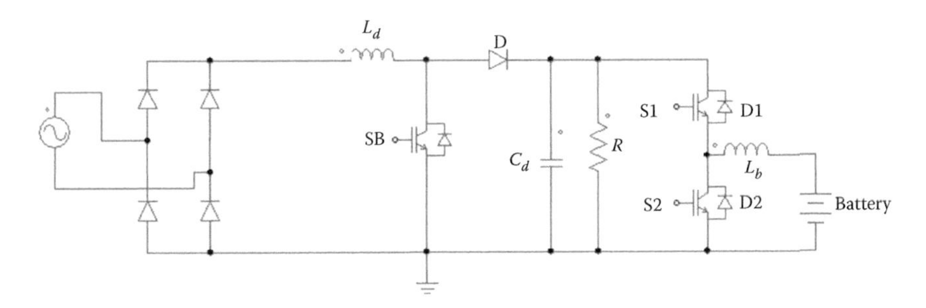

FIGURE 10.4
PFC boost converter + bidirectional converter.

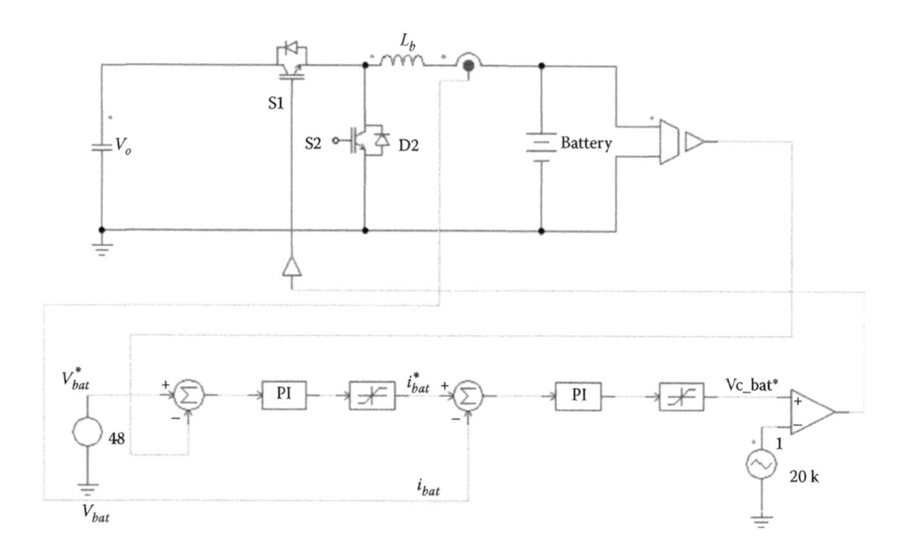

FIGURE 10.5
Bi-directional converter (charging mode).

The crossover frequency of this buck converter is chosen as 1 kHz. According to Figure 10.6, $\angle G_{ps\_bi}(s)\big|_{f_{ci\_bi}}$ is 0 degree at this crossover frequency. The small signal transfer function G_{ps_bi} has a gain $|G_{ps\_bi}(s)|_{fci\_bi}(s) = -20_{dB}$ in Figure 10.5. The P and I gains of the current controller G_{c_bi} for the bidirectional converter can be determined such that a desired phase margin of the loop transfer function is to be $\phi_{PM2} = 60°$, similarly in the PFC boost converter.

10.2.3 Simulation and Experimental Results

10.2.3.1 Simulation

The nominal values of parameters for the simulation are shown in Table 10.1. Powersim (PSIM) software is used for this simulation.

TABLE 10.1
Simulation Parameters

Parameter	Values
L_d (PFC)	1 mH
L_b (bidirectional)	500 μH
C_d	3000 μF
R_{load}	100 Ω (initial)
f_s	30 kHz
V_{bat} (battery)	48 V
V_s	120 $V_{s,rms}$, 60 Hz
V^*_{dc}	300 V

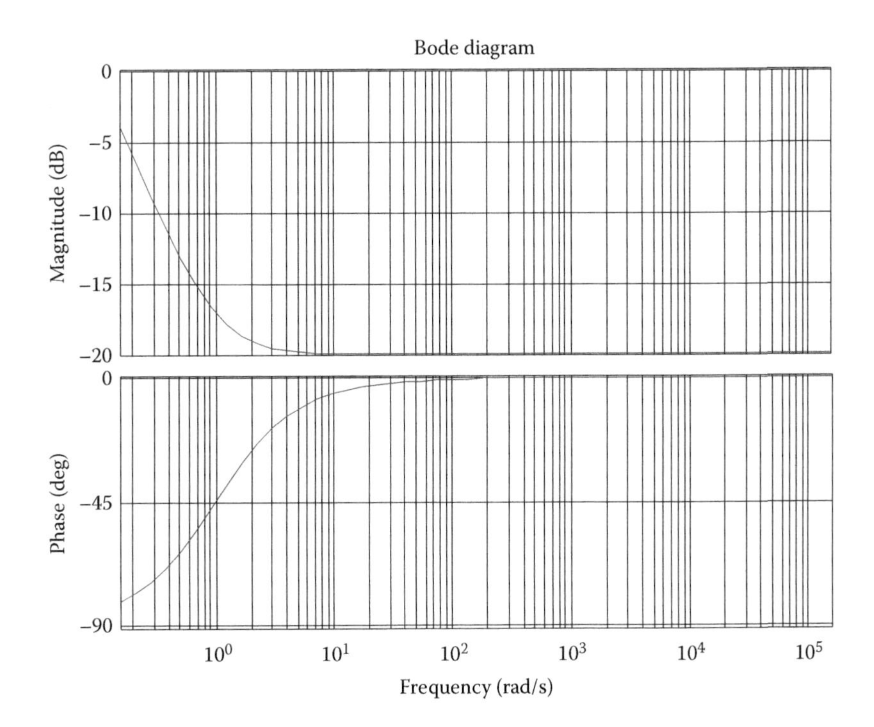

FIGURE 10.6
Bode plot of $\tilde{v}_o/\tilde{i}_L$.

In the simulation, a step load change is applied. The value of the initial load is set to 100 Ω, and at 0.4 s, the step load change occurs from 100 to 50 Ω. Figure 10.7 shows the PFC boost converter inductor current, its reference, and the scaled input voltage (0.04 V s). As shown in Figure 10.7, the inductor current and the input voltage are in phase because of the PFC control.

Figure 10.8 shows the DC link voltage during the discharging mode of the battery and Figures 10.9 and 10.10 show the battery current and its reference during the discharging mode. The DC link voltage is well regulated within 0.1 s under the load step change. Figure 10.11 shows the battery voltage during the charging mode. Figures 10.12 and 10.13 also show the battery current and its reference during the charging mode. As shown in Figures 10.10 through 10.13, with the proposed controller, the battery voltage and current are properly regulated under the load change.

10.2.3.2 Experiment

In the preliminary experiment setup of Figure 10.14, IRFP460A N-Type Power MOSFET, NTE5328—bridge rectifier, and the diode, VS-HFA50PA60CPBF are used to build a boost converter. The controller is realized with a 32-bit fixed-point DSP TMS320F2812 in Figure 10.14 and the PWM pulses are generated

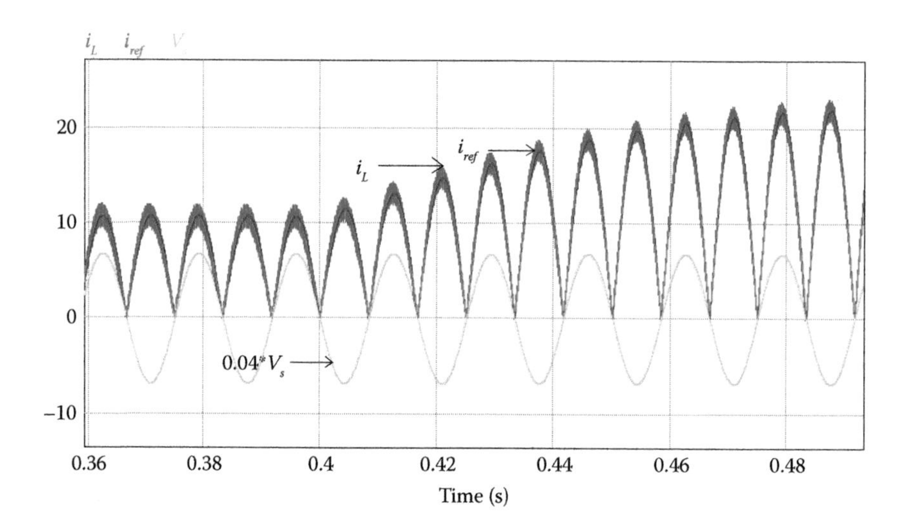

FIGURE 10.7
The current and the input voltage of the PFC boost converter.

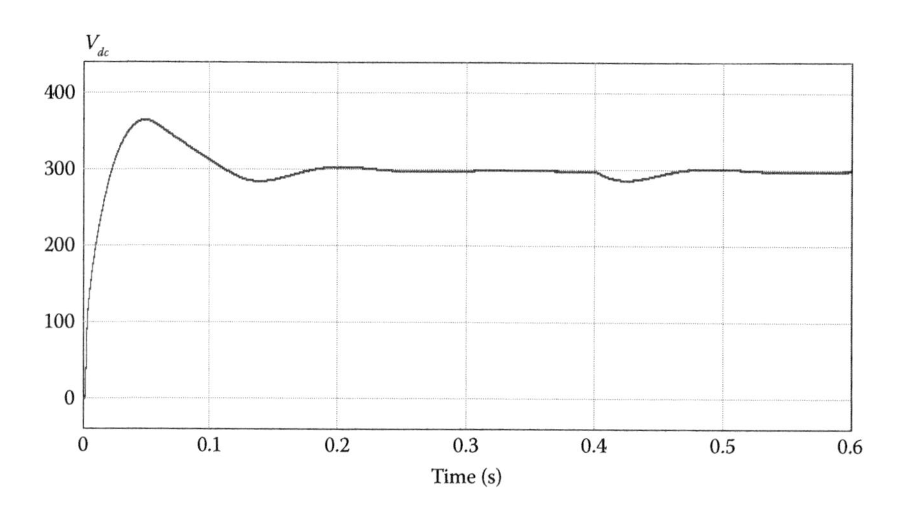

FIGURE 10.8
The DC link voltage during the discharging.

through the digital output of the DSP. The preliminary experimental result is shown in Figures 10.16 and 10.17. In Figure 10.18, the experiment results of the boost converter and PWM output are shown. The DC link voltage, V_d is regulated to 9.98 V shown in Figure 10.16a when the output voltage reference is 10 V, and the input is 5 V.

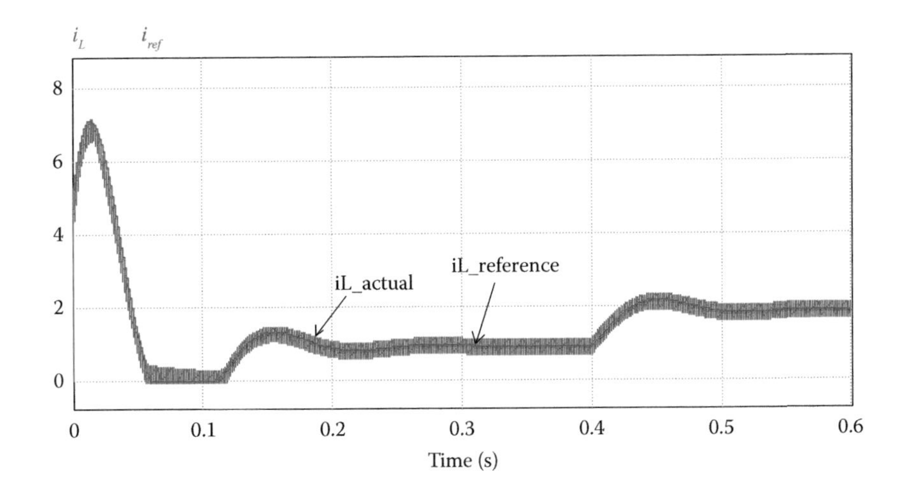

FIGURE 10.9
The inductor current and its reference of the bidirectional converter during the discharging mode.

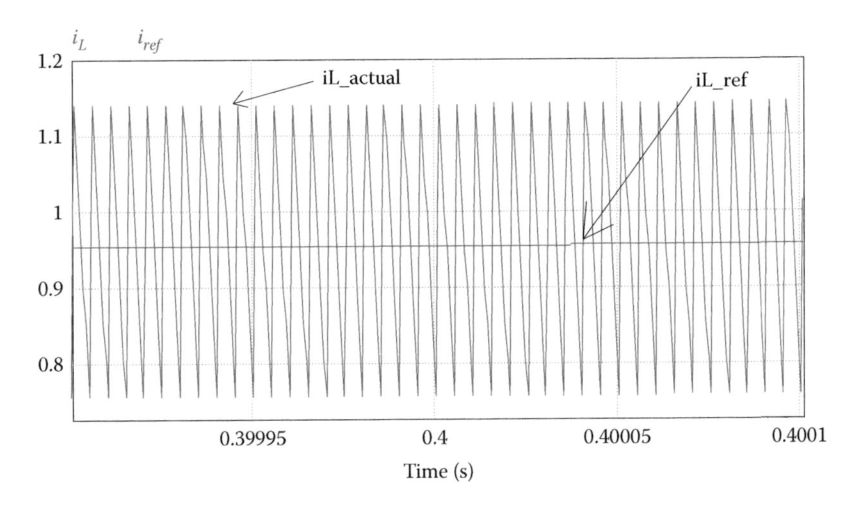

FIGURE 10.10
Zoom in the inductor current and its reference of the bidirectional converter during the discharging mode.

A Simulink model is built as shown in Figure 10.15 to run the control for the buck converter using the DSP. When the Simulink file is compiled, it generates a C code file in Code Composer and loads it into the DSP. When testing with the small-scale model with the input of 20 V, we could expect the output to be smaller than the input as a buck converter such as 10 V in the case of

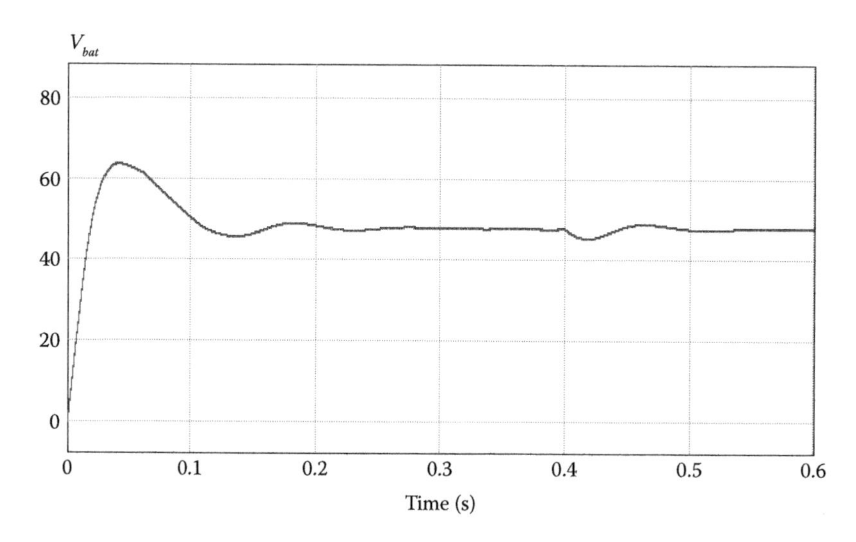

FIGURE 10.11
The battery voltage during the charging mode.

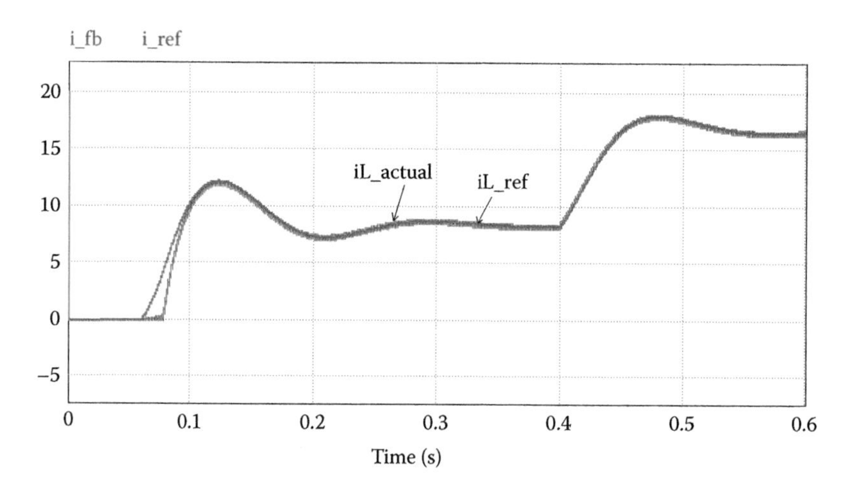

FIGURE 10.12
The inductor current and its reference of the bidirectional converter during the charging mode.

the duty cycle 50%. The outputs were very accurate as shown in Figure 10.15 when the duty cycle is 50% and 25%, respectively.

The same method described previously was used to calculate the PI controller's gains using a small-scale boost converter. After testing them in PSIM, a MATLAB–Simulink-based DSP program to run the boost converter shown in Figure 10.16 is built. This again was compiled into Code Composer to generate C code for the DSP to run.

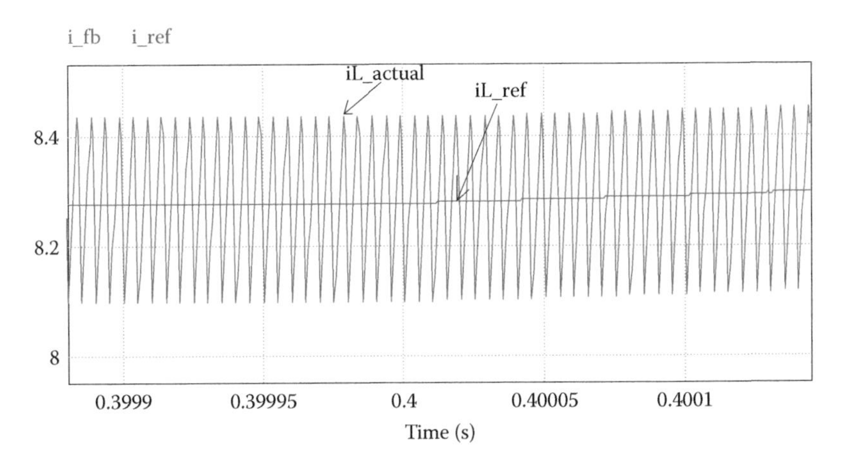

FIGURE 10.13
Zoom in the inductor current and its reference of the bidirectional converter during the discharging mode.

(a)
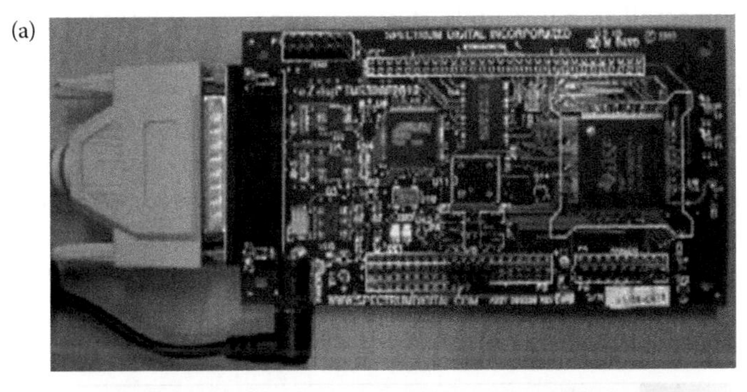

(b)

FIGURE 10.14
(a) TMS 320F2812 DSP board and (b) its experiment setup.

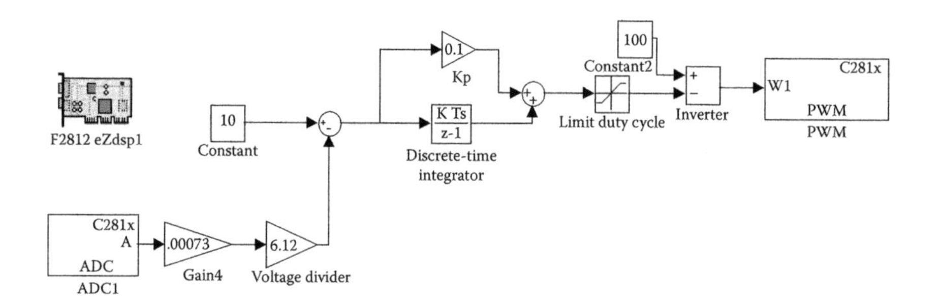

FIGURE 10.15
Boost converter control system in Simulink.

With a 5-V input into our system we can expect the output to be larger than that. As shown in Figure 10.17, the output voltages matched the constant reference value accurately as 10 V when the duty cycle is 50%.

10.2.4 Summary

Based upon small signal models, design procedures of linear current controllers for the PFC boost converter and the bidirectional converter for PHEV applications were explained in detail. Using the Bode plot analysis of the small signal transfer functions, $\tilde{i}_L/\tilde{d}$ and $\tilde{v}_o/\tilde{i}_L$, the gains of each current controllers can be determined. The design methods for the linear current controllers can be validated through the PSIM power electronics circuit simulation tool. Also it can be implemented using a simple MATLAB–Simulink-based DSP program, and was tested with a small-scale prototype without problems.

10.3 Sliding Mode Controller for Power Converters

Since each power converter is inherently operating with high-frequency switching devices, with easily above 100 kHz switching frequency in the case of using MOSFETs, highly nonlinear behaviors are observed during parameter variations and disturbances. SMC has been a good candidate in nonlinear control applications in terms of power electronics converters because an SMC technique is naturally suited for on–off switched controlled systems such as PWM-based power converters, particularly in a AC–DC rectifier, or DC–DC converter, or DC–AC inverter [8]. In addition, the main advantages of using the SMC technique [8] are the following:

1. The order of the system can be reduced.
2. Sensitivity under parameters variation and disturbances can be reduced.

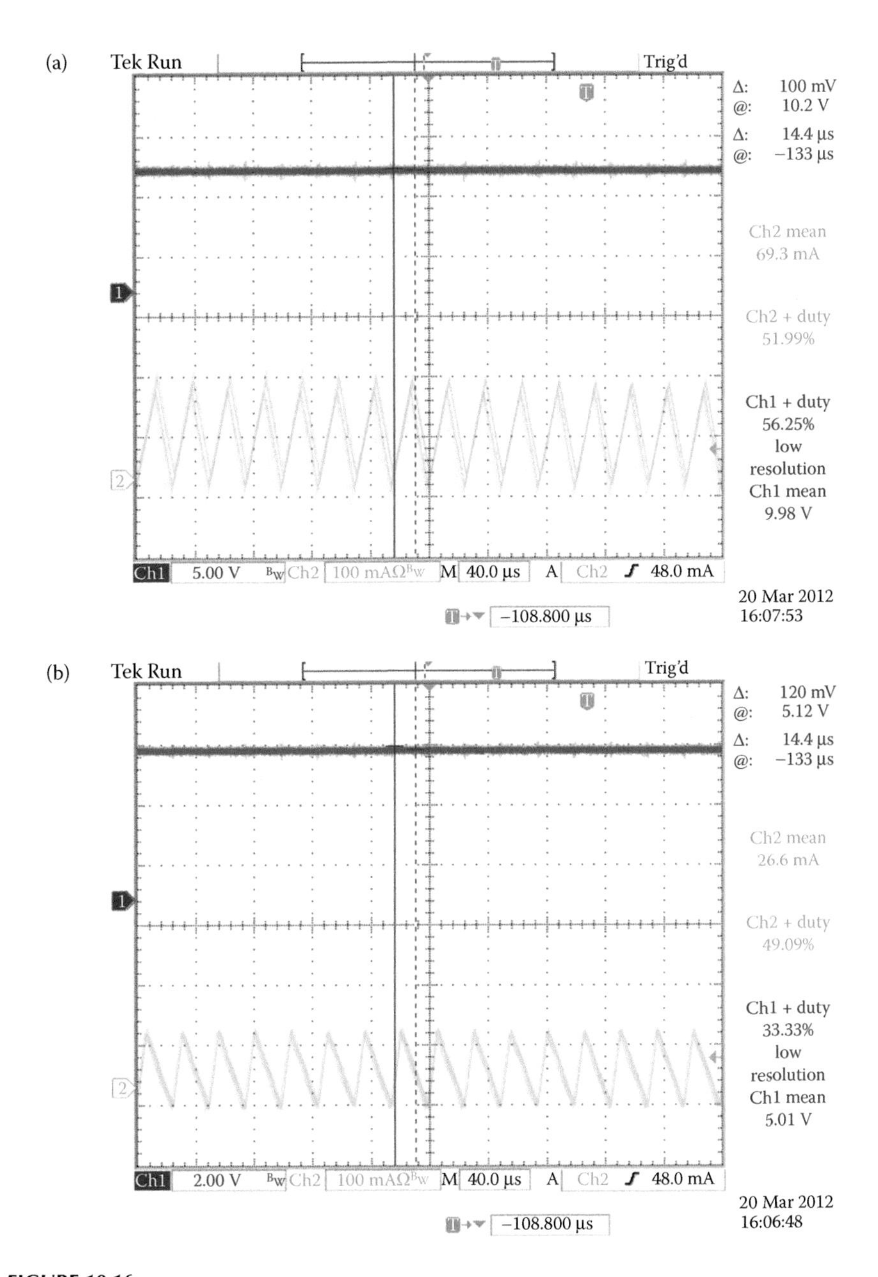

FIGURE 10.16
Results from the closed loop buck converter: Output voltage (upper) and inductor current (low) when the voltage reference is (a) 10 V and (b) 5 V.

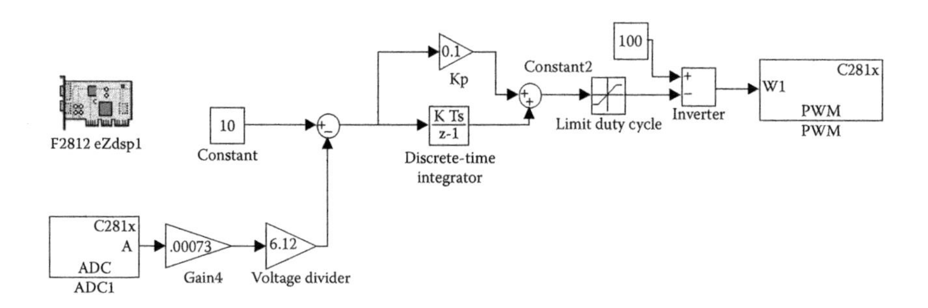

FIGURE 10.17
Boost converter control system in Simulink.

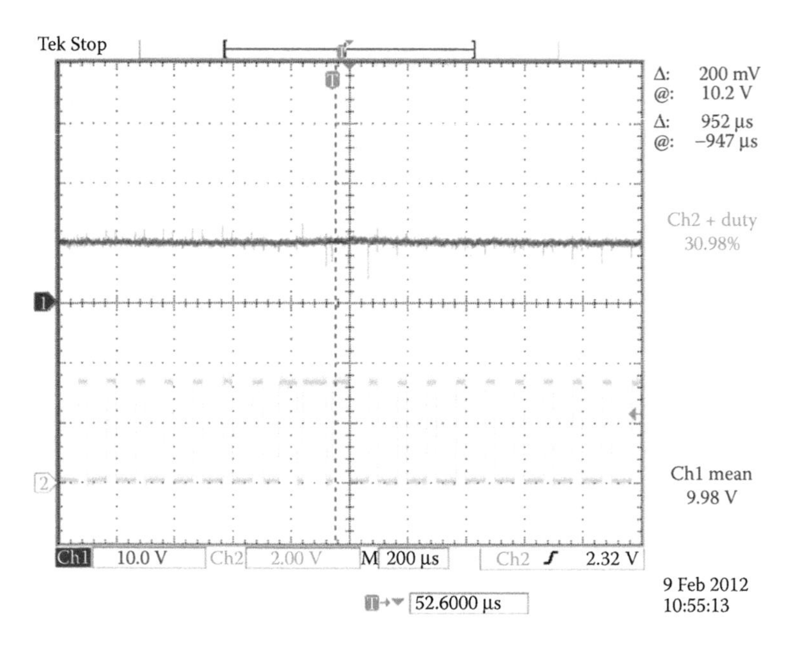

FIGURE 10.18
Results from the closed-loop boost converter. Output voltage (upper) and PWM pulse (lower).

In this section, designs of SMC for a PFC converter and bidirectional converter were explained.

10.3.1 SMC for a PFC Converter

SMC is designed such that the system can achieve the low sensitivity to system parameter variations from the high-frequency switching PFC boost converter. A dynamic model of the boost converter for a sliding mode current controller in the PFC boost converter, is to be built during the continuous conduction mode (CCM) using Kirchhoff voltage law (KVL) and Kirchhoff current law (KCL) as follows:

$$\dot{x}_1 = -(1-u)\frac{1}{L}x_2 + \frac{v_s}{L} \tag{10.7}$$

$$\dot{x}_2 = (1-u)\frac{1}{C}x_1 - \frac{1}{RC}x_2 \tag{10.8}$$

The state-space representation forms of Equations 10.7 and 10.8 are rewritten as

$$\begin{bmatrix} \dot{x}_1 \\ \dot{x}_2 \end{bmatrix} = \begin{bmatrix} 0 & -(1-u_1)\frac{1}{L_d} \\ (1-u_1)\frac{1}{C} & -\frac{1}{RC} \end{bmatrix} \begin{bmatrix} x_1 \\ x_2 \end{bmatrix} + \begin{bmatrix} \frac{1}{L} \\ 0 \end{bmatrix} v_s \tag{10.9}$$

where L_d is the inductor on the PFC converter, C, DC is the link capacitor value, R is the load resistance, v_s is the input voltage, $x_1 = i_{Ld}$ is the inductor current, $x_2 = V_d$ is the output voltage, and u_1 is the switching control function.

The switching function for sliding mode current control is defined as

$$s_1 = x_1 - x_1^* \tag{10.10}$$

where x_1^* is the reference inductor current and $x_1^* = I_d \sin \omega t$. Since it is assumed to be a lossless system, the input active power is equal to the output power under this R load [9], and then the following equation is derived:

$$\frac{1}{2}I_d V_s = \frac{V_d^2}{R} \tag{10.11}$$

Using Equation 10.11, I_d can be determined as

$$I_d = \frac{2V_d^2}{RV_s} \tag{10.12}$$

The switching control function for the current control loop, u_1 is then designed to enforce the sliding mode on the surface $s_1 = x_1 - x_1^* = 0$ [8].

$$u_1 = \frac{1}{2}(1 - sign(s_1)) \tag{10.13}$$

By solving the equation $\dot{s} = \dot{x}_1 = 0$, the equivalent control of u_{eq} becomes

$$u_{eq} = \frac{V_d - V_s \sin \omega t - \omega L_d I_d \cos \omega t}{V_d} \tag{10.14}$$

The sliding mode on the sliding surface $s_1 = 0$ exists when the condition $s_1 \dot{s}_1 < 0$ is satisfied.

For sliding mode to exist, $s_1 \dot{s}_1 < 0$ holds or $s_1 > 0(\ u_1 = 0)$, and $\dot{s}_1 < 0$.

One of the sliding mode conditions in this system is

$$\dot{s}_1 = \frac{V_s \sin \omega t}{L_d} - \omega I_d \cos \omega t - \frac{x_2}{L_d} < 0 \tag{10.15}$$

Rearranging Equation 10.15 we get

$$x_2 > V_s \sin \omega t - \omega L_d I_d \cdot \cos \omega t \tag{10.16}$$

and $x_2 > V_s$ because this is the boost converter operation.

The other sliding mode condition is that $s_1 < 0$ ($u_1 = 1$), and $\dot{s}_1 > 0$ to hold $s_1 \dot{s}_1 < 0$.

$$\dot{s}_1 = \frac{V_s \sin \omega t}{L_d} - \omega I_d \cos \omega t > 0 \tag{10.17}$$

Rearranging Equation 10.17 we get

$$\omega t = \arctan\left(\frac{\omega L_d I_d}{V_s}\right) > 0 \tag{10.18}$$

For a sliding mode to locally exist on $s_1 = 0$, the corresponding equivalent control of Equation 10.14 satisfies

$$0 < u_{eq1} = \frac{V_d - V_s \sin \omega t - \omega L_d I_d \cos \omega t}{V_d} < 1 \tag{10.19}$$

This implies that the sliding mode can be enforced as long as all the above sliding mode conditions (10.16) through (10.19) are satisfied including $x_2 > V_s$.

10.3.2 SMC for a Bidirectional Converter

The buck or boost converter can be realized with a two-cascade control structure as shown in Figure 10.19 [7,8]. For example, the buck (charging)

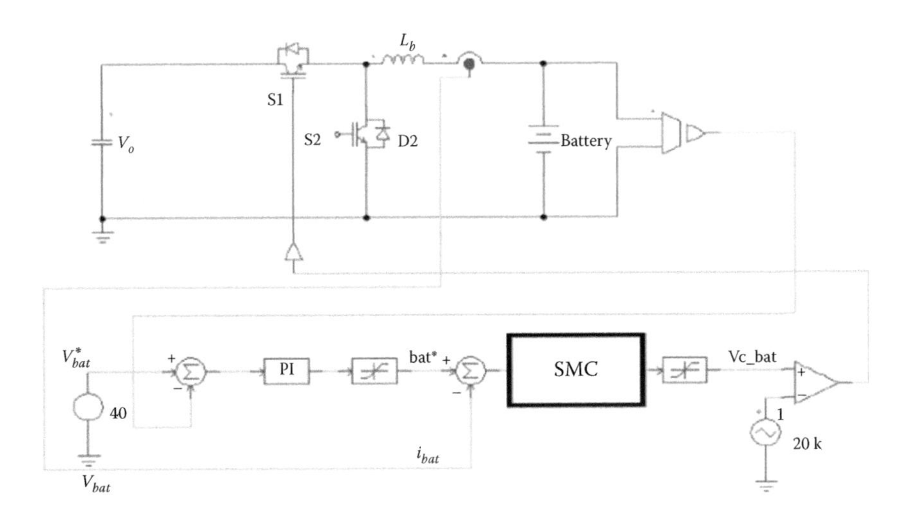

FIGURE 10.19
Bi-directional converter (buck mode).

mode can be achieved by utilizing switch S1 and the antiparallel diode D2 in Figure 10.19. Likewise, the boost mode is achieved utilizing switch S2 and the antiparallel diode D1 of switch S1 in Figure 10.19. The aforementioned two-cascaded control structure requires regulating the battery voltage to 48 V, for instance precisely with a good control performance. The error between the measured battery voltage, V_{bat} and the reference V^*_{bat} can be amplified to produce the battery or inductor current reference, i^*_{bat} which is compared with the measured inductor current.

10.3.2.1 Boost Mode

To design a SMC, the dynamic models of the boost converter and buck converter are to be built during the CCM using KVL and KCL.

$$\dot{x}_3 = -(1-u_2)\frac{1}{L}x_4 + \frac{V_{bat}}{L} \tag{10.20}$$

$$\dot{x}_4 = (1-u_2)\frac{1}{C}x_3 - \frac{1}{RC}x_4 \tag{10.21}$$

Equations 10.20 from KVL and Equation 10.21 from KCL are for the boost converter operation. Where L is the inductor value in the bidirectional converter, C is the DC link capacitor value, R is the load resistance, $x_3 = i_L$ is the inductor current, $x_4 = V_d$ is the output DC link voltage, and u_2 is the switching control function. The desired current is obtained from the outer voltage loop as [8]

$$x_3^* = \frac{V_d^*}{R \cdot V_{bat}} \tag{10.22}$$

where V_d^* is the desired voltage output.

From the SMC theory [8], the state variable error forms the sliding function to enforce the current to track the desired current x_3^* as given below

$$s_2 = x_3 - x_3^* \tag{10.23}$$

The switching control function for the inner control loop, u_1, is designed as to enforce the sliding mode surface $s_1 = 0$ [8]

$$u_2 = \frac{1}{2}(1 - sign(s_2)) \tag{10.24}$$

Also, by solving $\dot{s}_2 = \dot{x}_3 = 0$ [8], the equivalent control of u_2 [8] is achieved.

$$u_{eq2} = 1 - \frac{V_{bat}}{x_4} \tag{10.25}$$

The suitable SMC law is achieved by applying the conversion condition $s_2\dot{s}_2 < 0$ [8].

$$x_4 > V_{bat} \quad \text{or} \quad 0 < u_{eq2} = 1 - \frac{V_{bat}}{x_4} < 1 \tag{10.26}$$

This implies that the sliding mode can be enforced as long as the output DC link voltage is higher than the battery voltage.

10.3.2.2 Buck Mode

Likewise in (1) boost mode, the dynamic model of the buck converter is as follows during the CCM using KVL and KCL:

$$\dot{x}_3 = -\frac{1}{L}x_5 + u\frac{V_d}{L} \tag{10.27}$$

$$\dot{x}_5 = \frac{1}{C}x_3 - \frac{1}{RC}x_5 \tag{10.28}$$

where L is the inductor value, C is the DC link capacitor value, R is the load resistance, $x_3 = i_L$ is the inductor current, $x_5 = V_{bat}$ is the battery voltage, and

u_3 is the switching control function. The desired current is obtained from the outer voltage loop as [8]

$$x_3^* = \frac{V_{bat}^*}{R} \tag{10.29}$$

where V_{bat}^* is the desired battery voltage, in the simulation, 48 V is used for the desired battery voltage.

From the sliding mode control theory [8] that has exact tracking properties, a sliding function is enforced as

$$s_3 = x_3 - x_3^* = 0 \tag{10.30}$$

Similarly, the switching control function, is designed as Equations 10.24 and 10.25 to enforce the sliding mode surface $s_3 = 0$ [8].

$$u_3 = \frac{1}{2}(1 - sign(s_3)) \tag{10.31}$$

Also, by solving $\dot{s}_3 = \dot{x}_3 = 0$ [8], the equivalent control of u [8] is achieved.

$$u_{eq3} = \frac{x_5}{V_d} \tag{10.32}$$

The proper SMC law is achieved as by applying the conversion condition $s\dot{s} < 0$ [8].

$$0 < x_5 < V_d \quad \text{or} \quad 0 < u_{eq3} = \frac{x_5}{V_d} < 1 \tag{10.33}$$

This implies that the sliding mode can be enforced as long as the output DC link voltage is higher than the battery voltage as seen in the boost mode.

10.3.3 Simulation and Experimental Results

10.3.3.1 Simulation

The nominal values of parameters for the simulation are shown in Table 10.1. PSIM software is used for this simulation. SMC controller in the simulation is implemented with SR flip-flop, comparator, i + delta and i – delta input that indicates the reference current ± the hysteresis band in Figure 10.20. The reference current, i_reference in Figure 10.20 is generated from the voltage controller (Figures 10.21 and 10.22).

In the simulation, a step load change is applied. The value of the initial load is set to 100 Ω, and at 0.4 s in the PFC converter, the step load change

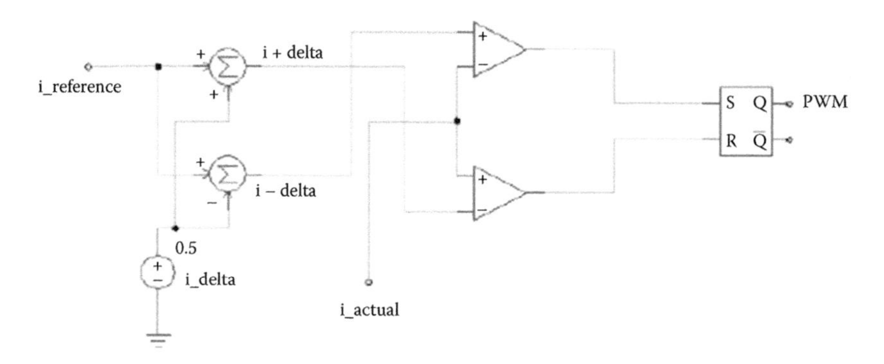

FIGURE 10.20
SMC controller in PSIM.

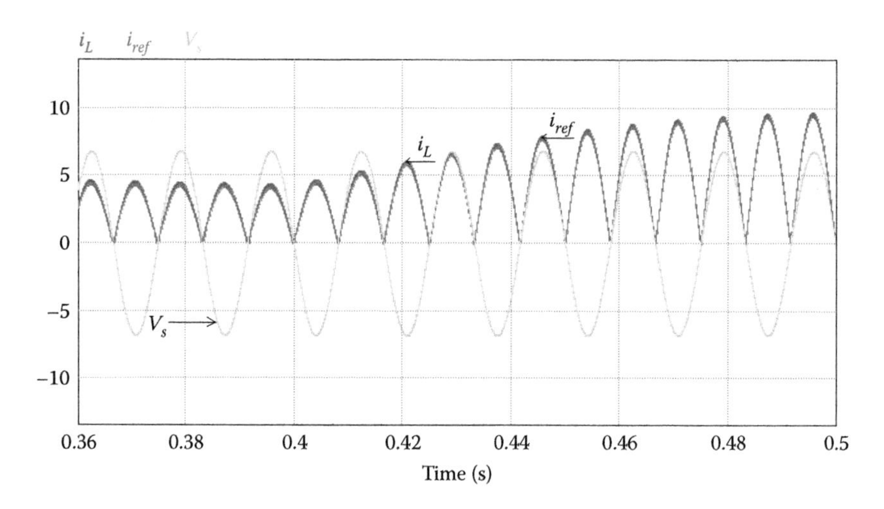

FIGURE 10.21
The inductor current and the input voltage of the PFC boost converter.

occurs from 100 to 50 Ω. And at 0.5 s, a step load change is enforced from 100 to 50 Ω in the bidirectional converter as well. Figure 10.23 shows the DC link voltage during the discharging mode of the battery (Figure 10.24). Figures 10.25 and 10.26 show the battery current, and its reference during the discharging mode, the DC link voltage is well regulated within 0.1 s under the load step change.

10.3.3.2 Experiment

The SMC is realized with a MATLAB–Simulink program using a 32-bit fixed-point DSP TMS320F2812 and the PWM pulses are generated through the digital output of the DSP. The Simulink code is shown in Figure 10.27.

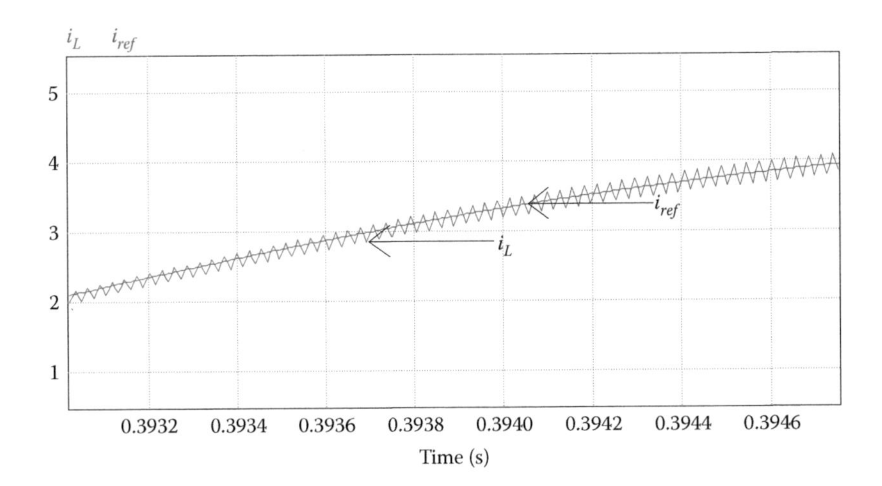

FIGURE 10.22
Zoom in the reference inductor current, i_{ref} and the actual inductor current i_L of the PFC boost converter.

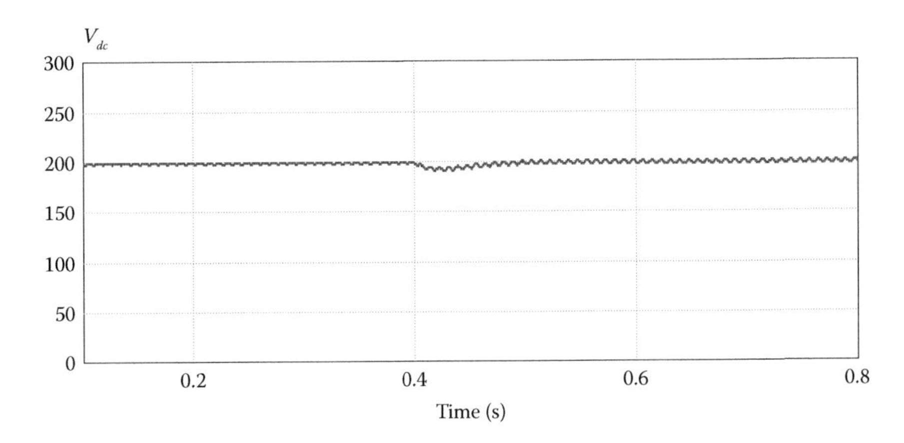

FIGURE 10.23
DC-bus voltage.

The experimental result is shown in Figures 10.28 and 10.29. In Figure 10.28, experiment results of the PFC boost converter and PWM output are shown. The DC link voltage, V_d and the inductor current are well regulated.

For the boost converter, V_d is regulated to 20 V as shown in Figure 10.29a when the output voltage reference is 20 V when the input is 5 V. Due to the nature of the SMC controller, inconstant switching frequency varies from 2 to 20 kHz, although the voltage is well controlled to 20 V, the inductor current shown in Figure 10.29b enters in the discontinuous conduction mode when the load resistance is 200 Ω.

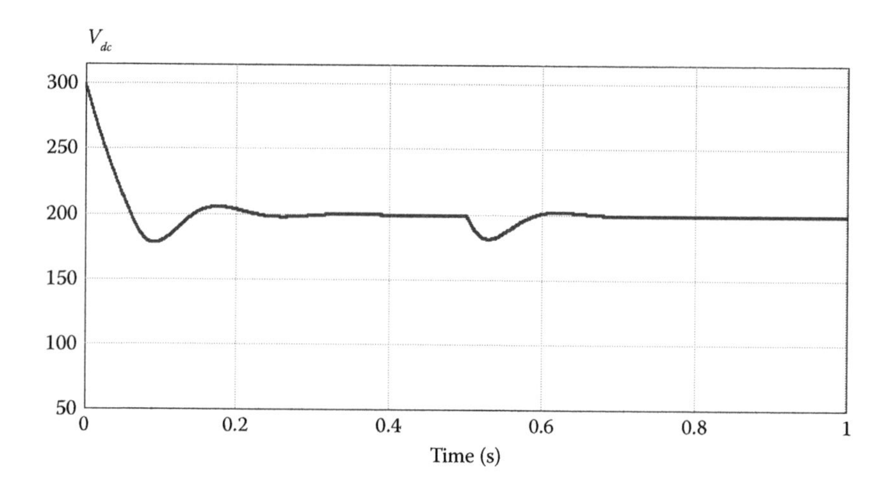

FIGURE 10.24
The output voltage of the boost converter.

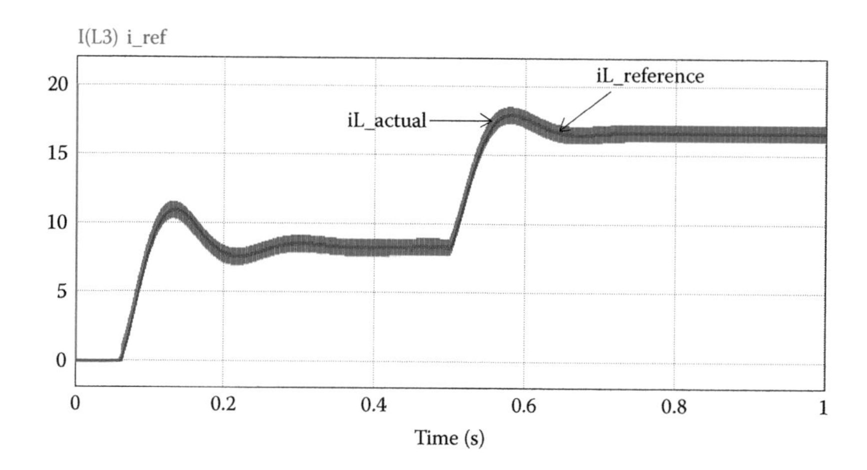

FIGURE 10.25
The inductor current and its reference of the boost converter.

10.3.4 Summary

An SMC based for EV and energy applications has been explained. By adopting sliding mode control, robust control performances are expected with a simple MATLAB–Simulink-based DSP in a real-time operation. An experimental test bed has been built using a Texas Instruments 32-bit fixed-point DSP TMS320F2812 and each mode of SMC operation such as the PFC, and bidirectional modes has been validated throughout the DSP test bed with a PSIM simulation.

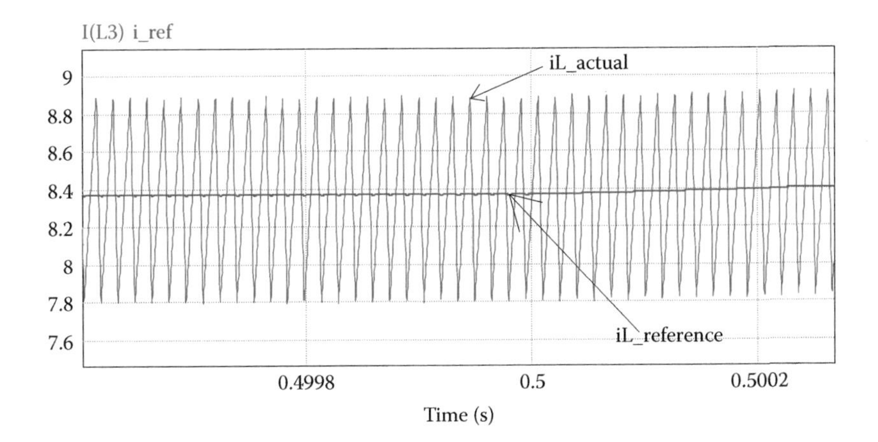

FIGURE 10.26
Zoom-in figure of Figure 10.25 around 5 s when a load change is enforced.

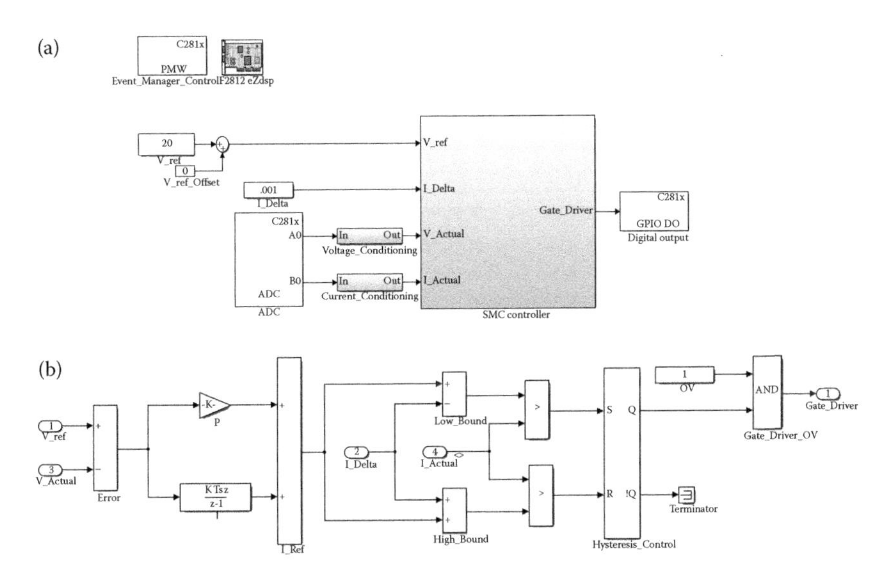

FIGURE 10.27
DSP real-time MATLAB–Simulink SMC program. (a) Overall program including a sensing block and (b) controller block including the SMC.

10.4 Predictive Controller for FCVs

In this section, a predictive current-controlled fuel cell boost converter, UC bidirectional converter, and brushless DC (BLDC) motor drive are used to provide a better dynamic response than a conventional PI controller-based

(a)

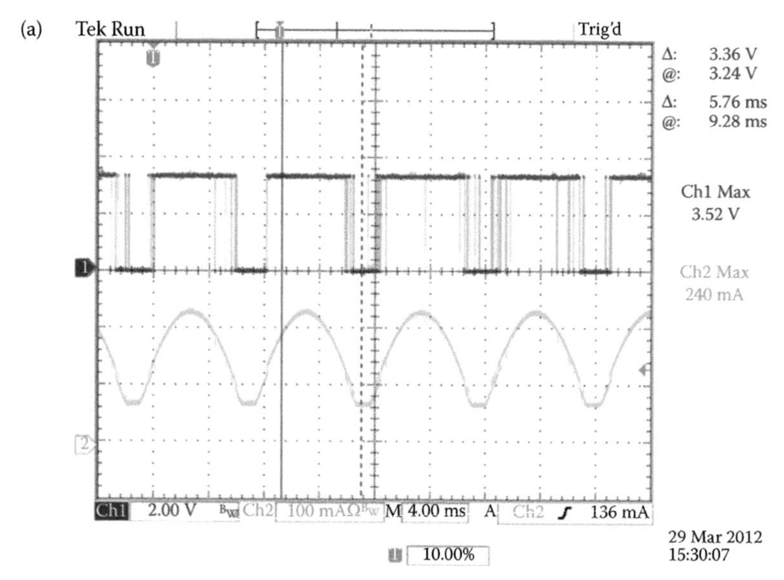

Ch1: PWM duty cycle for PFC
Ch2: Output of current probe

(b)

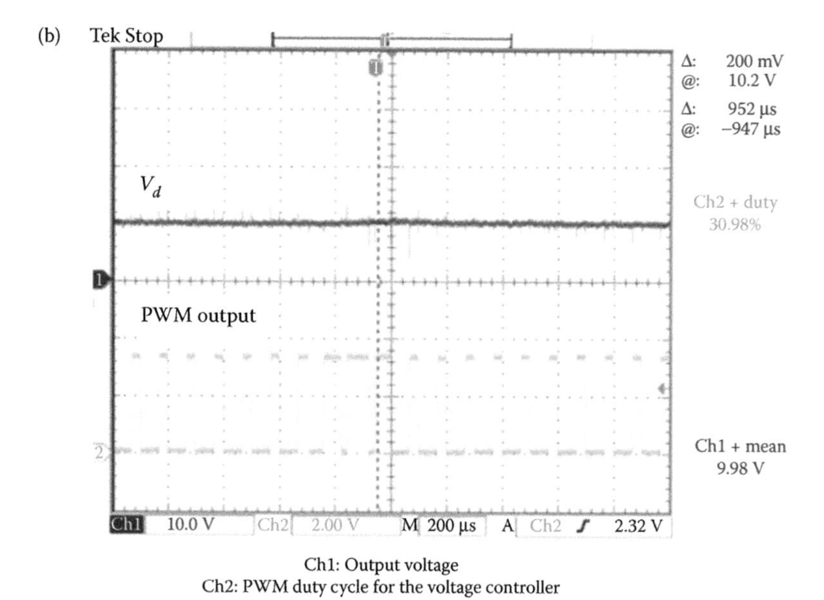

Ch1: Output voltage
Ch2: PWM duty cycle for the voltage controller

FIGURE 10.28
Experiment results of the PFC boost converter. (a) PWM and the current and (b) PWM duty cycle and the voltage.

(a)

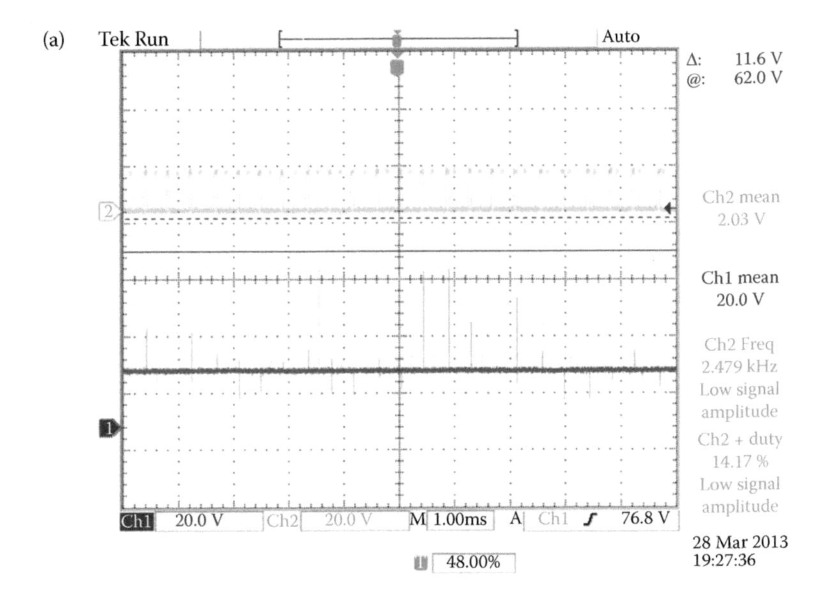

(b)

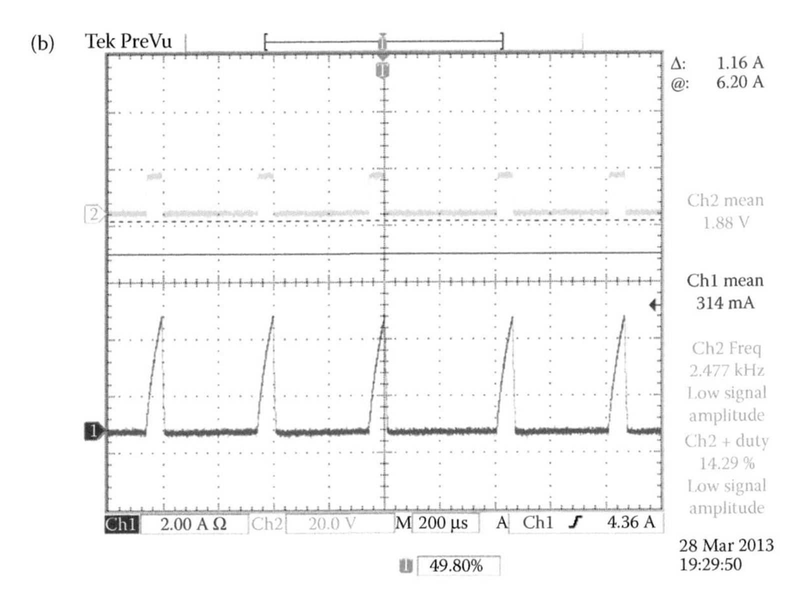

FIGURE 10.29
Experiment results of the boost converter. (a) PWM (upper) and output voltage (lower) and (b) PWM (upper) and inductor current (lower).

system. The concept of predictive control is introduced for hybrid FCVs in Reference 24, which focusses more on a prediction of the load power and an optimal system control based on a simulation analysis to minimize fuel consumption. In contrast, in this section, component level predictive current controllers are explained for the fuel cell boost converter, UC bidirectional converter, and BLDC motor drive to control them independently.

Normally, a fuel cell boost converter and UC bidirectional converter employ an outer voltage loop and an inner current loop, while the BLDC motor controller has only a torque (current) controller. In nature, all these current controllers are supposed to be faster than the outer loops, voltage loop, or speed loop. The following systems based on three independent predictive current controllers for the boost converter, UC manager, and motor drive provide a faster dynamic response on the DC-bus side, feeding the motor current for a faster torque response. In addition, the faster dynamic responses of the boost converter and UC prolong battery life reducing the burden of the battery due to less peaking current and harmonics. The details of the system configuration and its controller followed by the system modeling, simulation verifications are described in subsequent sections.

10.4.1 System Configuration and Modeling

Figure 10.30 illustrates the overall electrical system configuration of the fuel cell powered hybrid vehicle. The system is targeted for light FCVs such as

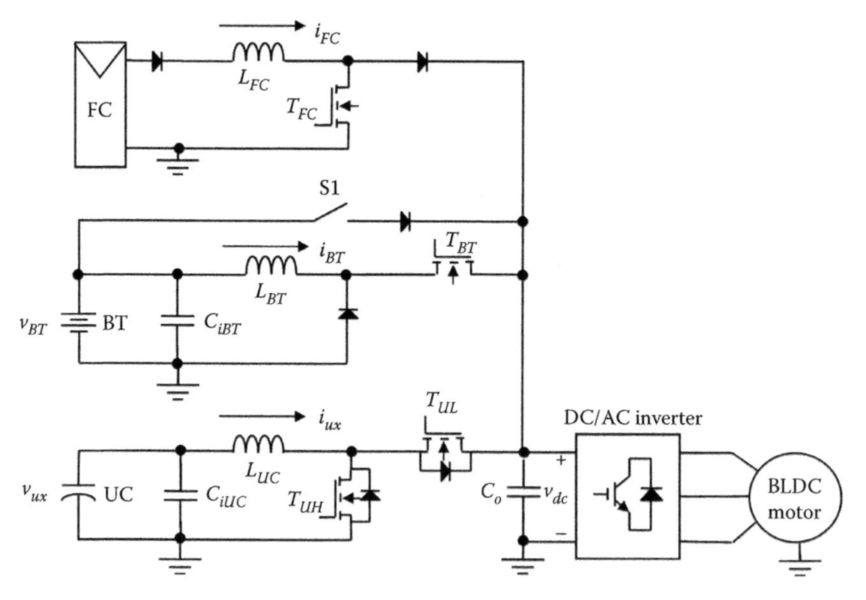

FIGURE 10.30
Overall system circuit configuration of the light FCV.

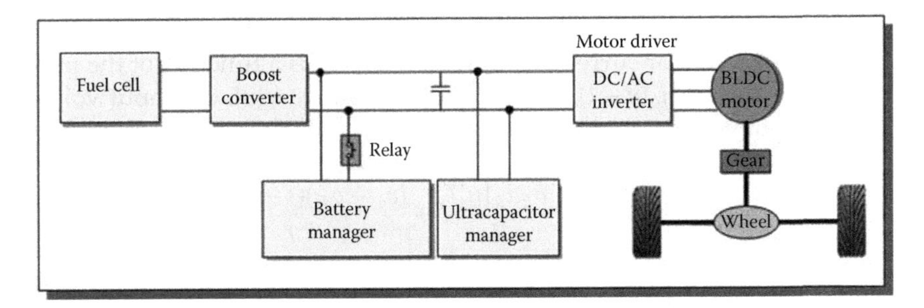

FIGURE 10.31
System structure of the fuel cell hybrid electric vehicle (for light vehicles). (Adapted from W. Na et al. *IEEE Transactions on Vehicular Technology*, 60(1), 2011, 87–97.).

motor cycles or scooters. There are five key components in the system: (1) fuel cell system (FCS); (2) motor driver (MD); (3) battery manager (BM); (4) boost converter (BC); and (5) UC manager.

For FCV applications, the DC-bus voltage needs to be boosted to increase the motor power density lowering the motor current for the same traction power. Therefore, a boost converter is typically employed between the fuel cell stacks and the DC bus. In the hybrid system, a UC can be used for faster response, and to protect the battery from over charging and over discharging currents. The DC-bus voltage of the system is quite low for light vehicles, and thus isolated converters are not considered for this application.

Figure 10.31 illustrates the overall system configuration of the light fuel cell hybrid electric vehicle. When the main switch is on, the fuel cell controller, boost converter, battery relay, UC manager, and BLDC MD modules are sequentially on to be ready for a driver command. When a driver shuts off the main power, the motor drive module is shut off first, turning off all inverter switches. Then, the UC, battery, boost converter modules, and fuel cell controller are shut off sequentially.

For the analysis of FCVs, a previous simple fuel cell circuit model in Chapter l is considered.

10.4.2 FCVs Based on the Predictive Controllers

In this section, the design methodologies of the proposed predictive controllers for the boost converter, UC bidirectional converter, and BLDC motor drive are explained.

10.4.2.1 Predictive Control for the Fuel Cell Boost Converter

The goal of this predictive current control (PCC) method [10–14] is to ensure that the inductor current of the boost converter follows the reference current predicting the next step dynamic response. By assuming that the fuel cell voltage, v_{fc}, and DC-bus voltage, v_{dc}, are slowly varying compared to the currents, they can be considered to be constant during the switching period. The sampled inductor current, $i_{fc}(n)$, at the time nT_s is a function of the previous sampled value, $i_{fc}(n-1)$, the applied duty ratio, $d_{fc}[n]$, the input voltage, v_{fc}, the output voltage, v_{dc}, the inductance, L, and the switching period, T_s as

$$i_{fc}(n) = i_{fc}(n-1) + \frac{v_{fc} d_{fc}[n] T_s}{L} + \frac{(v_{fc} - v_{dc})(1 - d_{fc}[n]) T_s}{L} \tag{10.34}$$

Equation 10.34 will be simplified as

$$i_{fc}(n) = i_{fc}(n-1) + \frac{v_{fc} T_s}{L} - \frac{v_{dc}(1 - d_{fc}[n]) T_s}{L} \tag{10.35}$$

At the next cycle

$$i_{fc}(n+1) = i_{fc}(n) + \frac{v_{fc} d_{fc}[n+1] T_s}{L} + \frac{(v_{fc} - v_{dc})(1 - d_{fc}[n+1]) T_s}{L} \tag{10.36}$$

Then, Equation 10.36 yields following equation by substituting Equation 10.34 into Equation 10.35

$$\begin{aligned} i_{fc}(n+1) = i_{fc}(n-1) &+ \frac{v_{fc} T_s}{L} - \frac{v_{dc}(1 - d_{fc}[n]) T_s}{L} \\ &+ \frac{v_{fc} d_{fc}[n+1] T_s}{L} + \frac{(v_{fc} - v_{dc})(1 - d_{fc}[n+1]) T_s}{L} \end{aligned} \tag{10.37}$$

Equation 10.37 can be reduced as follows:

$$i_{fc}(n+1) = i_{fc}(n-1) + 2\frac{v_{fc} T_s}{L} - \frac{v_{dc}(1 - d_{fc}[n]) T_s}{L} - \frac{v_{dc}(1 - d_{fc}[n+1]) T_s}{L} \tag{10.38}$$

Assuming that the next sampling current signal is equal to the reference command signal ($i_{fc}(n+1) = i_{fc}^*$), the next step duty ratio can be derived based on Equation 10.38 as follows:

$$d_{fc}[n+1] = 2 - d_{fc}[n] + \frac{L}{v_{dc} T_s}\left[i_{fc}^* - i_{fc}(n-1)\right] - 2\frac{v_{fc}}{v_{dc}} \tag{10.39}$$

According to Equation 10.39, the next step duty ratio $d[n+1]$ will be predicted based on the previous duty ratio, $d[n]$, and the inductor current, $i(n-1)$. The theoretical inductor current waveform is illustrated in Figure 10.32.

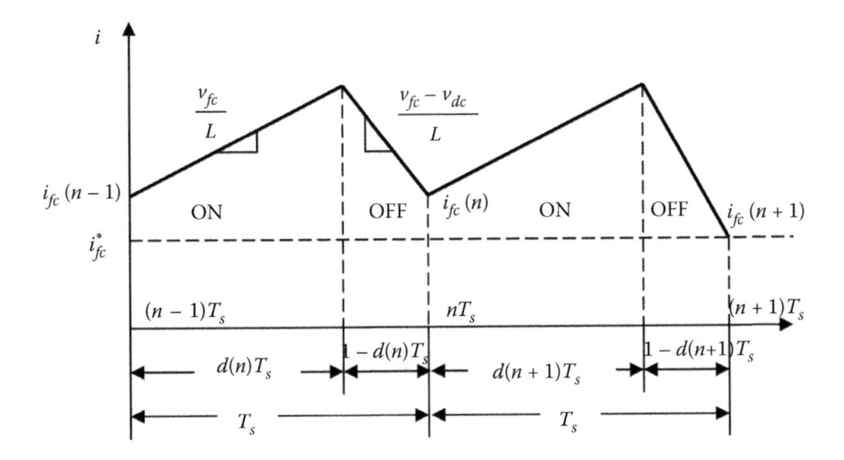

FIGURE 10.32
Inductor current waveform based on the PCC.

As depicted in Figure 10.32, it should be noted that the predictive current controller initially assumes that the error of $i_{fc}^* - i_{fc}(n-1)$ at the $(n-1)_{th}$ sampling will disappear at $(n+1)_{th}$ sampling time, and hence the predictive controller is inherently stable for all operating points [12].

10.4.2.2 Predictive Control for the UC Bidirectional Converter

During the boost mode from the UC to the main DC bus, the control principle is similar to that of the fuel cell boost converter. Thus the duty ratio of the UC bidirectional converter at the boost mode is

$$d_{uc\_up}[n+1] = 2 - d_{uc\_up}[n] + \frac{L}{v_{dc}T_s}\left[i_{uc}^* - i_{uc}(n-1)\right] - 2\frac{v_{uc}}{v_{dc}} \tag{10.40}$$

where i_{uc} is UC inductor current.

However, during the buck mode, the energy flows from the main DC bus to the UC yielding

$$i_{uc}(n) = i_{uc}(n-1) + \frac{(v_{dc} - v_{uc})d_{uc\_d}[n]T_s}{L} - \frac{v_{dc}(1-d_{fc}[n])T_s}{L} \tag{10.41}$$

Similarly, for the buck mode the duty ratio is obtained based on Figure 10.33 and Equations 10.39 through 10.41.

$$d_{uc\_down}[n+1] = -d_{uc\_down}[n] - \frac{L}{v_{dc}T_s}[i_{uc}^* - i_{uc}(n-1)] + 2\frac{v_{uc}}{v_{dc}} \tag{10.42}$$

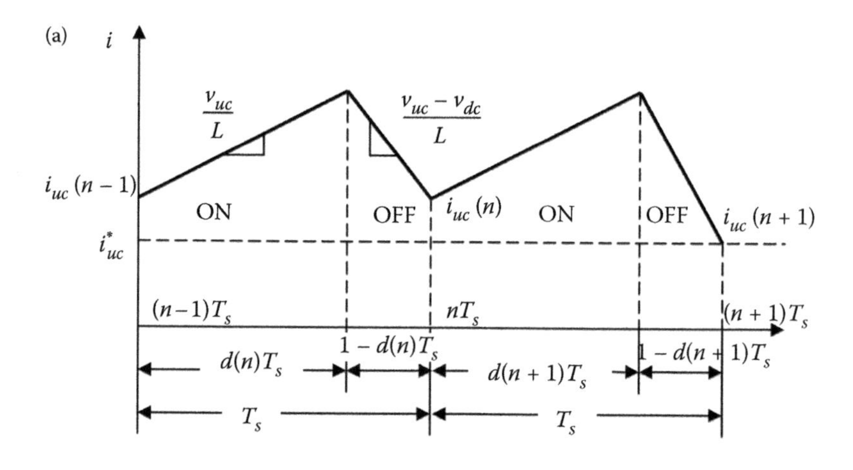

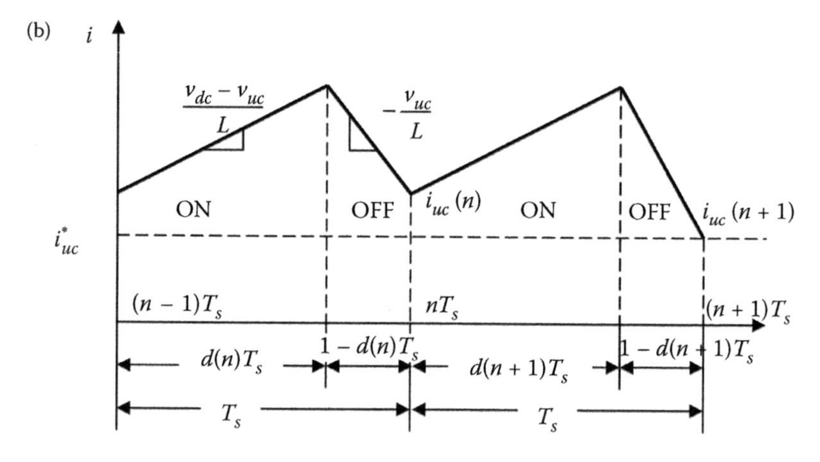

FIGURE 10.33
UC inductor waveform based on the PCC. (a) Boost mode and (b) buck mode.

The theoretical UC inductor current switching waveforms in the boost and buck mode are described in Figure 10.33.

10.4.2.3 Predictive Control of the BLDC Motor Drive

Figure 10.34 illustrates the system control structure including the BLDC motor control configuration. A typical BLDC motor drive system consists of the BLDC motor, power inverter, and digital controller. The voltage source inverter is connected to the DC bus. The digital controller energizes each phase of the BLDC motor in a sequence synchronizing with the rotor position in order to produce constant torque (120° current conduction case). Therefore, the system requires a mechanical position sensor (Hall sensor)

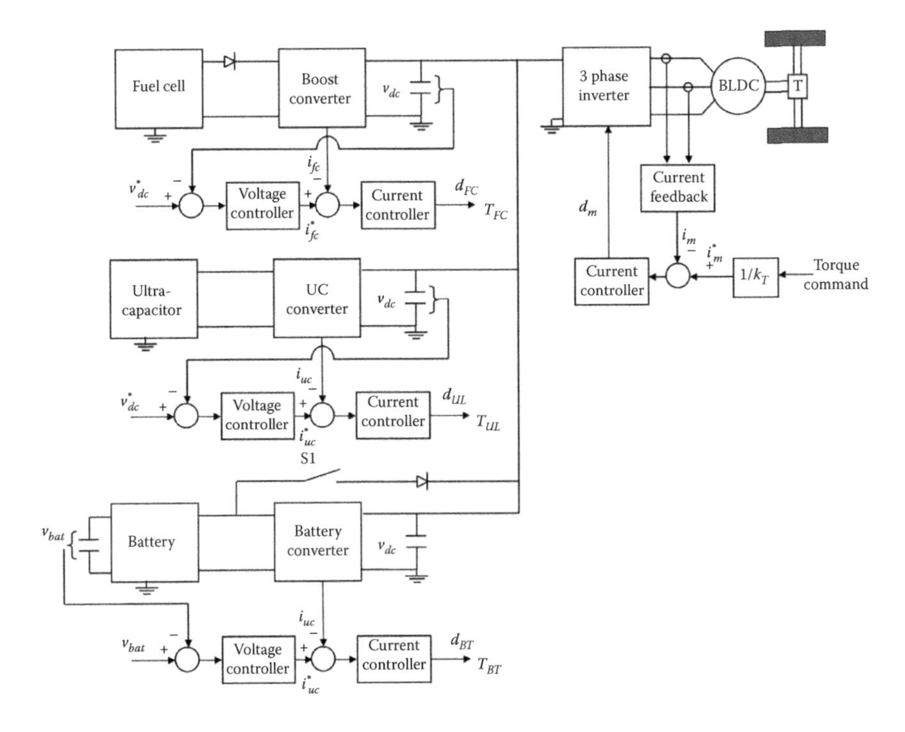

FIGURE 10.34
BLDC motor drive configuration in the fuel cell hybrid drive system.

that is usually attached inside the BLDC motor in order to provide rotor position feedback for the controller. The typical BLDC motor drive system with the position feedback [14] is shown in Figure 10.34 with the boost converter, BM, and UC converter.

In order to derive a PCC law, the motor voltage equation is considered during a switching cycle as

$$d_m \cdot v_{dc} - (1 - d_m) \cdot v_{dc} = Ri_m + L\frac{di_m}{dt} + E_{back} \tag{10.43}$$

where i_m is the motor line to line current, L and R are the motor inductor and resistance (line-to-line), respectively, and E_{back} is the back EMF, and d_m is the duty ratio. Then, the nth sampled motor current can be expressed based on the previous sampled value and the duty ratio as

$$i_m(n) = i_m(n-1) - \frac{E_{back}T_s}{(R + LT_s)} + \frac{2v_{dc}d_m[n]T_s}{(R + LT_s)} - \frac{2v_{dc}T_s}{(R + LT_s)} \tag{10.44}$$

where T_s is the sampling period. Similarly, the predicted duty ratio is calculated as follows:

$$d_m[n+1] = 2 - d_m[n] - \frac{(R + LT_s)}{2v_{dc}T_s}[i_m^* - i_m(n-1)] - \frac{E_{back}}{v_{dc}} \tag{10.45}$$

The concept behind Equation 10.46 is that the controller is to force the $(n + 1)_{th}$ sampled current to be the reference value such that the error between the desired current reference and the sampled current is reduced at the end of the switching cycle. To improve the current controller performance under the motor parameter inaccuracies and the nonideal inverter operation, the current error correction approach [15,16] has been applied. Then, the current reference, i_m^* in Equation 10.46 can be replaced with the newly updated current reference, $i_{m\_new}^*$ based on the following equation:

$$i_{m\_new}^*(s) = i_m^*(s) + \frac{i_e(s)}{sg_e} \tag{10.46}$$

where $i_{m\_new}^*$ is the new current reference; i_m^* is the old current reference; i_e is the current error ($i_e = i_m^* - i_m$); and g_e is an error correction gain. In the discrete time domain with a sampling time, T_s, Equation 10.46 can be approximated as

$$i_{m\_new}^*(n+1) = i_m^*(n+1) + \frac{T_s}{g_e}\sum_{j=0}^{n}\left(i_m^*(j) - i_m(j)\right) \tag{10.47}$$

Thus, the updated predictive duty ratio is written as

$$d_m[n+1] = 2 - d_m[n] - \frac{(R + LT_s)}{2v_{dc}T_s}[i_{m\_new}^* - i_m(n-1)] - \frac{E_{back}}{v_{dc}} \tag{10.48}$$

Unlike the predictive controllers in the boost converter and UC converter, an error correction gain is applied along with the proportional gain and the feed forward term related to the back EMF in the predictive BLDC motor controller. The detailed PCC block diagram for the BLDC motor is described in Figure 10.35.

10.4.3 Simulation Verification

The outer voltage control loops for the fuel cell boost converter and the UC bidirectional converter are designed to provide a maximized bandwidth considering high inner current loop dynamics. The crossover frequency of the outer loop is set to maximize the phase lead of the controller [7] as

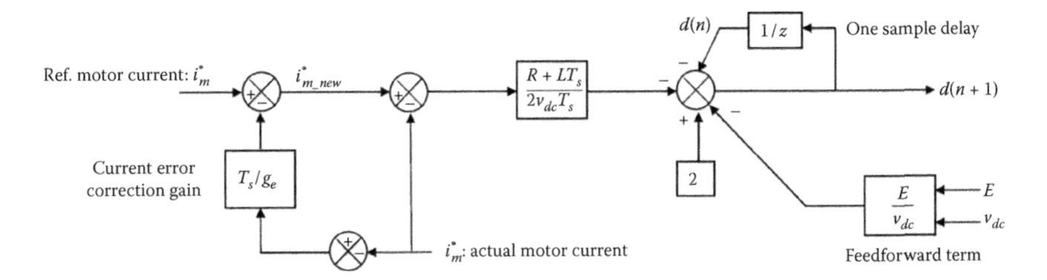

FIGURE 10.35
PCC block diagram for the BLDC motor.

$$f_x = \frac{1}{10 \cdot T_s} \tag{10.49}$$

The PI controller for the outer voltage control loop is designed to obtain a satisfactory phase margin of 45° [17] with the crossover frequency of 2 kHz. To compare the proposed PCC with a conventional PI current controller, the PI current controllers in the boost converter and UC converter are set to have the loop crossover frequencies of 10 kHz and the phase margin of 60° as recommended in Reference 17. For the BLDC motor controller, the step current reference is applied to evaluate the dynamic response of the controllers (conventional PI versus the proposed predictive controllers). To design an appropriate PI current controller in the BLDC motor for a comparison purpose, a simplified open-loop transfer function [18] is used as

$$G_{I,OL}(s) = \underbrace{\frac{k_i}{s}\left[1 + \frac{s}{k_i/k_p}\right]}_{PI-controller} k_{pwm} \frac{1/R}{1+(s/1/\tau_e)} \tag{10.50}$$

where k_{pwm} is the ratio between control signal and peak control signal; R is the motor phase resistance; and τ_e is the electrical time constant of the motor (L/R). To cancel the motor pole at $1/\tau_e$, the following condition is used [17]:

$$k_p = \tau_e k_i \tag{10.51}$$

Based on a the crossover frequency of 1 kHz for the PI current controller for the BLDC motor, the PI controller gains can be calculated to avoid interference with the switching frequency noise [17] as

$$k_i = \frac{\omega_{cl} R}{k_{pwm}} \tag{10.52}$$

TABLE 10.2
Simulation Parameters

Parameter	Values
L_d (PFC)	1 mH
L_b (bidirectional)	500 μH
C_d	3000 μF
R_{load}	100 Ω (initial)
f_s	30 kHz
V_{bat} (battery)	48 V
V_s	120 $V_{s,rms}$, 60 Hz
V_{dc}^*	200 V

where ω_{cl} is the crossover frequency. Table 10.2 shows system parameters, which include each crossover frequency f_{cl}, and phase margin φ in the controller. The switching frequency is set to 20 kHz. The system parameters of the PI controllers are shown in Table 10.3.

The simulation is performed and the results justify that the proposed predictive current controllers have advantages over the PI controller respective to battery protection and longer battery life because the fast current response of all other components can lead to less battery current drain. For the control performance comparison, the load current is step changed from 2 to 5 A at 0.3 s in Figure 10.36. The predictive controller demonstrates a faster current response and less overshoot compared to the PI controller case.

Figure 10.37 presents the boost converter current response and DC link voltage under the step load current change from 5 to 2 A. Similarly the predictive controller generates a less DC link voltage overshoot and a faster current response. Figure 10.38 illustrates the motor torque and phase current waveforms when the motor current command is step changed from 5 to 10 A. At the transient period, the discrepancy of the torque responses between the conventional PI controller-based system (PI current controllers are used for

TABLE 10.3
System Parameters

	Boost Converter	UC Converter	BLDC
Inner loop f_{cl}	10 kHz	10 kHz	1 kHz
Inner loop φ	60°	60°	90°
Outer loop f_{cl}	2 kHz	2 kHz	N/A
Outer loop φ	45°	45°	

Source: Adapted from W. Na et al. *IEEE Transactions on Vehicular Technology*, 60(1), 87–97, 2011.

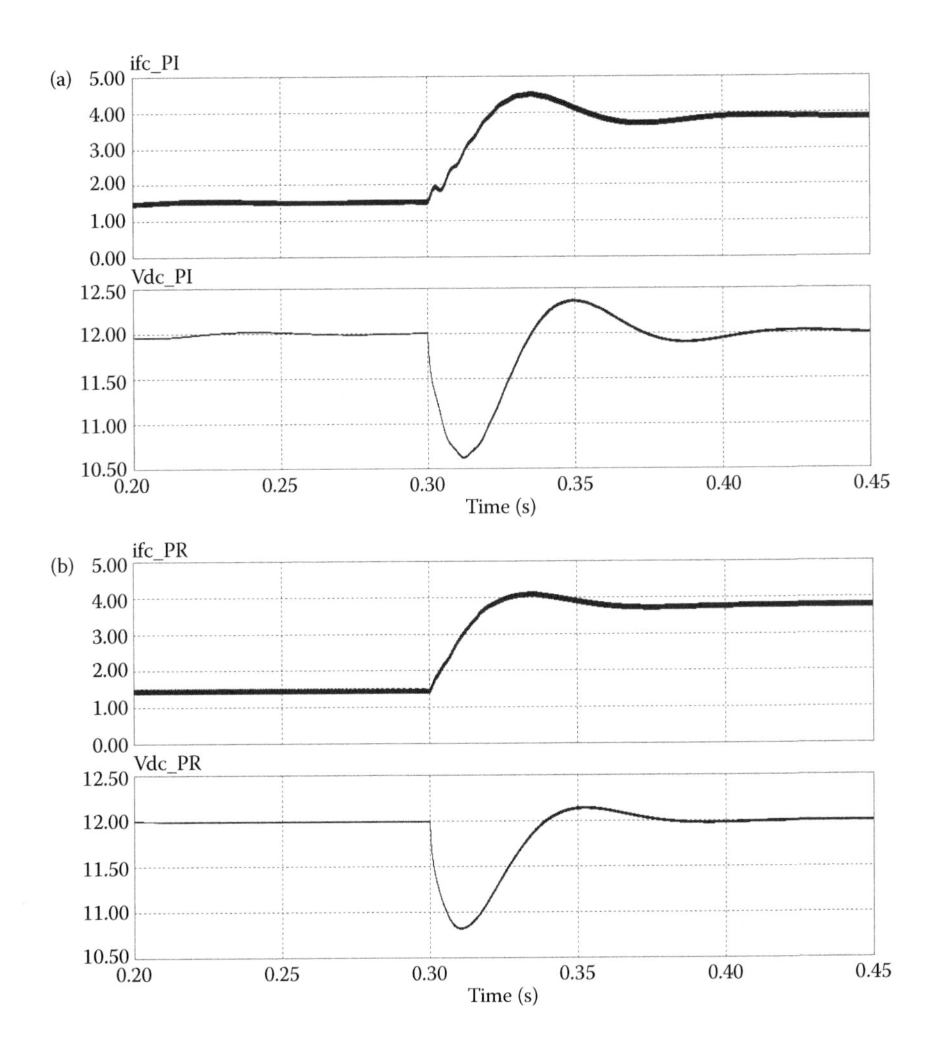

FIGURE 10.36
Simulation results: boost converter inductor current (upper) and DC link voltage (lower) under the step load change (2 → 5 A). (a) PI controller and (b) predictive controller.

boost, UC converters, and the BLDC motor drive) and the predictive controller-based system is recognizable (faster torque response is observed). According to Figures 10.36 through 10.38, the advantages of the predictive controller can be observed.

10.4.4 Summary

In this section, the predictive controllers for a fuel cell powered light hybrid vehicle have been discussed. The predictive controllers are implemented in

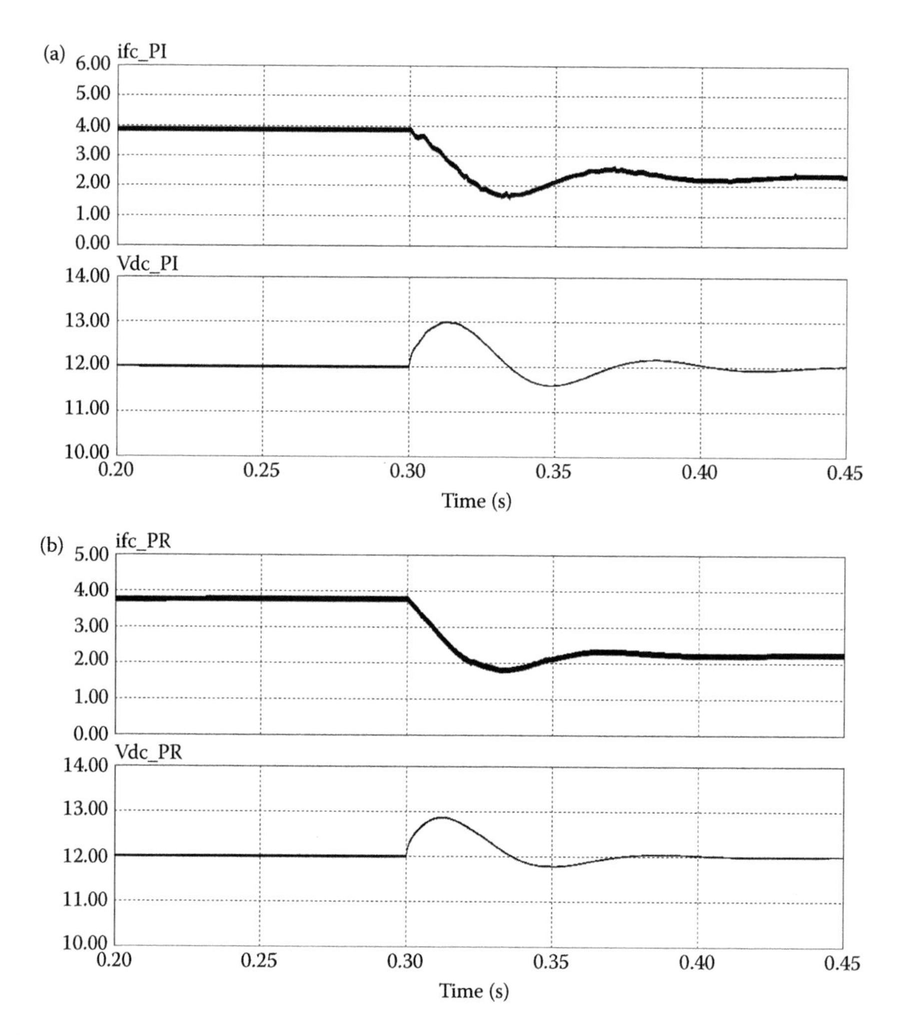

FIGURE 10.37
Simulation results: boost converter inductor current (upper) and DC link voltage (lower) under the step load change (5 → 2 A). (a) PI controller and (b) predictive controller. (Adapted from W. Na et al. *IEEE Transactions on Vehicular Technology*, 60(1), 87–97, 2011.)

the boost converter, UC manager, and BLDC motor drive. In that way, each component and traction drive has faster response and less current ripple. The faster responses from the boost converter, UC, and motor drive help decrease the battery current variation (di/dt). Therefore, improved battery protection and longer battery life are expected. Simulations have been performed and results verified using this predictive control approach. Details of the experiment set-up and its experimental verification using DSP for the predictive control approaches can be found in Reference 18.

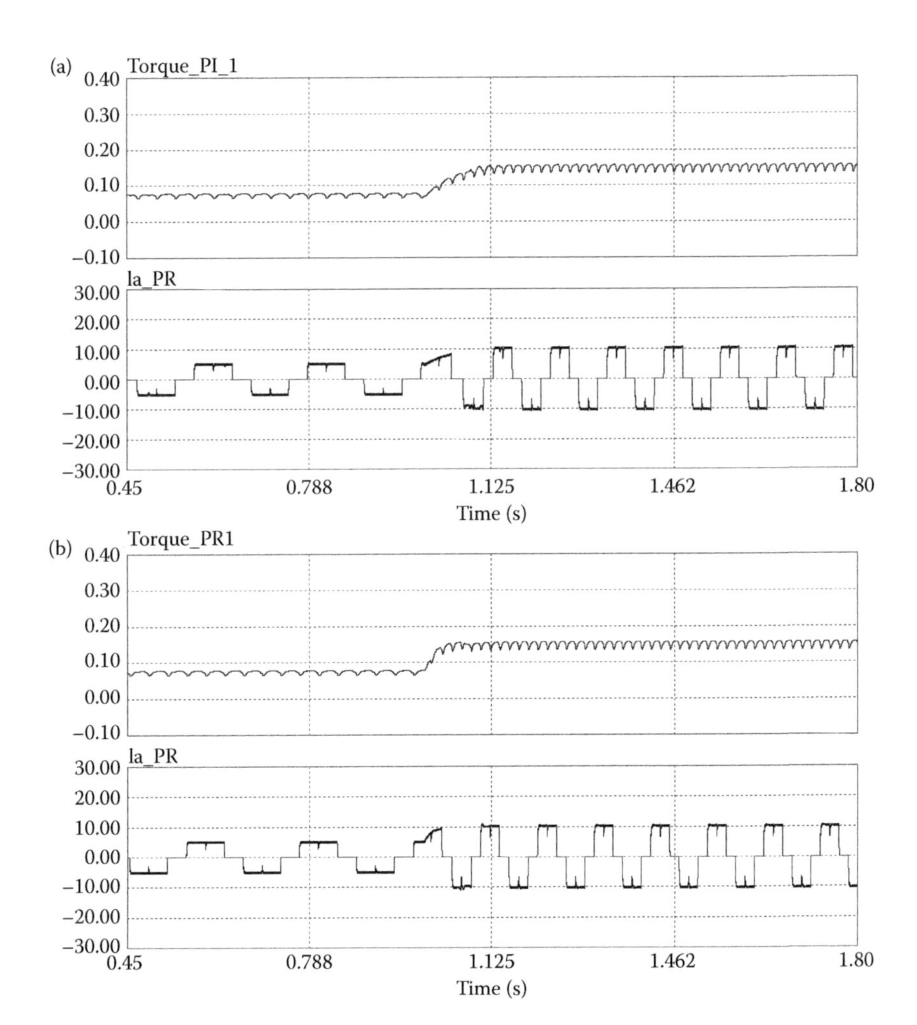

FIGURE 10.38
Simulation results: Motor torque (N m) and motor phase current (A) when the motor torque command is step changed (5 → 10 A). (a) PI controller and (b) predictive controller. (Adapted from W. Na et al. *IEEE Transactions on Vehicular Technology*, 60(1), 87–97, 2011.)

10.5 Conclusion

In this chapter, designs of linear controllers, SMCs for a PFC and bidirectional (buck/boost) converter have been explained. And one of the digital control algorithms, predictive controllers for FCVs applications have been described in detail. In order to validate each controller, computer simulations, and DSP-based power electronics tests have been performed.

References

1. W.D. Jones, Take this car and plug it, *IEEE Spectrum*, 42(7), 10–13, 2005.
2. A.F. Burke, Batteries and ultracapacitors for electric, hybrid, and fuel cell vehicles, *Proceedings of the IEEE*, 95(4), 806–820, 2007.
3. M.G. Egan, D.L. O'Sullivan, J.G. Hayes, M.J. Willers, and C.P. Henze, Power-factor-corrected single stage inductive charger for electric vehicle batteries, *IEEE Transactions on Industrial Electronics*, 54(2), 1217–1226, 2007.
4. B. Singh, B.N. Singh, A. Chandra, K. Al-Haddad, A. Padey, and D.P. Kothari, Review of single phase improved power quality AC–DC converters, *IEEE Transactions on Industrial Electronics*, 50, 962–981, 2003.
5. Y.-J. Lee, A. Khaligh, and A. Emadi, Advanced integrated bidirectional AC/DC and DC/DC converter for plug-in hybrid electric vehicles, *IEEE Transactions on Vehicular Technology*, 58(8), 2009.
6. F. Blaabjerg and Z. Chen, Power electronics as an enabling technology for renewable energy integration, *Journal of Power Electronics*, 3(2), 2003, 81–89.
7. N. Mohan, *Power Electronics: A First Course*, Wiley, Hoboken, New Jersey, 2012.
8. V. Utkin, J. Guldner, and J. Shi, *Sliding Mode Control in Electromechanical Systems*, Taylor & Francis, Philadelphia, Pennsylvania, 1999.
9. R. Revathy and N. Senthil Kumar, Design and evaluation of robust controller for AC-to-DC boost converter, In: *IEEE International Conference on Computer, Communication, and Electrical Technology*, Tamil Nadu, India, 2011, no. 3, May 2005, pp. 763–770.
10. R. Bartholomaeus, A. Fischer, and M. Klingner, Real-time predictive control of hybrid fuel cell drive trains, In: *Fifth IFAC Symposium on Advances in Automotive Control (2007), Advances in Automotive Control*, Pajaro Dunes, California, Vol. 5, Part 1.
11. P. Mattavelli, G. Spiazzi, and P. Tenti, Predictive digital control of power factor preregulators with input voltage estimation using disturbance observers, *IEEE Transactions on Power Electronics*, 20(1), 140–147, 2005.
12. J. Chen, A. Prodic, R.W. Erickson, and D. Maksimovic, Predictive digital current programmed control, *IEEE Transactions on Power Electronics*, 18(1), 411–419, 2003.
13. S. Jeong and M. Woo, DSP-Based active power filter with predictive current control, *IEEE Transactions on Industrial Electronics*, 1(3), 329–336, 1997.
14. T. Kim and J. Yang, Control of a brushless DC motor/generator in a fuel cell hybrid electric vehicle, *IEEE International Symposium on Industrial Electronics*, 1973–1977, 2009.
15. L. Hoang, K. Slimani, and P. Viarouge, Analysis and implementation of a real-time predictive current controller for permanent-magnet synchronous servo drives, *IEEE Transactions on Industrial Electronics*, 41(1), 110–117, 1994.
16. S. Bibian and H. Jin, High performance predictive Dead-Beat digital controller for DC power supplies, *IEEE Transactions on Power Electronics*, 7(3), 420–427, 2002.
17. N. Mohan, *Electric Drives*, MNPERE, Minneapolis, MN, 2003.
18. W. Na, T. Park, T. Kim, and S. Kwak, Light fuel cell hybrid electric vehicles based on predictive controllers, *IEEE Transactions on Vehicular Technology*, 60(1), 2011, 87–97.

11

A PEM Fuel Cell Temperature Controller*

11.1 Introduction

In general, there are several kinds of fuel cells such as PEMFCs, SOFCs, PAFCs, MCFCs, AFCs, DMFCs, ZAFCs, and PCFCs [1]. Among these fuel cells, PEMFC is suitable for fuel cell vehicles and distributed generators because it has high power density, solid electrolyte, a long cell and stack life, and low corrosion. Moreover, PEMFCs can easily operate at low temperatures (50–100°C), which enables them to have a faster startup time. However, in order to generate a reliable and efficient power response and to prevent membrane damage and oxygen depletion, a sophisticated control technique is crucial for the operation of fuel cell systems [2,3]. Several control approaches have been developed for PEM fuel cell systems [2–5] to achieve an optimal air supply, hydrogen flow rate, and pressure. The steady-state electrochemical model of PEMFCs has been developed in References 6, 7 and a new dynamic model has been introduced [8,9]. However, even though the temperature significantly affects the operation of fuel cell stack, a control strategy for temperature has mostly not been considered. In Reference 10, an optimal temperature control study has been conducted based upon the consideration that the humidity and temperature are limited in boundaries, instead of a detailed control analysis by checking Bode plots. There are many literature PEMFC studies regarding control approaches for PEMFCs [2–5], PEMFC modeling studies [6–9,11–14], and PEMFCs applications regarding power electronics, power systems, and auxiliary system control [12,15–21]. In References 2–22, because temperature has a slow dynamic compared with other parameters such as air, hydrogen flow, and pressure, the stack temperature is assumed to be constant so that a simplified control-oriented dynamic model was derived. However, in reality, the change of the fuel cell stack temperature dramatically affects the output current as well as the output power of the fuel cells system. Thus, this chapter is focused on the fuel cell temperature model based upon an electrical aspect to design a better

* This chapter was mainly prepared by Dr. Woonki Na, California State University, Fresno and Dr. Bei Gou, Smart Electric Grid, LLC, USA.

accurate fuel cell temperature controller. In this study, the thermal equivalent circuit of PEMFC described in Chapter 6 is used to design a temperature controller, which makes it easier to develop a temperature control algorithm for PEMFC. With the PEMFC thermal equivalent circuit model, the fuel cell system can be viewed as one of electrical systems, and the design procedure of the controller is almost the same as the design for the conventional power converter's controller explained in Reference 23. Therefore, the analysis of the fuel cell controller can be simply done by checking the phase margin and magnitude of the transfer function of Bode plots.

This chapter is organized as follows. Section 11.2 addresses the design of PEMFC temperature controller based on the thermal equivalent circuit. Sections 11.3 and 11.4 provide analysis results of the PEMFC thermal transfer functions and its simulation data verification, and Section 11.5 concludes the chapter.

11.2 PEMFC Temperature Controller Design

In PEMFCs, the temperature $T_s(t)$ is a function of the fuel cell voltage V_{st}, the fuel cell current I_{fc}, the cooling pump control variable, u_{cl} described in Chapter 6, and other temperature values such as ambient temperature T_{amb}, assumed to be perturbed in the range of ±5%. In order to obtain a stable temperature under any disturbances, it is necessary to adjust u_{cl} in the cooling/heat exchanger system. The typical open-loop PEMFC for temperature control is illustrated in Figure 11.1. Because the voltage depends on the current in the fuel cell system, only two disturbances, i_{fc} and t_{amb}, are considered by ignoring $v_{st}(t)$. To adjust the u_{cl} automatically under any circumstance, the feedback control system in Reference 23 is utilized as presented in Figure 11.2. The output stack temperature, $t_s(t)$ is measured

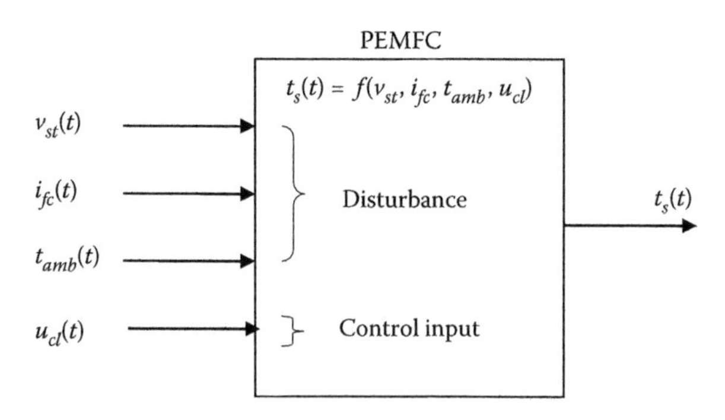

FIGURE 11.1
Open-loop PEMFC for temperature control.

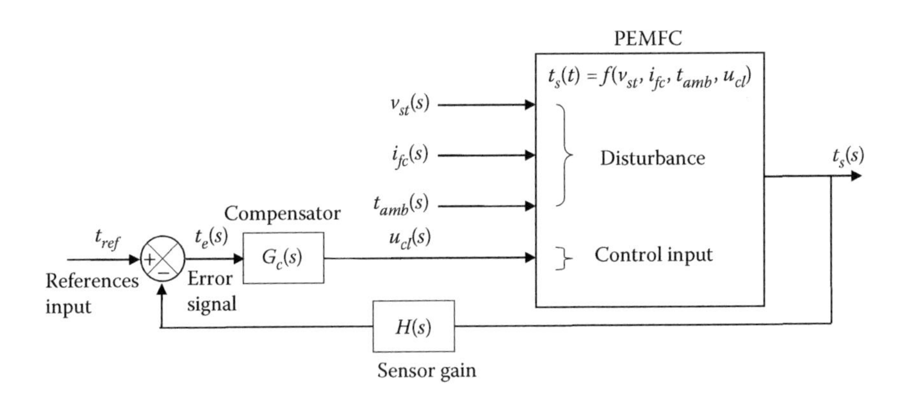

FIGURE 11.2
Feedback loop PEMFC for temperature control.

using a temperature sensor with gain $H(s)$. The sensor circuit is usually a voltage divider consisting of high precision resistors. The sensor output signal $H(s)t_s(s)$ is compared with the reference input temperature, $t_{ref}(s)$ to make $H(s)t_s(s)$ being equivalent to $t_{ref}(s)$ so that $t_s(s)$ follows $t_{ref}(s)$ during disturbances in the system.

The small signal ac transfer functions of a thermal equivalent circuit can be derived assuming that the variation of $\hat{t}_s(t)$, $\hat{i}_{fc}(t)$, $\hat{t}_{amb}(t)$, and $\hat{u}_{cl}$ are relatively small [23]. According to the transient energy balance equation (6.1) and Figure 11.1, the stack temperature variation can be expressed as follows:

$$\hat{t}_s(s) = G_{ti}(s)\hat{i}_{fc}(s) - G_{tu}(s)\hat{u}_{cl}(s) + G_{tt}(s)\hat{t}_{amb}(s) \tag{11.1}$$

where $G_{ti}(s)$ is the fuel cell current to output thermal transfer function as

$$G_{ti}(s) = \left. \frac{(N/2F)\Delta H - \hat{v}_{st}(s)}{sC_t + (1/R_t)} \right|_{\substack{\hat{u}_{cl}=0 \\ \hat{t}_{amb}=0}} \tag{11.2}$$

where ΔH is the enthalpy change for hydrogen (285.5 kJ mol/s), N is the number of the fuel cell, F is the Faraday constant, R_t is the thermal resistance, and C_t is the thermal capacitance.

For details of those parameters in Equation 11.2, please refer to Chapters 6 and 7.

$G_{tu}(s)$ is the control input to output transfer function as

$$G_{tu}(s) = G_d(s)/(sC_t + (1/R_t))\Big|_{\substack{i_{fc}=0 \\ \hat{t}_{amb}=0}} \tag{11.3}$$

where

$$G_d(s) = k_c/(1+\lambda_c s) \tag{11.4}$$

where τ_c is the time delay constant, 70 s, and k_c the conversion factor, 1.5 are assumed.

The ambient temperature to output transfer function $G_{tt}(s)$ as

$$G_{tt}(s) = (\hat{t}_{amb}(s)/R_t)/(sC_t + (1/R_t))\Big|_{\substack{\hat{i}_{fc}=0 \\ \hat{u}_{cl}=0}} \tag{11.5}$$

To analyze this system, the reference and error of the stack temperature perturbed are defined as follows:

$$\begin{aligned} t_{ref}(t) &= T_{ref} + \hat{t}_{ref}(t) \\ t_e(t) &= T_e + \hat{t}_e(t) \end{aligned} \tag{11.6}$$

Using the small signal ac transfer functions of a thermal equivalent circuit, the PEMFC block in Figure 10.2 can be redrawn by adding an FC thermal block described by Equation 11.1 (Figure 11.3).

The output stack temperature variation, $\hat{t}_s$, can be reconstructed as

$$\hat{t}_s = -\hat{t}_{ref}\frac{G_cG_{tu}}{1+HG_cG_{tu}} + \hat{i}_{fc}\frac{G_{ti}}{1+HG_cG_{tu}} + \hat{t}_{amb}\frac{G_{tt}}{1+HG_cG_{tu}} \tag{11.7}$$

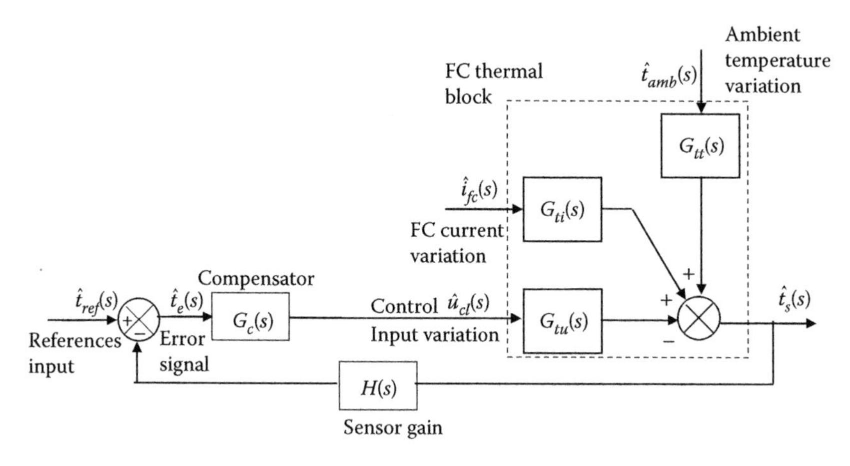

FIGURE 11.3
Complete feedback loop PEMFC for temperature control.

Equation 11.11 can be rewritten assuming that $L = HG_cG_{tu}$ as

$$\hat{t}_s = -\hat{t}_{ref}\frac{1}{L}\frac{L}{1+L} + \hat{i}_{fc}\frac{G_{ti}}{1+L} + \hat{t}_{amb}\frac{G_{tt}}{1+L} \tag{11.8}$$

where $L = L(s) = H(s)G_c(s)G_{tu}(s)$. The loop gain $L(s)$ in Equation 11.8 is an important quantity to identify the system performance when a controller is adopted. Further detailed analysis of the loop gain is described in the following section.

11.3 Analysis of the PEMFC Thermal Transfer Function

To analyze the transfer function a Bode plot is used. First of all, the loop gain $L(s)$ is constructed by the transfer functions, $H(s)$, $G_c(s)$, and $G_{tu}(s)$, to draw the Bode plot. The loop gain $L(s)$ is

$$L(s) = \frac{0.15}{\tau_s\tau_c C_t}\frac{(sk_p + k_i)}{s(s+(1/\tau_s))(s+(1/\tau_c))(s+(1/R_tC_t))} \tag{11.9}$$

where $H(s) = 0.1/(1 + \tau_s s)$ represents the sensor gain function with the sensor time delay 70 s, which means that in the case of 100°C stack temperature, 10 V is used as the input of the controller ranging 0–10 V depending on the stack temperature. The lag compensator, G_c which is a PI controller with gains k_i and k_p ($G_c(s) = (k_i + k_p s)/s$) is used in Equation 11.9, since the lag compensator is required to increase the low-frequency loop gain, which leads to a rejection of low-frequency disturbance. The variation of temperatures usually exists in the low-frequency region. The Bode plots of $L(s)$ are illustrated in Figure 11.6 when the k_i is varying from 0.1 to 50, and k_p from 1 to 50. As presented in Figure 11.4, the small gains ($k_i = 0.1$, $k_p = 1$) are desired to make the system stable, because the phase margin φ_L of the loop gain $L(s)$ is to be positive and the $L(s)$ contains no right half plane pole.

Then, the phase margin can be calculated as

$$\varphi_L = 180° + \angle L(2\pi f_c) \tag{11.10}$$

where f_c is the crossover frequency where the magnitude of the loop gain is unity:

$$\|T(j2\pi f_c)\| = 1 \Rightarrow 0\ \text{dB} \tag{11.11}$$

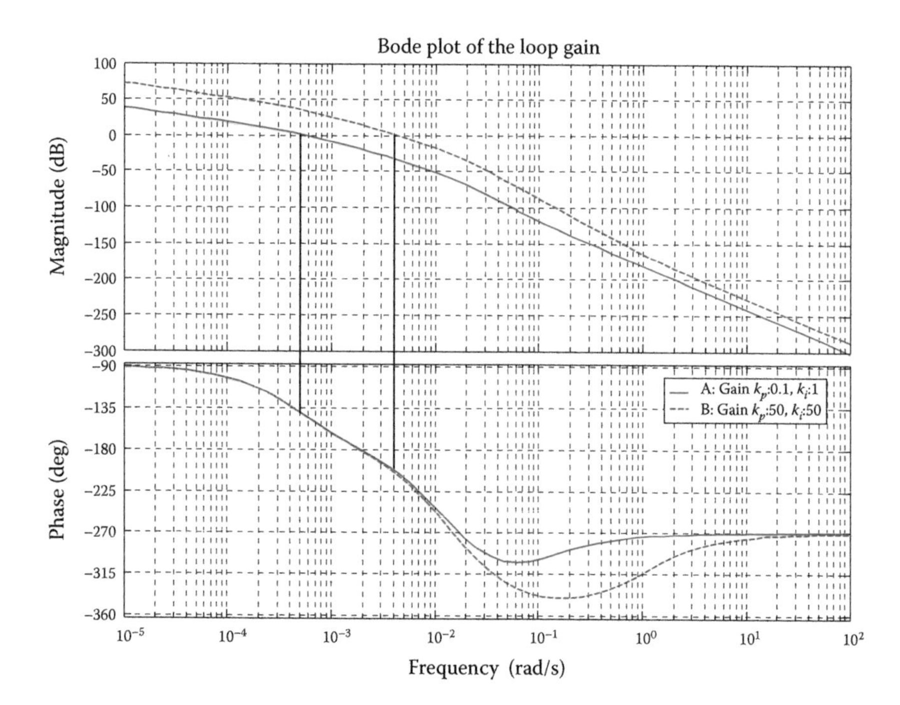

FIGURE 11.4
The Bode plots of loop gain $L(s)$ with lead compensator.

Therefore, the phase margin of $L(s)$ is approximately 40° in the case of the small gains of the lag compensator, that is, $k_i = 0.1$ and $k_p = 1$. If the gains were selected with high values (i.e., $k_i = 50$, $k_p = 50$), the phase margin would be negative, and hence the system would be unstable, which is absolutely not recommended.

11.4 Verification of the PEMFC Temperature Controller

For the model verification, the experimental results (Ballard MK5-E) [11] are compared with our simulation results during the load step up condition. As shown in Figure 11.5 at 0 s, the load step up change occurs based on the assumption that the start-up times of the fuel cell is not considered, but in the proposed works, 2500 s after thermal time constant which is mentioned in Chapter 2, the load step up is imposed. For this step up condition, since the temperature is increased from 40°C to 60°C in Figure 11.5, and the explained temperature controller is designed so that operating fuel cell system temperature is within 80°C, this case does not need to activate a temperature controller. As shown in Figure 6.1, at 72°C, a better fuel cell V–I polarization

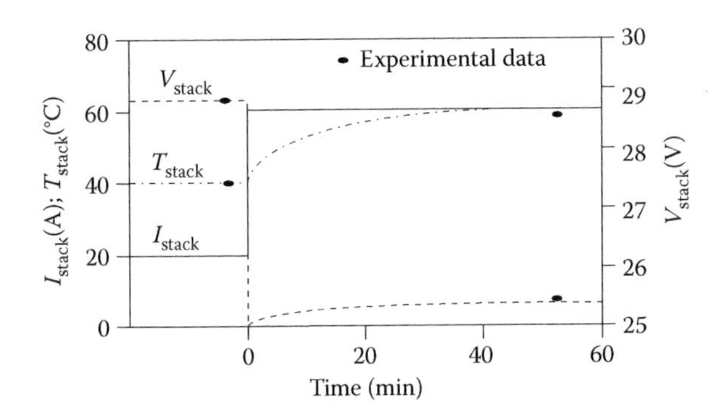

FIGURE 11.5
Load step up condition. (Adapted from Hamelin, J. et al. *International Journal of Hydrogen Energy*, 26, 625–629, 2001.)

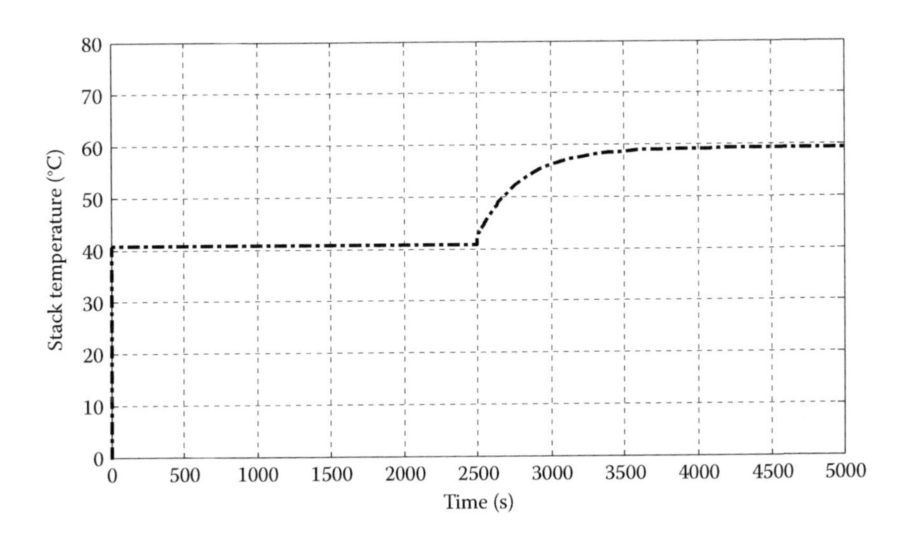

FIGURE 11.6
Temperature changes based on the proposed model.

curve can be achieved. However, at high temperature, cooling system (temperature controller) is required to prevent degradation of the fuel cell characteristics from the high thermal stresses.

In Figure 11.6, the simulation results match with the experimental results (<5% error). Especially the thermal time constant (2059 s) is almost same.

The simulated stack current changes in Figure 11.7 also matches well with the experimental results [11]. However, in the experimental result, the greater stack voltage drops are observed (Figure 10.5) than the simulated voltage in Figure 10.8 since the practical fuel cell system has little more losses than the fuel cell model in the simulation (Figure 11.8).

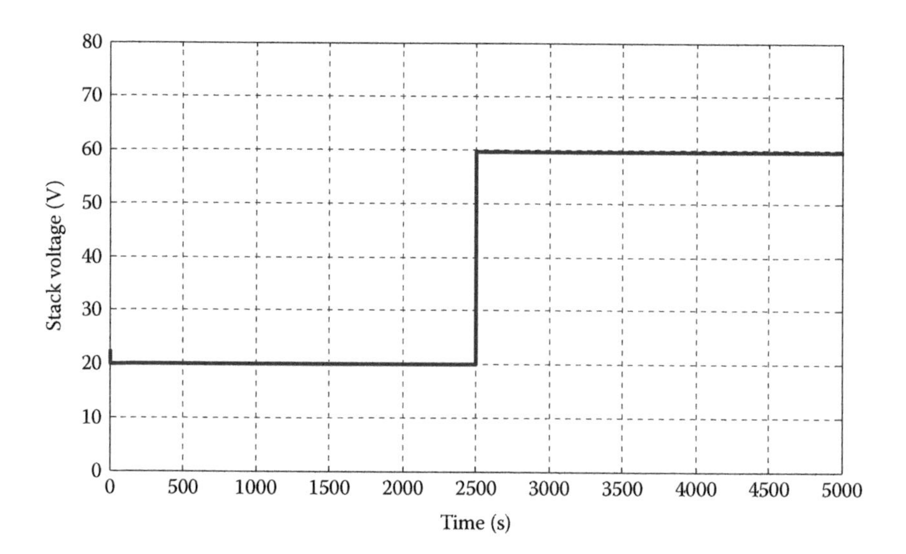

FIGURE 11.7
Stack current changes based on the proposed model.

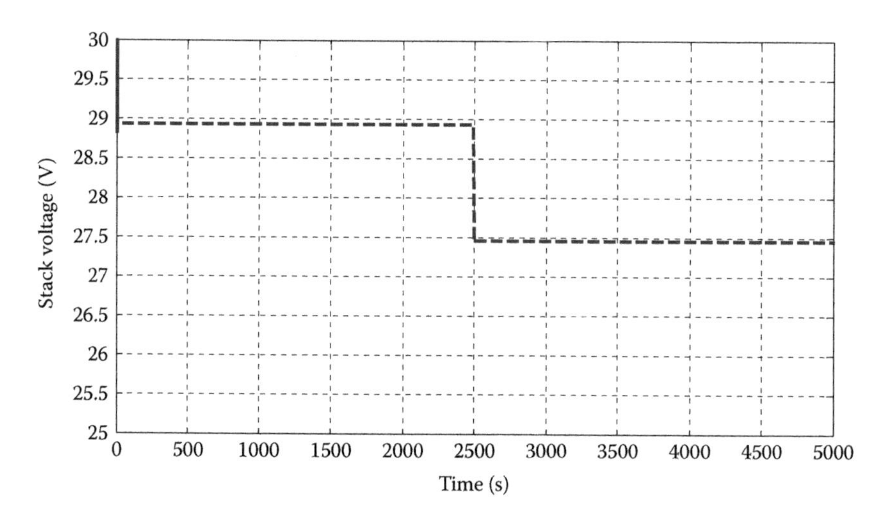

FIGURE 11.8
Stack voltage changes by using proposed model.

Up to date data related to the fuel cell temperature controller are not commercially available. Thus, based on the simulation results, the validity of the described temperature controller can be evaluated.

Figure 10.9 illustrates the temperature change due to the sudden load change (20–190 A) at 2500 s in Figure 11.10. As shown in Figure 10.9, the dotted line shows the temperature change with no controller, whereas the solid

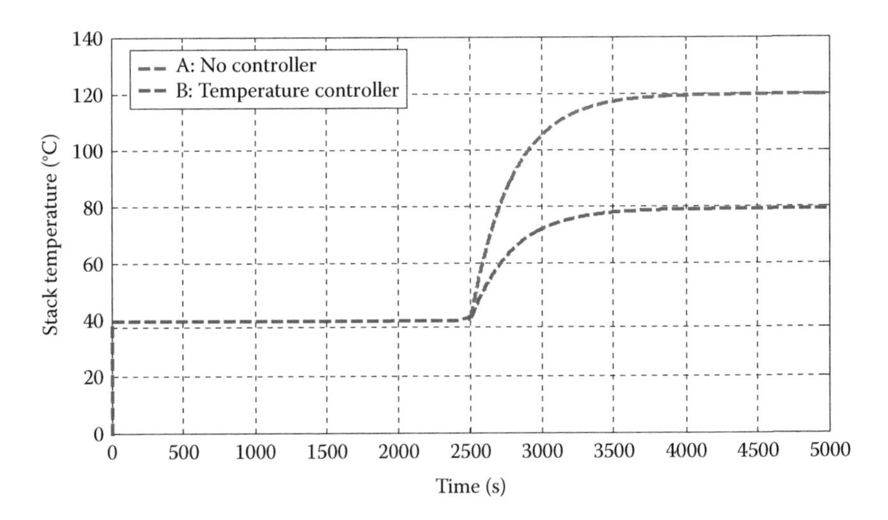

FIGURE 11.9
Temperature changes comparison.

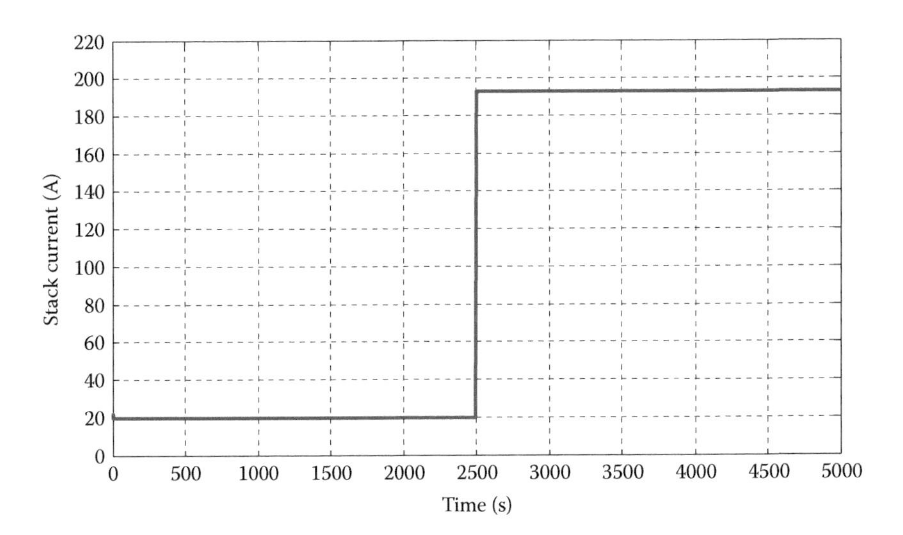

FIGURE 11.10
Stack current change from 20 to 190 A.

line indicates the temperature change based on the temperature controller explained (Figure 11.9).

With the temperature controller, the operating fuel cell temperature can be limited to 80°C. However, without the controller the temperature is increased up to 120°C.

When the load change from 20 to 190 A is enforced in Figure 11.10, the voltage drop can be observed from 29 to 25 V at 2500 s in Figure 10.11 (Figure 11.11).

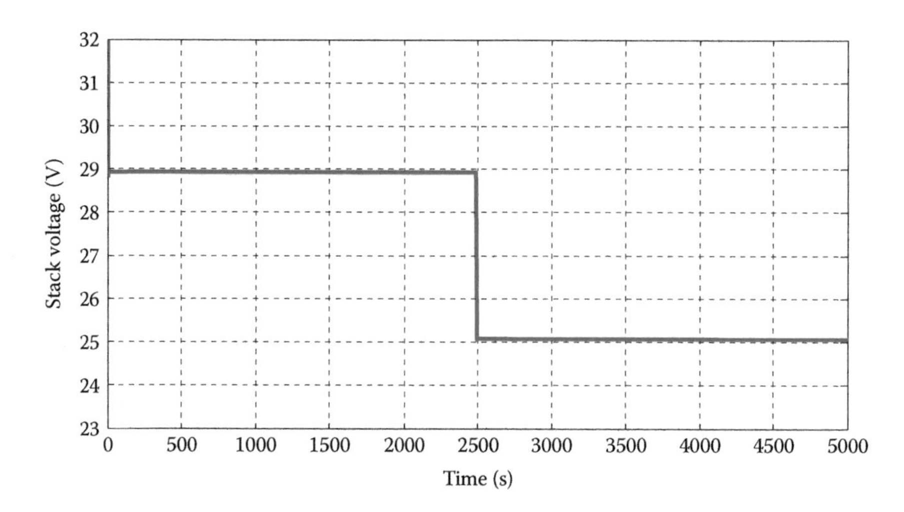

FIGURE 11.11
Stack voltage changes.

11.5 Conclusion

This chapter explains the design of a fuel cell temperature controller based on the thermal equivalent circuit in Chapter 6. The analysis for the design of the PEMFC temperature controller is performed using Bode plots of the transfer functions using the thermal equivalent circuit. The Bode plots indicate that the lag compensator with small PI gains is an appropriate option for the design of the temperature controller for PEMFC. Based on this temperature controller, the temperature of PEMFC is controlled for limiting the temperature less than 80°C. The analytical results in this chapter provide useful guidelines for the design of a temperature controller for PEM fuel cell systems.

References

1. J. Larminie and A. Dicks, *Fuel Cell Systems Explained*, Wiley, New York, 2002.
2. J.T. Pukrushpan, A.G. Stefanopoulou, and H. Peng, Control of fuel cell breathing, *IEEE Control System Magazines*, 24(2), 30–46, 2004.
3. J. Purkrushpan and H. Peng, *Control of Fuel Cell Power Systems: Principles, Modeling, Analysis and Feedback Design*, Springer, Germany, 2004.
4. W. Na and B. Gou, Feedback linearization based nonlinear control for PEM fuel cells, *IEEE Transactions on Energy Conversion*, 23(1), 179–190, 2008.

5. J. Purkrushpan, A.G. Stefanopoulou, and H. Peng, Modeling and control for PEM fuel cell stack system, *Proceedings of the American Control Conference*, Anchorage, Alaska, pp. 3117–3122, 2002.
6. J.C. Amphlett, R.M. Baumert, R.F. Mann, B.A. Peppy, and P.R. Roberge, Performance modeling of the Ballard Mark IV solid polymer electrolyte fuel cell, *Journal of Electrochemical Society*, 142(1), 9–15, 1995.
7. R.F. Mann, J.C. Amphlett, M.A. Hooper, H.M. Jesen, B.A. Peppy, and P.R. Roberge, Development and application of a generalized steady-state electrochemical model for a PEM fuel cell, *Journal of Power Sources*, 86, 173–180, 2000.
8. M.J. Khan and M.T. Labal, Dynamic modeling and simulation of a fuel cell generator, *Fuel Cells*, 1, 97–104, 2005.
9. M.J. Khan and M.T. Labal, Modeling and analysis of electro chemical, thermal, and reactant flow dynamics for a PEM fuel cell system, *Fuel Cells*, 4, 463–475, 2005.
10. L. Riascos and D.D. Pereira, Optimal temperature control in PEM fuel cells, *Industrial Electronics, IECON '09, 35th Annual Conference of IEEE*, Porto, Portugal, 2009.
11. J.C. Amphlett, R.M. Baumert, R.F. Mann, B.A. Peppy, P.R. Roberge, and A. Rodrigues, Parametric modeling of the performance of a 5-kW proton exchange membrane fuel cell stack, *Journal of Power Sources*, 49, 349–356, 1994.
12. P. Famouri and R.S. Gemmen, Electrochemical circuit model of a PEM fuel cell, *IEEE Power Engineering Society General Meeting*, 3, 13–17, 2003.
13. L.Y. Chiu, B. Diong, and R.S. Gemmen, An improved small-signal mode of the dynamic behavior of PEM fuel cells, *IEEE Transactions on Industry Applications*, 40(4), 970–977, 2004.
14. F. Grasser and A. Rufer, A fully analytical PEM fuel cell system model for control applications, *IEEE Transactions on Industry Applications*, 43(6), 586–595, 2006.
15. M.Y. El-Sharkh, A. Rahman, M.S. Alamm, A.A. Sakla, P.C. Byrne, and T. Thomas, Analysis of active and reactive power control of a stand-alone PEM fuel cell power plant, *IEEE Transactions on Power Delivery*, 19(4), 2022–2028, 2004.
16. P. Almeida and M. Godoy, Neural optimal control of PEM fuel cells with parametric CMAC network, *IEEE Transactions on Industry Applications*, 41(1), 237–245, 2005.
17. J. Sun and V. Kolmannovsky, Load governor for fuel cell oxygen starvation protection: A robust nonlinear reference governor approach, *IEEE Transactions on Control Systems Technology*, 3(6), 911–913, 2005.
18. J. Hamelin, K. Abbossou, A. Laperriere, F. Laurencelle, and T.K. Bose, Dynamic behavior of a PEM fuel cell stack for stationary application, *International Journal of Hydrogen Energy*, 26, 625–629, 2001.
19. M. Tanrioven and M.S. Alam, Modeling, control, and power quality evaluation of a PEM fuel cell-based power supply system for residential use, *IEEE Transactions on Industry Applications*, 42(6), 1582–1589, 2006.
20. C. Wang, M.H. Nehrir, and H. Gao, Control of PEM fuel cell distributed generation systems, *IEEE Transactions on Energy Conversion*, 21(2), 586–595, 2006.
21. Agbossou, Y. Dube, N. Hassanaly, K.P. Adzakpa, and J. Ramousse, Experimental validation of a state model for PEMFC auxiliary control, *IEEE Transactions Industry Applications*, 45(6), 2098–2103, 2009.
22. A. Sakhare, A. Davari, and A. Feliachi, Fuzzy logic control of fuel cell for standalone and grid connection, *Journal of Power Sources*, 135(1–2), 165–176, 2004.
23. R.W. Erickson and D. Maksimovic, *Fundamentals of Power Electronics*, Kluwer Academic Publishers, Boulder 2000.

12

Implementation of Digital Signal Processor-Based Power Electronics Control[*]

12.1 Introduction

Today, digital signal processors (DSPs) are widely used in many areas, such as aerospace, communication, automatic vehicles, power systems, etc. DSPs have superior abilities in terms of automatic control and signal process. One of the advantages is that DSPs can handle complex control algorithms. In this chapter, the description for controlling DSPs in power electronics applications are based on the case of 32-bit TMS320X28xxx series DSPs (Texas Instruments) which provide advanced capabilities, processing a 32 × 32 bits multiplication or two 16 × 16 bits multiplication within one clock period. With the help of the short interrupt response time, 28xxx DSPs can protect key registers in time and accurately respond to asynchronous events.

Generally, in order to implement DSP-controlled systems, designers need to carefully select proper DSPs by considering their analog-to-digital converter (ADC) sampling rate, system clock frequency, resolution of high-frequency PWM signals, etc. In addition, the physical characteristics of the desired systems as well as the software development environment, its interface between the signal process systems, and power process systems must be taken in account. This chapter intends to provide a practical guideline for utilizing the C2000 series DSPs in power electronics applications. The implementation of DSP-controlled power electronics for Sliding Mode Control is briefly mentioned in this chapter.

[*] This chapter was mainly prepared by Dr. Woonki Na, California State University, Fresno, USA. Many thanks to former graduate student, Pengyuan Chen, California State University, Fresno, USA for setting up the DSP software, and preparing a DSP-controlled power electronics experimental test-bed for this chapter.

12.2 Overview of DSP

DSPs are the processors that can promptly process digital signals converted from analog signals. Based on an enhanced Harvard architecture and combined multiple types of digital arithmetic and signal process hardware, DSPs are extremely faster than traditional microprocessor units (MPUs) in terms of speed of computation. Currently, DSPs are the fundamental elements contributing to areas of communication, computation, automatic control, commercial electric product, and so on.

12.2.1 Development of DSP

The development of DSP can be generally categorized into three stages: invention of DSP concepts in the 1970s, popularization in the 1980s, and prompt improvement in the 1990s. Before DSPs were invented, the processing of digital signals had been only relying on the traditional microprocessor unit (MPU), but the low-processing speed of MPUs could not meet the requirement of high-speed real-time control. Until the 1970s, the early stage of fundamental DSP theories and algorithms had been developed. At that time, DSP technologies were premature because the concepts of DSPs were only introduced in textbooks. Even though some DSP systems were created at that time, those DSP systems consisting of separated parts were used only for military and aerospace purposes.

As the technologies of large-scale integrated circuits were heavily developed, the first DSP in the world was created in 1982. Such DSPs were manufactured using N-channel MOSFET transistor (NMOS) technologies. Although the power consumption of the first-generation DSPs was relatively high, the processing speed of the DSPs in 1982 was 10 times faster than that of the traditional microprocessor unit. Therefore, the first-generation DSPs were widely used for voice synthesizers, coders, and decoders in sound applications requiring a high speed of processing speed of the programs.

In the 1990s, as DSP technology improved, the fourth and fifth generation DSPs were popularized in commercial markets. The DSPs that are widely being used today can be categorized as fifth generation DSPs. For the fifth generation DSP product, the DSP core and peripheral parts are heavily integrated on a single chip.

12.2.2 Characteristics of DSP

Although DSPs are utilized in various areas for different objectives, most of the DSP's inner architectures are very similar. A DSP normally consists of one processor core, instruction buffers, data memory, ROM, I/O (input/output) interfaces, program address bus, program data bus, address bus, data bus, etc. The processor cores have the following characteristics:

- The architecture of DSPs is an enhanced Harvard architecture [2]. In the architecture, there are two buses, data bus and program bus. The data section and program section are separated. Both of them have their respective address bus and data bus so that the operation of obtaining instructions and operation of reading data can be handled simultaneously.
- Execution of each instruction is divided into several consecutive parts: reading instruction, decoding, reading data, execution, etc. The whole flow operations are concurrently handled by multiple function units.
- In each instruction period, a DSP handles one or multiple multiply-accumulate (MAC) operations.
- In a DSP, there are multiple address generating units supporting circular addressing and bit-reversed instructions so that in processing of FFT or convolution, the speed of addressing, sorting, and computing are considerably increased. For example, the typical computation time of 1024-bit FFT is less than 1 μs.
- A DSP consists of one or more independent DMA control logics. As a result, the bandwidth of data input/output can be improved.
- DSPs support iterative computations for avoiding consuming unnecessary time caused by circular operations.
- DSPs provide multiple serial/parallel I/O interfaces for achieving specified data process and control.

12.2.3 Selection of DSPs

DSPs can be used to implement various systems for the functioning of various areas, while they face their limitations in terms of clocking frequency, capabilities of noise-isolation, power consumption, etc. Hence, before practically constructing a designed system, the physical characteristics of the system, the cost, and the development period must be taken into account. The general considerations for selecting a desired DSP are as follows:

- *Objectives:* Every DSP is specialized in high performance in one or several specified areas, so the selection of a DSP can be determined by the objectives of the system to be designed. As examples of DSPs manufactured by Texas Instruments, C2000 series DSPs provide multiple peripheral parts required by control systems. Hence, the C2000 series can be used for control systems and power electronics systems. The C5000 series have advantages regarding high computation speed, low power consumption, and low cost. In consequence, C5000s are normally used for mobile devices and consumer electronics. Compared to the C2000 and C5000 series, the C6000 series have an extreme fast computation speed and high data precision, thus C6000s are suitable for contributing areas of communication and image processing. In other word, based on the objective of the system to be designed, designers may choose appropriate DSPs for their applications.

- *Algorithm format:* The digital signal process involves multiple algorithms. Various systems and algorithms require different algorithm format and real-time processing accuracy. Float-point arithmetic is a kind of complex arithmetic and adopts float-point data for establishing wide ranges of dynamic data. In fact, DSPs with float-point arithmetic systems do not involve concerns of handling dynamic range and dynamic precision. Such DSPs are more suitable for implementing a high-level programming language. Compared to fixed-point DSPs, float-point DSPs are very user-friendly in terms of high-performance software programming, although the cost and power consuming of float-point DSPs are higher than those of fixed-point DSPs.
- *System accuracy:* The demand for system accuracy directly determines the algorithm format to be adopted as well as the data width. Now, developers can use low data width processors to achieve high-data width operations, for example, users can use 16-bit DSP processor for executing 64-bit data exchanges that may need a lot of efforts in case of using an MCU.
- *Processing speed:* Processing speed should be considered as the first priority of DSPs. The processing speed is determined by the instruction period, and reflects the operation period of various core functions such as finite impulse response and infinite impulse response. Some DSPs contain very long instruction word (VLIW) architecture so that they can implement multiple instructions in a single instruction period. Since the processing speed is closely related to the system clock, the higher system clock normally can accelerate processing speed.
- *Power consumption:* DSPs are widely used in mobile devices such as cell phones and personal assistants (PDAs). Power consumption is always a significant concern for designers.
- *Multiprocessor support:* In certain applications, interconnected multiple processors are required for large volume data exchanging. Compatibility of DSPs will become an important concern in terms of system internal communication and time delays.

12.3 Texas Instruments DSPs

Texas Instruments manufactured the first-generation DSP products that can make a modem implement 5,000,000 instructions in 1 s, which was

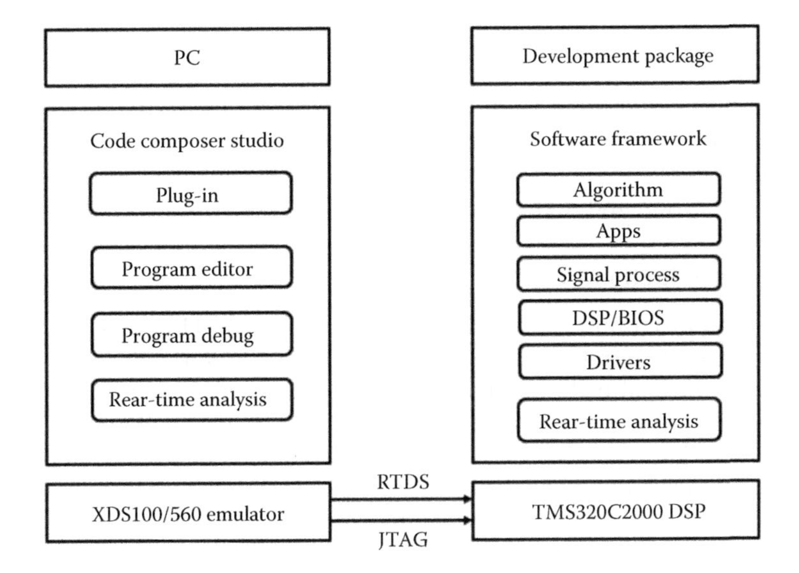

FIGURE 12.1
C2000 DSP development tools and software.

the great breakthrough for real-time signal processing. From TMS32010 to TMS320C2000/5000/6000 products, the latest TI DSPs have advantaged architectures and faster speed compared to old generation DSPs. The environment of the development of DSPs has become very comprehensive. For helping designers quickly construct their desired DSP systems, Texas Instruments provides multiple development tools including coding, debugging, and code optimization adjustment. Texas Instruments development tool, *Code Composer Studio IDE* is provided for accelerating the progressing of DSP development. Figure 12.1 presents the development tools and software environment of C2000 DSP using Code Composer Studio (CCS).

DSP development generally involves four steps: Apps design, code editing and compiling, testing, and debugging and analysis.

- *Apps design:* In this stage, the designer needs to consider the structure of the program (application level). Also, the designer should decide how many ADC channels, interrupts, timers, GPIOs, and PWM generators should be used for the implementation as well as the proper flow chart of the control algorithm.
- *Code editing and compiling:* At this stage, the designer should select proper software framework and code editor.
- *Testing and debugging:* At this stage, the designer may make a lot of effort to test the functionalities of the code in terms of grammar errors and logic errors. Multiple common problems may be seen in this stage,

for example, undeclared variables and functions, using local variables as external variables, inappropriate value assignment, etc.

- *Debugging and analysis:* In addition, different code and data structure can lead to various results, therefore the designer should consider the operating time for each subroutine, especially in cases of communication and high-frequency signal control.

12.3.1 Introduction to CCS

The official programing environment for C2000 series TI DSP is the CCS. In this chapter, the introduction of the CCS v3.3 is described. If the DSP application engineer is already familiar with programing environments (based on PC operation systems) such as Microsoft Visual Studio or Borland C++, the designer will find that the CCS programming environment is similar to other integrated development environment (IDE). The difference between the DSP development and general PC-based apps development is that the PC-based apps developments do not require that the users link the programming environments to the hardware systems because most PC-based apps only operate in the OS level such as Microsoft Windows. In the C2000 DSP real-time development programming environment, users need to use CCS to generate instructions and executive files in PC operation systems and then users should download the executive files to DSP real-time hardware for debugging purposes. For some simple projects, users may not necessarily download the executive files to the DSP simulator, which means that users can also run simple projects with simulators in the OS level. For each stages of the DSP development, the tools and their functions are summarized in Table 12.1.

12.3.2 First Step of CCS

After the user installs the CCS v3.3 in the Windows environment, two icons will be automatically created on the desktop. They are "Setup CCStudio v3.3"

TABLE 12.1

Development Tools and Related Functionalities

Application Design	Code and Build	Debug	Analyze and Tune
Select target	Code generating tools	Debugger	Real-time analysis
DSP/BIOS setting	Project management	Simulation	Description
Arithmetic standard	Editor	Processor	Visual data
Power management	Power supply management	Data exchange RTDX	Code manipulation
		High priority task trigger	High-speed buffer analysis
		Script editor	
		Connect/disconnect	

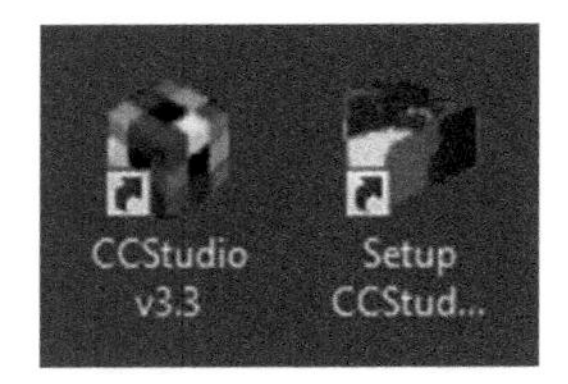

FIGURE 12.2
Icons of CCStudio v3.3 and Setup CCStudio v3.3.

and "CCStudio v3.3," respectively. These icons are shown in Figure 12.2. Setup CCStudio v3.3 app is used to configure the target DSP and hardware settings. CCStudio v3.3 is used to startup the CCS IDE development environment for coding and debugging purposes. To build a reliable connection between the programming environment and the target hardware throughout emulators, the corresponding hardware drivers must be installed in PC operation systems, for example, if users want to program using the TI F28335 DSP, the Flash driver and header files of F28335 must be installed in the Windows OS before the user begins to program into the F28335 DSP. Note that different driver files support various software/hardware systems. For detail, please refer to Texas Instruments official website: http://www.ti.com/tool/ccstudio-msp?keyMatch=ccs&tisearch=Search-EN-Everything.

12.3.3 Configuration of the Target System

This section will mainly introduce the configuration of the single processor and IDE. CCS setup app allows users to configure the software settings referring to their DSP boards. Before executing the CCS IDE, the system and emulator must identify the exact type of DSP board, and related information such as emulator and JTAG cables. The steps to configure the target system are as follows:

1. Click the icon of CCS setup app, the main window will be similar to Figure 12.3.
2. Before configuring the settings, designers can click *File* shown on the top menu, and select the option, *Remove all*, for clearing the previous settings (see Figure 12.4).
3. Search the column, *Available Factory Boards*, for finding the target board and emulator to be used in the implementation. For conveniently finding the target board, the user can use the filters on the right top menu. The approach is shown in Figure 12.5.
4. Select system hardware from the system configuration, and right click for checking the attributions (for details see Figure 12.6).
5. After checking the board configurations and properties, click the button, *Save & Quit*. This will lead the CCStudio Setup to be closed, and will automatically start the CCS IDE v3.3 (see Figures 12.7 and 12.8).

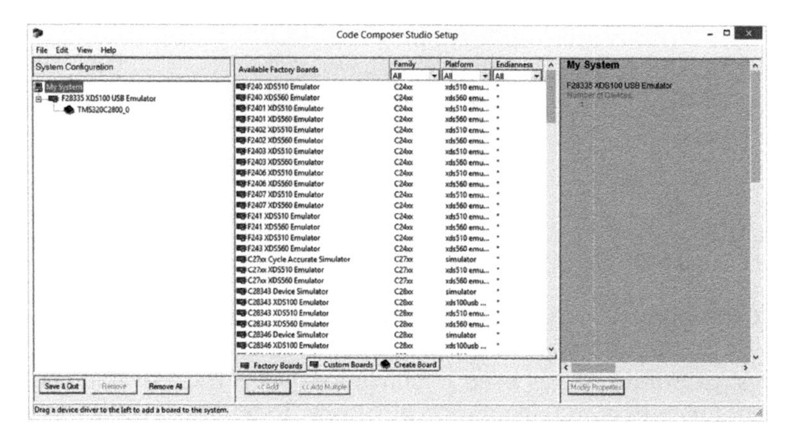

FIGURE 12.3
The main window of the CCStudio Setup v3.3.

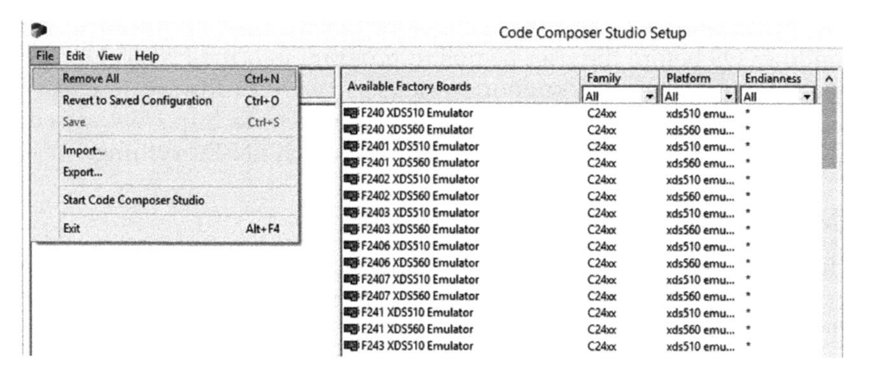

FIGURE 12.4
Clear the previous configuration.

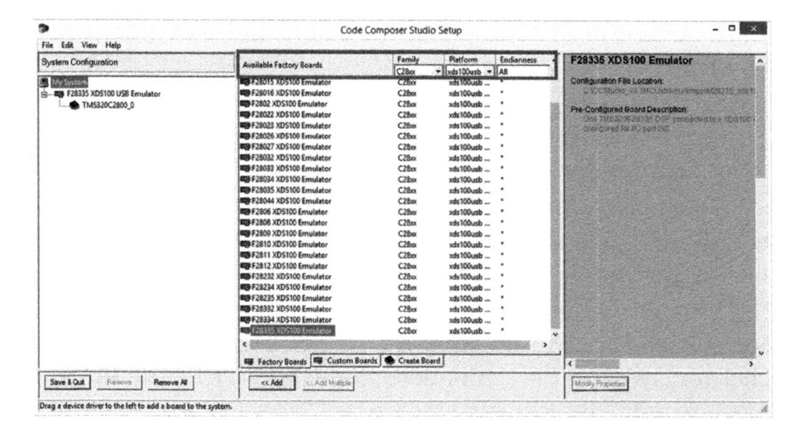

FIGURE 12.5
Use filters to find the target board.

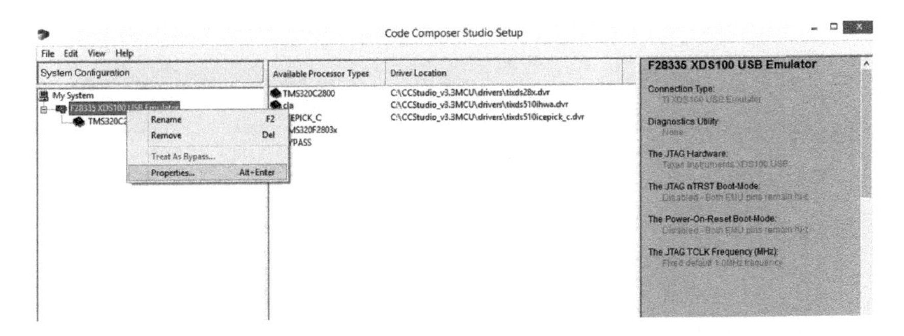

FIGURE 12.6
Board configurations and properties.

6. If the screen shown in Figure 12.8 appears on the PC screen after the CCStudio Setup v3.3 is automatically terminated, then the designer should be on the right track. The DSP board used for demonstration in this chapter is TI F28335, and the emulator is XDS100 USB. If the previous steps are properly followed, the main screen of the CCStudio IDE v3.3 should be similar to what is shown in Figure 12.9.
7. Note that the information highlighted by the rectangular box shown in Figure 12.9 is "DISCONNECTED (UNKNOWN)." This information means that the PC programming environment has not connected to the DSP hardware system. So, the user needs to click the option, *Debug*, on the top menu of the CCStudio IDE v3.3. And then select the option, *Connect*. This procedure is presented by Figure 12.10. If the programing environment is successfully connected to the DSP board, a result similar to Figure 12.11 can be seen.

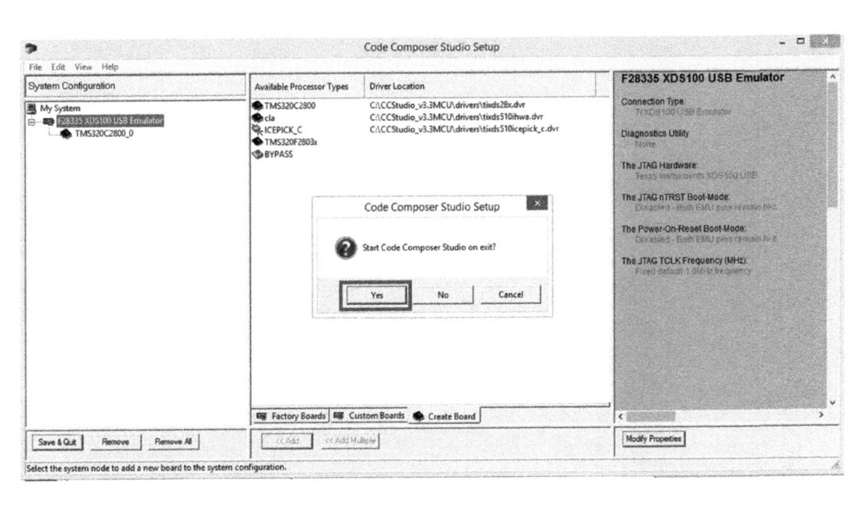

FIGURE 12.7
Click the Save & Quit button to automatically startup the CCStudio v3.3.

FIGURE 12.8
The startup screen of the CCStudio IDE v3.3.

8. Now, the user can download the executive files to the target DSP board and run the program. The names of executive files for TI DSP are normally with extension name, ".out." Assuming that the user already has a complied executive file and wants to implement the program throughout the DSP board, the first step is to download the executive file to the DSP, which is exhibited in Figures 12.12 and 12.13. The user can click the option, *File*, on the top menu and select the suboption, *Load Program*. To run the program, the user can click the blue button highlighted by the rectangular box shown in Figure 12.14.

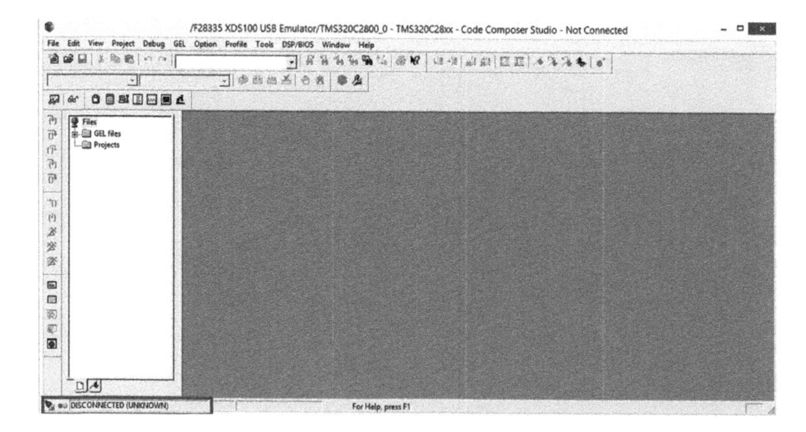

FIGURE 12.9
The initial main window of CCStudio IDE v3.3.

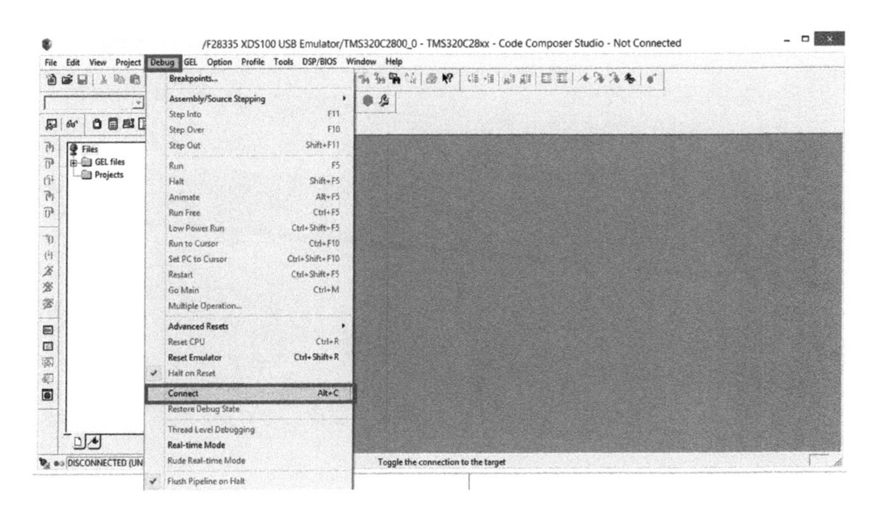

FIGURE 12.10
Connect to the DSP board.

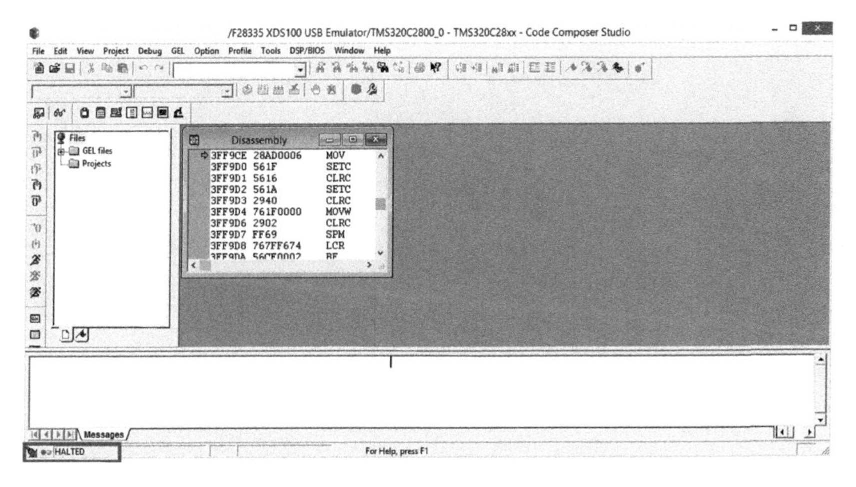

FIGURE 12.11
Demonstration for successful connection between CCStudio v3.3 to board.

In this section, the steps for configuring the programing environment and downloading the executive file to DSP boards are introduced. The following buttons that are generally used for coding and debugging are explained with simple statements:

Only compile the files that were changed

Rebuild all files

Set breakpoints

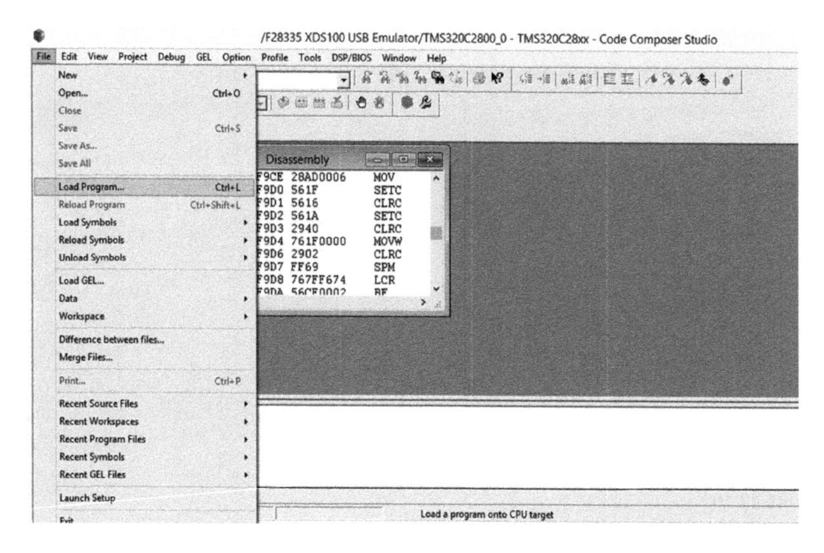

FIGURE 12.12
Download the program to DSP board.

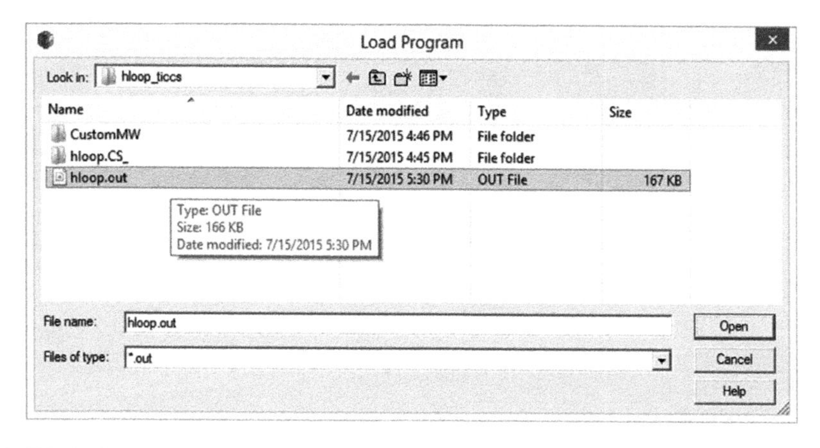

FIGURE 12.13
Select the executive files.

Stop program

Continuously run program

Discontinuously run program, the program pointer will stop at breakpoints

Single step, the observer pointer will go into the functions

Single step, the observer pointer will not go into the functions

- The register window
- The memory window

For viewing the data stored in data registers and control registers during the process of executing the program, the user can easily observe those register values by clicking the options listed in the top menu. The details are demonstrated in Figure 12.15.

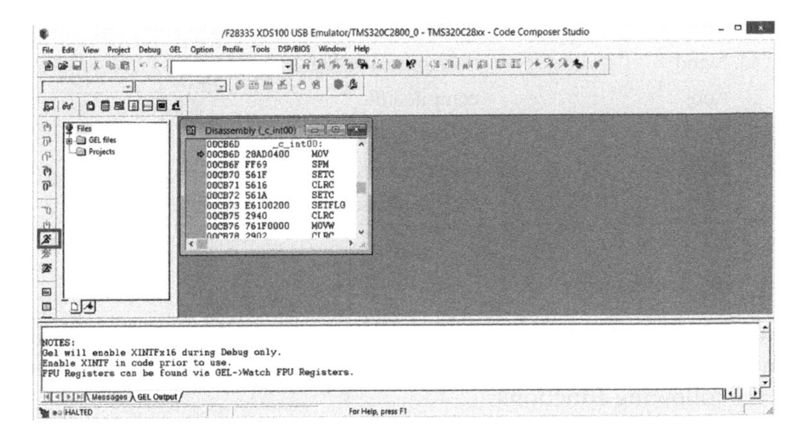

FIGURE 12.14
Run the program.

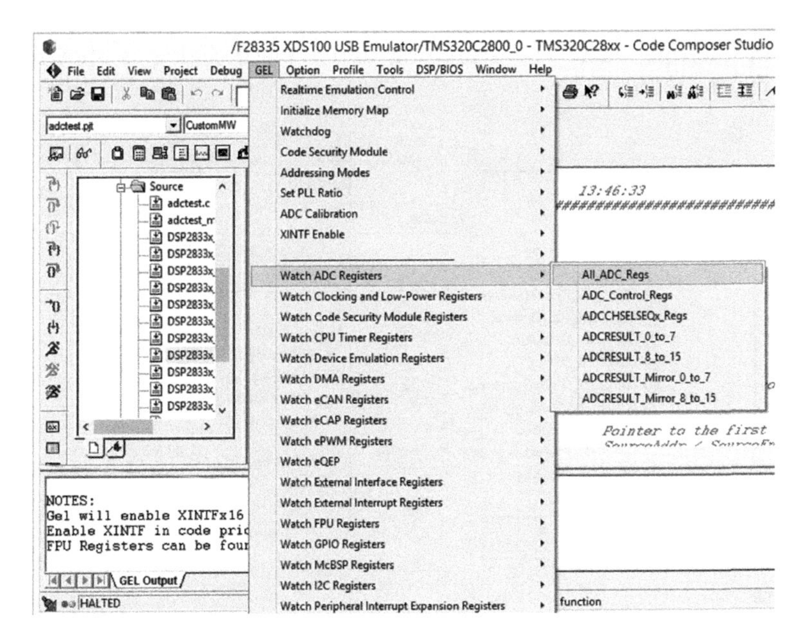

FIGURE 12.15
Menu for viewing data registers and control registers.

In CCStudio IDE v3.3, the designer will be familiar with the following types of files:

*.pjt	CCS project file
*.c	C program source file
.a or *.s*	assembly language file
*.h	C program header file
*.lib	library file
*.cmd	link command file
*.obj	compiled file
*.out	executive file for DSP boards
*.wks	programming environment setting file
*.cdb	CCS database file supporting DSP/BIOS API

12.3.4 Edit Source File

CCStudio IDE v3.3 can be used for editing "*.txt" files, C/C++ code, and assembly language code. In the CCS environment, the embedded editor supports the following functions:

1. *Distinct highlight display:* Key words, annotations, strings, and assembly language instructions are highlighted by various colors.
2. *Search and replace:* Users can search/replace strings in single file or multiple files. The searching and replace dialog is shown in Figure 12.16.

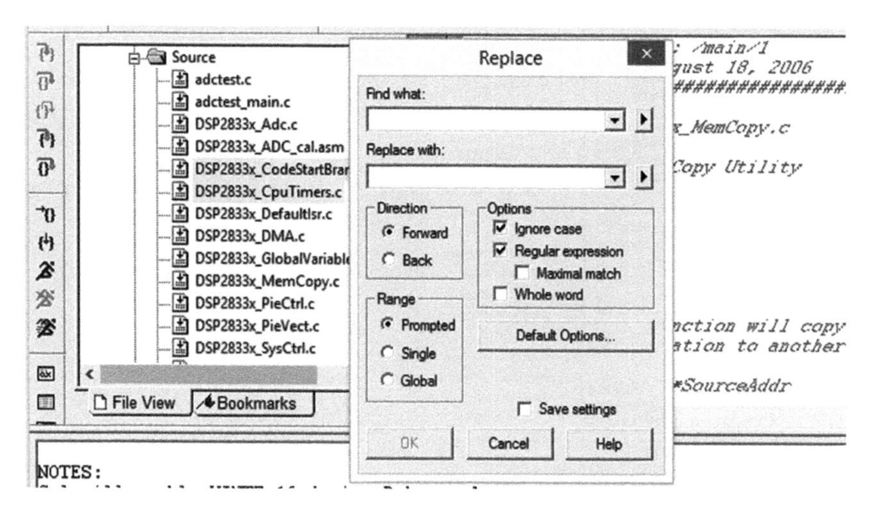

FIGURE 12.16
The dialog for finding/replace strings.

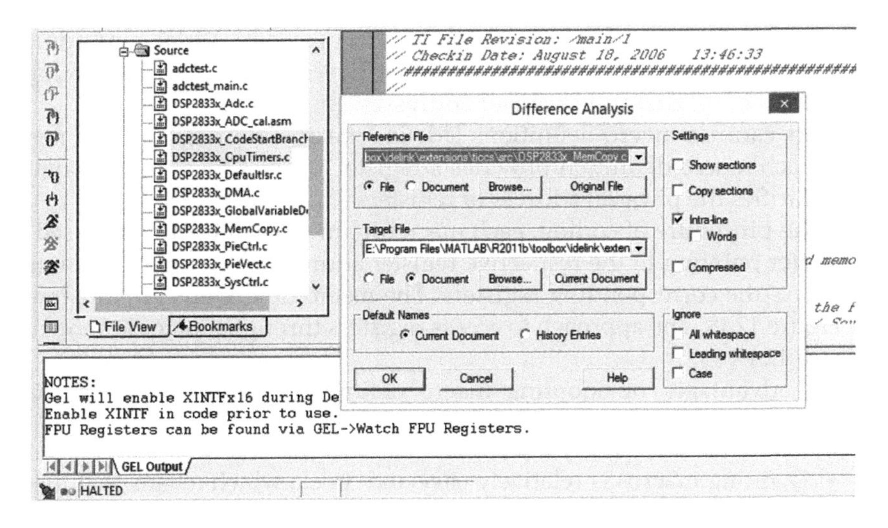

FIGURE 12.17
The dialog for comparing difference between files.

3. *Files comparison:* Users can check the differences between two files by using the function, *Difference between files,* in the option, *File.* The dialog is shown in Figure 12.17.
4. *Grammar error automatic check:* The embedded C program can automatically check grammatical errors while the user are coding.
5. *Hot key:* With using the hot keys, the user can shorten the development time.

12.3.5 TMS320X28335 C/C++ Programming

TMS320X28xx and TMS320X28xxx series DSPs are mainly used in embedded control systems. For simplifying coding procedures and improving the efficiency of C/C++ language implementation, Texas Instruments provides Hardware Abstract Layer (HAL) for simplifying the procedures for accessing peripheral registers. This HAL adopts register files structure and bit-defined formats for easily accessing registers and specified bits of registers. The implementation of this approach is simpler than the way of macro defined formats. In this section, the principles for programming C2000 DSP will be introduced.

12.3.5.1 Traditional Macro Definition

In view points of traditional C/C++ programing, users normally access hardware registers by using macros. Macro definitions are always seen in C/C++ header files, and every definition starts with "#define." For illustrating the approach-macro definition, an example of programing serial communication interface (SCI) is provided as follows. Table 12.2 [3] presents the functionalities of registers and register addresses.

Users can use macro definitions (#define macros) for defining SCI registers. Each macro definition provides a register address identifier or pointer and clarifies the physical address of registers.

In the procedure of coding, each macro definition can be directly used as a pointer pointing to the respective register address. Users can use pointers to access the corresponding registers. The macro definitions are illustrated in Figure 12.18. The approach to access registers through macros is shown in Figure 12.19.

The advantages of adopting macro definitions for accessing registers include:

- Defining macros is relatively easy and quick, which means that the coding is easy for users.
- Variable names can be directly denoted by using register names for programing purposes.

TABLE 12.2

SCI-A and SCI-B Register Files and Register Addresses

SCI-A Registers			SCI-B Registers		
Register	**Address**	**Function Description**	**Register**	**Address**	**Function Description**
SCICCRA	0x7050	Communication control register	SCICCRB	0x7750	Communication control register
SCICTL1A	0x7051	Control register1	SCICTL1B	0x7751	Control register1
SCIHBAUDA	0x7052	Baud rate (MSB) setting register	SCIHBAUDB	0x7752	Baud rate (MSB) setting register
SCILBAUDA	0x7053	Baud rate (LSB) setting register	SCILBAUDB	0x7753	Baud rate (LSB) setting register
SCICTL2A	0x7054	Control register2	SCICTL2B	0x7754	Control register2
SCIRXSTA	0x7055	RX status register	SCIRXSTB	0x7755	RX status register
SCIRXEMUA	0x7056	RX emulator data buffer	SCIRXEMUB	0x7756	RX emulator data buffer
SCIRXBUFA	0x7057	RX data buffer	SCIRXBUFB	0x7757	RX data buffer
SCITXBUFA	0x7059	TX data buffer	SCITXBUFB	0x7759	TX data buffer
SCIFFTXA	0x705A	FIFO RX register	SCIFFTXB	0x775A	FIFO RX register
SCIFFRXA	0x705B	FIFO TX register	SCIFFRXB	0x775B	FIFO TX register
SCIFFCTA	0x705C	FIFO control register	SCIFFCTB	0x775C	FIFO control register
SCIPR1A	0x705F	Priority control	SCIPR1B	0x775F	Priority control

Source: Adapted from TMS320x2833x, 2823x Serial Peripheral Interface (SPI) Reference Guide (SPRUEU3A).

```
/*****************************************************************
 This incompleted header file is only used for demostration
*****************************************************************/

#define Uint16 unsigned int
#define Uint32 unsigned long

#define SCICCRA (volatile Uint16 *)0x7050
#define SCICTL1A (volatile Uint16 *)0x7051
#define SCIHBAUDA (volatile Uint16 *)0x7052
#define SCILBAUDA (volatile Uint16 *)0x7053
#define SCICTL2A (volatile Uint16 *)0x7054
#define SCIRXSTA (volatile Uint16 *)0x7055
#define SCIRXEMUA (volatile Uint16 *)0x7056
#define SCIRXBUFA (volatile Uint16 *)0x7057
#define SCITXBUFA (volatile Uint16 *)0x7059
#define SCIFFTXA (volatile Uint16 *)0x705A
#define SCIFFRXA (volatile Uint16 *)0x705B
#define SCIFFCTA (volatile Uint16 *)0x705C
#define SCIPR1A (volatile Uint16 *)0x705F
```

FIGURE 12.18
Macro definition table.

```
/*****************************
 access register through macros
*****************************/

SCICTL1A= 0x0003; // write control register
SCICTL1B|=0x0001; // enable RX
```

FIGURE 12.19
The approach for using macros.

The disadvantages of tradition macro definitions are

- *Inconvenient operation:* In case of traditional macros operations, users always need to read/write the whole register for obtaining/sending bit information.
- Bit information can be totally revealed in the CCS environment.
- Incomplete implementation of CCS development environment.
- Inconvenient cooperation with peripheral devices.

12.3.5.2 Bit Definition and Register File

Compared to the traditional macros definitions, bit definition and register file structure can improve the efficiency of programing and be easy to use.

1. Register file structure
   Register file structure is the approach to define certain bit information which will be accessed by external devices in files. Registers are

categorized into different groups in C/C++ files. This is the so-called register file structure. Each register file structure can directly map external registers to their corresponding memory locations during compiling, and such mapping allows the compiler to use CPU Data Page Pointer (DP) to access registers.

2. Bit definition
   Bit definition is to allocate function bits of registers with corresponding names and data width. Hence, users can use bit-defined name to directly access the target bits of registers.

The operations of register file structures and bit definition are explained by the following SCI example. The operations involved in the following examples are

- Create new data structure for purposes of utilizing SCI registers
- Map the register file structure variable to the first register address
- Bit-define the SCI register
- Create a community for accessing certain bits of register or accessing whole registers
- Re-edit register file structure for register to add bit definitions and community definition

12.3.5.2.1 Define Register File Structures

As the macros definition is described above, users can access the registers of external devices throughout register file structure. Table 12.3 [3] shows the address-offset of the SCI registers.

TABLE 12.3

SCI-A and SCI-B Register Files

Register	Bit Length	Address-Offset	Function
SCICCR	16	0	Control register
SCICTL1	16	1	Control register1
SCIHBAUD	16	2	Baud-rate control register (MSBs)
SCILBAUD	16	3	Baud-rate control register (LSBs)
SCICTL2	16	4	Control register 2
SCIRXST	16	5	RX status register
SCIRXEMU	16	6	RX emulator data buffer register
SCIRXBUF	16	7	RX data buffer register
SCITXBUF	16	9	TX data buffer register
SCIFFTX	16	10	TX FIFO register
SCIFFRX	16	11	RX FIFO register
SCIFFCT	16	12	FIFO control register
SCIPR1	16	15	Priority control register

Source: Adapted from TMS320x2833x, 2823x Serial Peripheral Interface (SPI) Reference Guide (SPRUEU3A).

```
/**********************************************************
 * SCI Header Filer
 * Creat Register file strcutrue for SCI registers
 * Note that the following code are only used for demostrateion
 **********************************************************/

#define Uint16 unsigned int
#define Uint32 unsigned long

struct SCI_REGS{
Uint16 SCICCR_REG SCICCR;        //genral communication control reg
Uint16 SCICTL1_REG SCICTL1;      //control reg1
Uint16 SCIHBAUD;                 //Baud-rate control reg, high-bit
Uint16 SCILBAUD;                 //Baud-rate control reg, low-bit
Uint16 SCICTL2_REG SCICTL2;      //control reg2
Uint16 SCIRXST_REG SCIRXST;      //RX status reg
Uint16 SCIRXEMU;                 //RX Emulator data buffer reg
Uint16 SCIRXBUF_REG SCIRXBUF;    //RX data buffer reg
Uint16 rsvd1;                    //reserved space
Uint16 SCITXBUF;                 //TX data buffer reg
Uint16 SCIFFTX_REG SCIFFTX;      //TX FIFO reg
Uint16 SCIFFRX_REG SCIFFRX;      //RX FIFO reg
Uint16 SCIFFCT_REG SCIFFCT;      //FIFO control reg
Uint16 rsvd2;                    //reserved space
Uint16 rsvd3;                    //reserved space
Uint16 SCIPRI_REG SCIPRI;        //FIFO priority  control reg
}
```

FIGURE 12.20
Create new structure for register files.

Figure 12.20 presents that one can adopt C/C++ structure-define for grouping SCI registers, the identifier of a register with a lower address is normally allocated at the position ahead to that of a register with a higher address. Given that some address spaces should be reversed for the CPU so that one can use specified variable names for defining those spaces such as rsvd1, rsvd2, rsvd3, etc. The bit-widths occupied by registers can be defined by different data type, for example, Uint16 (16-bit) and Uint32 (32-bit).

In Figure 12.20, a new structure, SCI_REGS is created without defining variables. In the following example shown in Figure 12.21, the approach to define structure variable is provided.

```
/*********************************
Create variables for SCI registers
*********************************/


volatile struct SCI_REGS SCIaRegs; // Create variable for SCI-A regs
volatile struct SCI_REGS SCIbRegs; // Create variable for SCI-B regs
```

FIGURE 12.21
Creating variables for SCI registers.

12.3.5.2.2 Memory Allocation for Register File Structure

The complier generates new data modules and code modules which can be repositioned, and those modules are called segments. Those segments are allocated to address spaces specified by system configurations. The detailed method for allocating each segment is defined in the *.cmd file. In reserved modes, the compiler will allocate global variables and local variables to the *.ebss or *.bss segment, such as SCIaRegs and SCIbRegs. If the HAL is adopted, register file variables will be allocated to spaces named with extension names (.ebss or .bss) throughout the statements, for example, "# pragma DATA_SECTION." In C program, the coding method for realizing the above syntax (# pragma DATA_SECTION) is that:

```
# pragma DATA_SECTION(symbol, "section name");
```

In C++ program, the according coding is that:

```
# pragma DATA_SECTION("section name");
```

Figure 12.22 provides an example where users can use syntax "# pragma DATA_SECTION" to allocate variables SCIaRegs and SCIbRegs to data segments respectively named by "SciaRegsFile" and "ScibRegsFile." Then users can directly map those two data segments into the memory spaces occupied by SCI registers.

Throughout the methodologies shown in Figure 12.22, link command files will map data files to corresponding memory spaces. Table 12.2 [3] shows that the start address of SCI-A registers is 0x7050. With the allocated data segment, the variable, "SCIaRegs" will be linked to memory space, the starting address of which is 0x7050. The according link command file is given by Figure 12.23.

Users can simply change the structure names for mapping the register file structure to external register addresses, and directly use those variables to access registers in the C/C++ environment. For example, to write data to SCICCR just adopt the following steps:

SciaRegs.SCICCR = SCICCRA_MASK;

SciaRegs.SCICCR = SCICCRB_MASK;

12.3.5.2.3 Bit Definition

When users attempt to access certain bits of a register of an external device, the approach of bit definition provides a direct way to read/write bit information, which means that users do not necessarily read/write the whole register. In the case of macro definition, if users want to check/change certain bits of a register, the users will have to read/write the whole register. To define bit-space in C/C++ structures, users can type the name of the defined bit spaces and add a colon ":" at the end of the name. Then, users can give a

```
/*****************************************************
* use # pragma allocates variable to data segement

*C and C++ program adopts different types with respect to
# pragma

*During compling a C++ file, complier will automatically
define __cplusplus

*****************************************************/

#ifdef __cplusplus
#pragma DATA_SECTION("SciaRegsFile")
#else
#pragma DATA_SECTION(SCIaRegs,"SciaRegsFile");
#endif

volatile struct SCI_REGS SCIaRegs;

//-----------------------------------------------------

#ifdef __cplusplus
#pragma DATA_SECTION("ScibRegsFile")
#else
#pragma DATA_SECTION(SCIbRegs,"ScibRegsFile");
#endif

volatile struct SCI_REGS SCIbRegs;
```

FIGURE 12.22
Allocate variables to their according data segments.

```
/*****************************************************
* memory lineker.cmd

*link the SCI register file structure to memory space
*****************************************************/

MEMORY
{
PAGE 1;
SCIA : origin = 0x007050, length = 0x000010 /* SCI-A registers*/
SCIB : origin = 0x007750, length = 0x000010 /* SCI-B registers*/
}
SECTIONS
{
SciaRegsFile:>SCIA, PAGE=1
ScibRegsFile:>SCIB, PAGE=1
}
```

FIGURE 12.23
Memory space mapping.

```
/*****************************************************************
* SCI header file
*****************************************************************/
// SCICCR general communication control Reg definition
struct SCICCR_BITS{          // bit-function
Uint16 SCICHAR:3;            //2~0  byte length control
Uint16 ADDRIDLE_MODE:1;      //3    ADDR/IDLE mode control
Uint16 LOOPBKENA:1;          //4    loop self-check mode selection
Uint16 PARITYENA:1;          //5    polar enable control
Uint16 PARITY:1;             //6    odd/even selection
Uint16 STOPBITS:1;           //7    stop-bit length
Uint16 rsvd1:8;              //15~8 reserved
}
```

FIGURE 12.24
Bit definition.

number behind the colon ":" for denoting how much bit spaces should be occupied by the defined bit-space.

The operations can be illustrated by Figure 12.24.

12.3.5.3 Community

Bit definition allows user to directly program function bits, while users have to operate whole register for some reasons. Therefore, bits of a register can be treated as a whole value and be further operated throughout the approach named "community." The illustration is provided by Figure 12.25.

As long as the union is defined in C/C++, SCI register file structure and bit definition can be both used for coding. The following syntax can provide ideas about union operations:

```
/*****************************************************************
* SCI header file
*****************************************************************/
union SCICCR_REG{
                        Uint16 all;
                        struct SCICCR_BITS bit;
};
union SCICTL1_REG{
                        Uint16 all;
                        struct SCICTL_BITS bit;
};
```

FIGURE 12.25
Union definition.

```
SCICCR_REG.all = # Value;
```

This can be used for rewriting value of the whole register.

```
Variable Name = SCICCR_REG.all;
```

This can be used for obtaining value of the whole register.

```
SCICCR_REG.bit.STOPBITS = # Value;
```

This can be used for rewrite the stop-bit length of SCICCR where users can directly access the function bit using such syntax.

```
Variable Name = SCICCR_REG.bit.STOPBITS;
```

This can be used for obtaining value of the stop-bit length of SCICCR where users can directly access the function bit using such syntax.

Note that, for other register read/write operations, some function bits are only served as read-only, or write-only so that the practical utilization of union should be referred to specifications.

In this section, the fundamental operations for register read/write and coding in CCStudio v3.3 environment have been introduced. Programing DSP is a procedure to design a proper logic flow for fulfilling the users' objectives. The basic operations for DSP control is to read/write data registers and control registers to realize functionalities of each module embedded into the DSP board. Table 12.4 provides multiple technical documents supporting TI F28335 DSP development. Readers can fetch enough information from those documents, for programming ADCs, ePWMs, eCAPs, interrupt, etc. In the rest of sections of this chapter, the approaches for using ADCs, ePWMs, and Interrupt will be mainly explained.

12.4 Interrupt and Related Operations

For achieving periodical control and case-dependent control, there are two traditional approaches, polling and interruption. In the case of real-time control, the polling approach may consume plenty of time resources for checking the status flags of target events and result in inefficient program structures. Therefore, interrupts are widely used for real-time control projects.

C28x DSP core have 16 interrupt buses consisting of two nonmaskable interrupt (RESET and NMI) and other 14 maskable interrupts. Maskable interrupts can be configured throughout interrupt controller for enabling/disenabling. The overview of interrupt buses is shown in Figure 12.26.

TABLE 12.4
Technical Support Documents

Document Title	Literature No.
C2000 Guide to Reference Manuals	SPRU566L
Serial Peripheral Interface (SPI)	SPRUEU3A
System Control and Interrupts	SPRUFB0D
Enhanced Quadrature Encoder	SPRUG05As
Serial Communication Interface (SCI)	SPRUFZ5A
Multi-Channel Buffered Serial Port (McBSP)	SPRUFB7B
Enhanced Controller Area Network (CAN)	SPRUEU1
TMS320C28x Floating Point Unit and Instruction Set Reference Guide	SPRUEO2Bs
TMS320C28x DSP CPU and Instruction Set Reference Guide	SPRU430F
Direct Memory Access Controller (DMA)	SPRUFB8D
Enhanced Pulse-Width Modulator (PWM)	SPRUG04A
High Resolution Pulse-Width Modulator (HRPWM)	SPRUG02B
Analog-to-Digital Converter	SPRU812A
Inter-Integrated Circuit (I2c)	SPRUG03B
Enhanced Capture	SPRUFG4A
External Interface (XINTF)	SPRU949D
Boot ROM	SPRU963A
DSP Peripherals Reference	SPRU566L
C28x Assembly Language Tools	SPRU513H
C28x C/C++ Compiler	SPRU514H
TMS320C28x DSP/BIOS 5.x Application Programming Interface (API) Reference Guide	SPRU625L

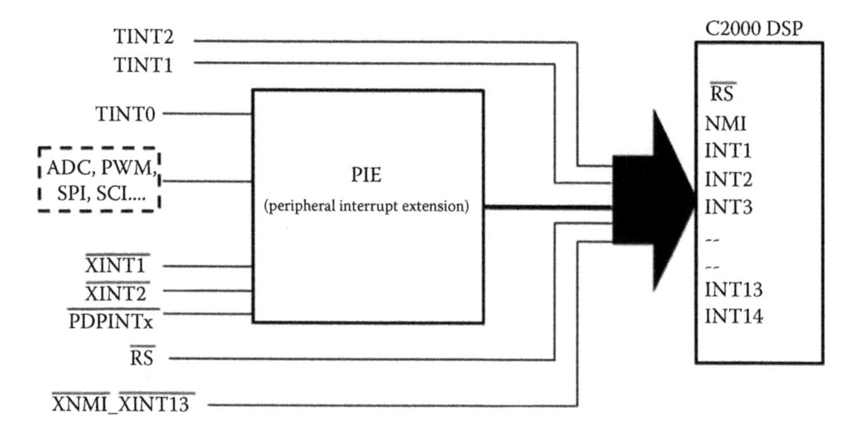

FIGURE 12.26
Overview of interrupt source and links.

C2000 series DSPs have timer1 and timer2 reserved for real-time system control, and according interrupts are allocated to INT14 and INT13. Two nonmaskable interrupts RESET and NMI possess their specified interrupts. Other 12 maskable interrupts are directly linked to the peripheral INT module for supporting external interruption and CPU inner peripheral units.

12.4.1 Peripheral Interrupt Extension

C2000 series processors integrate multiple peripherals. Each peripheral device generates one or multiple interruptions. Given that CPU does not have capabilities to handle all CPU level interruptions so that C2000 series CPU contains a peripheral interrupt extension (PIE) for arbitrating peripheral interruptions. Interrupt arbitration can know the positions of interrupt service routines (ISRs) referring to the addresses of ISRs where the ISR addresses are stored in the PIE vector table. The structure of PIE is shown in Figure 12.27 [4].

All interrupts are connected to according interrupt buses throughout PIE units. The detailed interrupt connections are illustrated in Table 12.5 [4].

12.4.2 Peripheral Interrupts

When the peripheral device generates interrupt, the corresponding of the interrupt flag (IF) will be set to 1. If the according bit of interrupt enable (IE) is also set to enable, then the generated interrupt will interrupt request to PIE controller. If the peripheral interrupt is disenabled, the value of the IF will not be changed unless software clears the flag. If the interrupt is enabled, and if the IF has not been cleared, the generated interrupt will keep sending requests to PIE. Note that IFs of peripheral registers must be clear throughout the software.

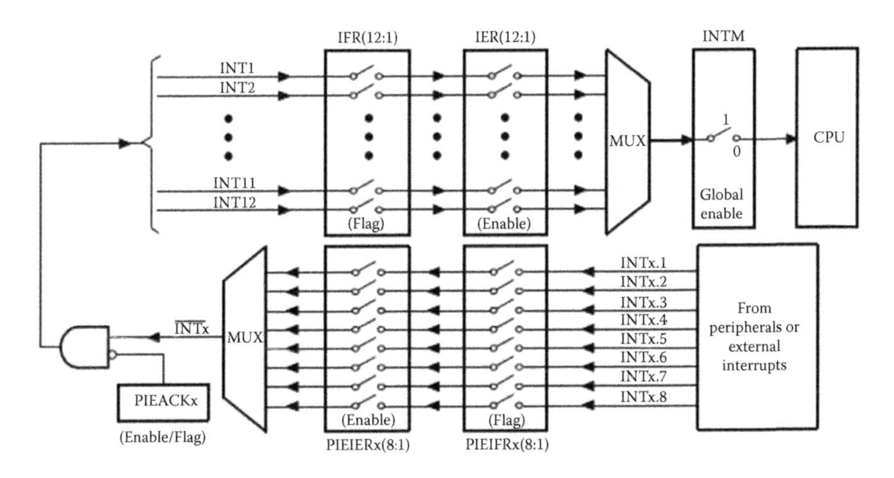

FIGURE 12.27
The structure of PIE. (Adapted from TMS320x2833x, 2823x System Control and Interrupts [SPRUFB0D].)

TABLE 12.5
Interrupt Connections

	INTx.8	INTx.7	INTx.6	INTx.5	INTx.4	INTx.3	INTx.2	INTx.1
INT1	WAKEINT	TINT0	ADCINT	XINT2	XINT1		PDPINTB	PDPINT
INT2		T1OFINT	T1UFINT	T1CINT	T1PINT	CMP3INT	CMP2INT	CMP1INT
INT3		CAPINT3	CAPINT2	CAPINT1	T2OFINT	T2UFINT	T2CINT	T2PINT
INT4		T3OFINT	T3UFINT	T3CINT	T3PINT	CMP6INT	CMP5INT	CMP4INT
INT5		CAPINT6	CAPINT5	CAPINT4	T4OFINT	T4UFINT	T4CINT	T4PINT
INT6			MXINT	MRINT			SPITXINTA	SPIRXINTA
INT7								
INT8								
INT9			ECAN1INT	ECAN0INT	SCITXINTB	SCIRXINT	SCITXINTA	SCIRXINTA
INT10								
INT11								
INT12								

Source: Adapted from TMS320x2833x, 2823x System Control and Interrupts (SPRUFB0D).

12.4.3 PIE Interruptions

The PIE module sends interrupt requests to the CPU throughout multiplexes 8 external interrupt pins. Those interruptions can be divided into 12 groups. Each group sends interrupt requests to CPU throughout one interrupt signal. For example, PIE 1st group multiplexes INT1, PIE 12th group multiplexes INT12. For the nonmultiplexed interrupts, the PIE passes the request directly to the CPU.

In each group of the PIE, there are IF bit (PIEIFRx.y) and interrupt enable bit (PIEIERx.y). Besides, each group has one acknowledge bit (PIEACK). Figure 12.28 [4] illustrates the flow operations with different PIEIFR settings and PIEIER settings.

As long as the PIE controller generates interruptions, corresponding acknowledge bits (PIEIFRx.y) will be set to 1. And if corresponding interrupt

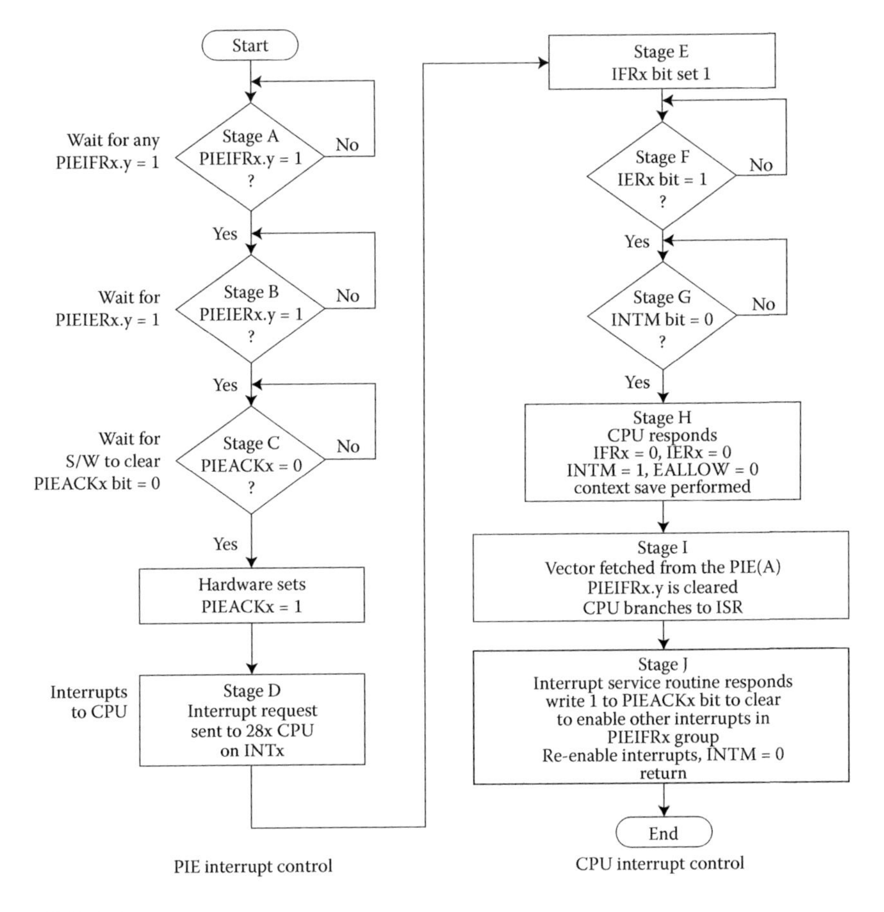

FIGURE 12.28
Typical PIE/CPU interrupt response—INTx.y. (Adapted from TMS320x2833x, 2823x System Control and Interrupts [SPRUFB0D].)

enable bits are set to 1, the PIE will check corresponding acknowledge bits for checking that the CPU will prepare to respond to interrupts or not. If the PIEACKx bits are cleared (0), the PIE will send interrupts to the CPU; if PIEACKx is set to 1, the PIE will keep waiting and send requests to the CPU after the PIEACKx is cleared (0).

12.4.4 CPU Interruptions

As long as the CPU receives the interrupt requests, the CPU level interrupt flag bits of interrupt flag register (IFR) will be set to 1. The information of IF bits will be updated into flag registers. The interrupt request can be responded to if only the corresponding enable bits of interrupt enable register (IER), debug interrupt enable register (DBGIER), and interrupt mask registers (INTM) are enabled.

12.4.5 Interrupt Vectors

The PIE supports 96 interrupts, each interrupt has specified interrupt vectors stored in the RAM. Those vectors are used to build the interrupt vector table shown in Figure 12.29 [4]. Users can adjust the interrupt vector table as required. The CPU can automatically fetch interrupt vectors from the interrupt vector table, where the CPU will spend nine CPU clock periods on fetching interrupt vectors and saving registers. The polarity of interruptions can be controlled through software/hardware. Every interruption can be enabled and be disenabled in the PIE module.

	INTx.8	INTx.7	INTx.6	INTx.5	INTx.4	INTx.3	INTx.2	INTx.1
INT1.y	*WAKEINT* (LPM/WD) 0xD4E	*TINTO* (TIMER 0) 0xD4C	*ADCINT* (ADC) 0xD4A	*XINT2* Ext. int. 2 0xD48	*XINT1* Ext. int. 1 0xD46	Reserved - 0xD44	*SEQ2INT* (ADC) 0xD42	*SEQ1INT* (ADC) 0xD40
INT2.y	Reserved - 0xD5E	Reserved - 0xD5C	*EPWM6_TZINT* (ePWM6) 0xD5A	*EPWM5_TZINT* (ePWM5) 0xD58	*EPWM4_TZINT* (ePWM4) 0xD56	*EPWM3_TZINT* (ePWM3) 0xD54	*EPWM2_TZINT* (ePWM2) 0xD52	*EPWM1_TZINT* (ePWM1) 0xD50
INT3.y	Reserved - 0xD6E	Reserved - 0xD6C	*EPWM6_INT* (ePWM6) 0xD6A	*EPWM5_INT* (ePWM5) 0xD68	*EPWM4_INT* (ePWM4) 0xD66	*EPWM3_INT* (ePWM3) 0xD64	*EPWM2_INT* (ePWM2) 0xD62	*EPWM1_INT* (ePWM1) 0xD60
INT4.y	Reserved - 0xD7E	Reserved - 0xD7C	*ECAP6_INT* (eCAP6) 0xD7A	*ECAP5_INT* (eCAP5) 0xD78	*ECAP4_INT* (eCAP4) 0xD76	*ECAP2_INT* (eCAP3) 0xD74	*ECAP2_INT* (eCAP2) 0xD72	*ECAP1_INT* (eCAP1) 0xD70
INT5.y	Reserved - 0xD8E	Reserved - 0xD8C	Reserved - 0xD8A	Reserved - 0xD88	Reserved - 0xD86	Reserved - 0xD84	*EQEP2_INT* (eQEP2) 0xD82	*EQEP1_INT* (eQEP1) 0xD80
INT6.y	Reserved - 0xD9E	Reserved - 0xD8C	*MXINTA* (McBSP-A) 0xD9A	*MXINTA* (McBSP-A) 0xD98	*MXINTB* (McBSP-B) 0xD96	*MRINTB* (McBSP-B) 0xD94	*SPITXINTA* (SPI-A) 0xD92	*SPIRXINTA* (SPI-A) 0xD90
INT7.y	Reserved - 0xDAE	Reserved - 0xDAC	*DINTCH6* (DMA6) 0xDAA	*DINTCH5* (DMA5) 0xDA8	*DINTCH4* (DMA4) 0xDA6	*DINTCH3* (DMA3) 0xDA4	*DINTCH2* (DMA2) 0xDA2	*DINTCH1* (DMA1) 0xDA0
INT8.y	Reserved - 0xDBE	Reserved - 0xDBC	*SCITXINTC* (SCI-C) 0xDBA	*SCITXINTC* (SCI-C) 0xDB8	Reserved - 0xDB6	Reserved - 0xDB4	*I2CINT2A* (I2C-A) 0xDB2	*I2CINT1A* (I2C-A) 0xDB0
INT9.y	*ECAN1INTB* (CAN-B) 0xDCE	*ECAN0INTB* (CAN-B) 0xDCC	*ECAN0INTA* (CAN-A) 0xDCA	*ECAN0INTA* (CAN-A) 0xDC8	*SCITXINTB* (SCI-B) 0xDC6	*SCIRXINTB* (SCI-B) 0xDC4	*SCITXINTA* (SCI-A) 0xDC2	*SCIRXINTA* (SCI-A) 0xDC0
INT10.y	Reserved - 0xDDE	Reserved - 0xDDC	Reserved - 0xDDA	Reserved - 0xDD8	Reserved - 0xDD6	Reserved - 0xDD4	Reserved - 0xDD2	Reserved - 0xDD0
INT11.y	Reserved -	Reserved -	Reserved -	Reserved -	Reserved -	Reserved -	Reserved -	-
INT12.y	LUF (FPU) 0xDDE	LVF (FPU) 0xDFC	Reserved - 0xDFA	*XINT7* Ext. Int. 7 0xDF8	*XINT6* Ext. Int. 6 0xDF6	*XINT5* Ext. Int. 5 0xDF4	*XINT4* Ext. Int. 4 0xDF2	*XINT3* Ext. Int. 3 0xDF0

FIGURE 12.29
PIE MUXed peripheral interrupt vector map. (Adapted from TMS320x2833x, 2823x System Control and Interrupts [SPRUFB0D].)

For accelerating the responding speed of the CPU, the reader can obtain more information from the technical document, *SPRUFB0D*. To clarify the procedure of interrupt settings and help readers to quickly develop DSPs, two examples are provided.

The ADC channels, ADC sampling rate, and ISRs can be configured in CCStudio IDE v3.3 through configuring the respective ADC registers, IER, and IFR. In the following example, ADC interrupt configuration is stressed. Several function registers are involved in the interrupt settings.

12.4.5.1 IER: CPU Register

The IER is a 16-bit CPU register. The IER contains enable bits for all the maskable CPU interrupt levels (INT1–INT14, RTOSINT, and DLOGINT). Neither NMI nor XRS is included in the IER; thus, IER has no effect on those interrupts. As presented in Figure 12.26, INT1–INT14 is an available interrupt source for designers. Setting bits of IER can enable/disenable the corresponding interrupts: setting bit 0 of the IER to 1 can enable interrupt 1 (INT1); setting bit 1 of the IER to 0 can disenable interrupt 2 (INT2). Figure 12.30 [4] shows the structure of IER.

12.4.5.2 CPU Interrupt Flag Register

The CPU IFR is a 16-bit CPU register and is used to identify and clear pending interrupts. The IFR contains flag bits for all the maskable interrupts at the CPU level. If a maskable interrupt is requested, the flag bit in the corresponding bit of IFR will be set to 1. This indicates that the interrupt is pending or waiting for acknowledgement. The structure of the IFR is presented in Figure 12.31 [4].

15	14	13	12	11	10	9	8
RTOSINT	DLOGINT	INT14	INT13	INT12	INT11	INT10	INT9
R/W-0	R/W-0	R/W-0	R/W-0	R/W-0	R/W-0	R/W-0	R/W-0

7	6	5	4	3	2	1	0
INT8	INT7	INT6	INT5	INT4	INT3	INT2	INT1
R/W-0	R/W-0	R/W-0	R/W-0	R/W-0	R/W-0	R/W-0	R/W-0

Legend: R/W = Read/Write; R = Read only; -n = value after reset

FIGURE 12.30
Interrupt enable register (IER)—CPU register. (Adapted from TMS320x2833x, 2823x System Control and Interrupts (SPRUFB0D).)

15	14	13	12	11	10	9	8
RTOSINT	DLOGINT	INT14	INT13	INT12	INT11	INT10	INT9
R/W-0	R/W-0	R/W-0	R/W-0	R/W-0	R/W-0	R/W-0	R/W-0

7	6	5	4	3	2	1	0
INT8	INT7	INT6	INT5	INT4	INT3	INT2	INT1
R/W-0	R/W-0	R/W-0	R/W-0	R/W-0	R/W-0	R/W-0	R/W-0

Legend: R/W = Read/Write; R = Read only; -n = value after reset

FIGURE 12.31
Interrupt flag register (IFR)—CPU register. (Adapted from TMS320x2833x, 2823x System Control and Interrupts [SPRUFB0D].)

12.4.5.3 PIE Vector Table

Accessing the interrupt throughout the PIE vector table is relatively easy in cases of developing C2000 series DSP. To link the ISRs to corresponding interrupt IDs, users only need to assign the start address of the ISRs to the interrupt vectors. As an example, such assignment can be written as follows:

```
PieVectTable.ADCINT = &adc_isr;
```

The above syntax is to link the ADC interrupt callback function to the ADC interrupt vector. As long as the ADC interrupt is enabled, at the end of each ADC conversion, an interruption will be generated. And the ADC interrupt callback function will be called until the CPU is idle.

```
PieCtrlRegs.PIEIER1.bit.INTx6 = 1;
```

The function of the syntax ("PieCtrlRegs.PIEIER1.bit.INTx6 = 1") is to enable the PIE group1 register. For helping reader to understand the syntax and related value assignment, Figure 12.32 [4] is provided.

Another significant control register is involved for implementing ADC interrupt control is the ADC control register2 (ADCTRL2). The following syntax is generally used for ADC initialization.

```
AdcRegs.ADCTRL2.bit.INT_ENA_SEQ1 = 1;
```

The bit, ADCTRL2.bit.INT_ENA_SEQ1 can be set to 1 for enabling the SEQ interrupt. On the contrary, this bit can be set to 0 for disenabling the SEQ interrupt, where the SEQ is the abbreviation of "sequencer." The ADC sequencer consists of two independent 8-state sequencers (SEQ1 and SEQ2) that can also be cascaded together to form one 16-state sequencer (SEQ). For gathering more details, readers can refer to the technical document, SPRU812A. An example related to use ADC interrupt is provided as follows:

PIE group 1 vectors—MUXed into CPU INT1							
INT1.1	32	0x0000 0D40	2	SEQ1INT	(ADC)	5	1 (Highest)
INT1.2	33	0x0000 0D42	2	SEQ2INT	(ADC)	5	2
INT1.3	34	0x0000 0D44	2	Reserved		5	3
INT1.4	35	0x0000 0D46	2	XINT1		5	4
INT1.5	36	0x0000 0D48	2	XINT2		5	5
INT1.6	37	0x0000 0D4A	2	ADCINT	(ADC)	5	6
INT1.7	38	0x0000 0D4C	2	TINT0	(CPU-Timer0)	5	7
INT1.8	39	0x0000 0D4E	2	WAKEINT	(LPM/WD)	5	8 (Lowest)

FIGURE 12.32
Descriptions of bits of PIEIER1. (Adapted from TMS320x2833x, 2823x System Control and Interrupts [SPRUFB0D].)

```
//###############################################################
//Example of ADC interrupt utilization
//###############################################################
#include "DSP28x_Project.h" // Device Headerfile and Examples
Include File
// Prototype statements for functions found within this file.
__interrupt void adc_isr(void);
main()
{
// Step 1. Initialize System Control:
  InitSysCtrl();
  EALLOW;
  #if (CPU_FRQ_150MHZ) // Default - 150 MHz SYSCLKOUT
    #define ADC_MODCLK 0x3 // HSPCLK = 25.0 MHz
  #endif
  #if (CPU_FRQ_100MHZ)
    #define ADC_MODCLK 0x2 // HSPCLK=25.0 MHz
  #endif
  EDIS;
  // Define ADCCLK clock frequency ( less than or equal to 25 MHz )
  // Assuming InitSysCtrl() has set SYSCLKOUT to 150 MHz
  EALLOW;
  SysCtrlRegs.HISPCP.all = ADC_MODCLK;
  EDIS;
//Step 2. Clear all interrupts and initialize PIE vector table:
//Disable CPU interrupts
  DINT;
//Initialize the PIE control registers to their default state.
  InitPieCtrl();
//Disable CPU interrupts and clear all CPU interrupt flags:
  IER = 0x0000;
  IFR = 0x0000;
//Initialize the PIE vector table with pointers to the shell
Interrupt
//Service Routines (ISR)..
  InitPieVectTable();
//Interrupts that are used in this example are re-mapped to
//ISR functions found within this file.
  EALLOW; // This is needed to write to EALLOW protected register
  PieVectTable.ADCINT = &adc_isr;
  EDIS; // This is needed to disable write to EALLOW protected
  registers
//Step 3. Initialize all the Device Peripherals:
  InitAdc(); // For this example, init the ADC
//Enable ADCINT in PIE
  PieCtrlRegs.PIEIER1.bit.INTx6 = 1;
  IER |= M_INT1; // Enable CPU Interrupt 1
  EINT; // Enable Global interrupt INTM
  ERTM; // Enable Global realtime interrupt DBGM
//Configure ADC
```

```
  AdcRegs.ADCMAXCONV.all = 0x0000; // Setup 1 conv's on SEQ1
  AdcRegs.ADCCHSELSEQ1.bit.CONV00 = 0x0; // Setup ADCINA3 as
  1st SEQ1 conv.
  AdcRegs.ADCTRL2.bit.EPWM_SOCA_SEQ1 = 1;
//Enable SOCA from ePWM to start SEQ1
  AdcRegs.ADCTRL2.bit.INT_ENA_SEQ1 = 1; // Enable SEQ1
  interrupt (every EOS)
//Assumes ePWM1 clock is already enabled in InitSysCtrl();
  EPwm1Regs.ETSEL.bit.SOCAEN = 1; // Enable SOC on A group
  EPwm1Regs.ETSEL.bit.SOCASEL = 4; // Select SOC from from CPMA
  on upcount
  EPwm1Regs.ETPS.bit.SOCAPRD = 1; // Generate pulse on 1st event
  EPwm1Regs.CMPA.half.CMPA = 0x0080; // Set compare A value
  EPwm1Regs.TBPRD = 0xFFFF; // Set period for ePWM1
  EPwm1Regs.TBCTL.bit.CTRMODE = 0;        // count up and start
}
__interrupt void adc_isr(void)
{
      {
      /*******************************************************
      Users can type code here to gather ADC conversion results.
      Note that the ADC_isr can be also used for implementing
      the digital controller.
      *******************************************************/
      }
  //Reinitialize for next ADC sequence
  AdcRegs.ADCTRL2.bit.RST_SEQ1 = 1; // Reset SEQ1
  AdcRegs.ADCST.bit.INT_SEQ1_CLR = 1; // Clear INT SEQ1 bit
  PieCtrlRegs.PIEACK.all = PIEACK_GROUP1; // Acknowledge
  interrupt to PIE
  return;
}
```

In the above example, the ePWM module is used to generate triggers for starting ADC conversions. In the ePWM module, there is a counter comparing the counting number to the set threshold value. Each time the counting number exceeds the threshold value, the ePWM will trigger the ADC module for starting a new round conversion. After every ADC conversion finishes, ADC module will generate an interrupt and send an interrupt request to the CPU. As results, the CPU will handle the ISR as long as it is idle. And the program pointer will be automatically linked to the starting address of the interrupt callback function which is the ISR function in our case.

The next key element for implementing real-time control for most electrical systems is the ADC. Users normally implement digital compensators throughout DSPs. There will be various requirements for ADC sampling rate and data precision, given objectives of practical designed systems. The fundamental control of ADCs of C2000 series DSP will be explained in the following section by using the examples of TI F28335 ADCs.

12.5 Analog-to-Digital Conversion

ADCs of C2000 series DSPs convert analog signals to digital signals. The converted digital signals are generally used to generate digital control signals for purposes of signal filtering and feedback control, especially for power electronics applications. For describing how to use ADCs of TI DSPs, this section mainly focusses on the TI F28335 ADC modules.

12.5.1 Features of Analog-to-Digital Conversion for TI F28335

The ADC module has 16 channels which are divided into two independent 8-channel sequencers. The two independent sequencers can be cascaded to perform a 16-channel module. Although there are multiple input channels and two sequencers, there is only one converter in the ADC module [6]. Figure 12.33 [6] shows the block diagram of the ADC module.

Functions of the ADC module include:

- 12-bit ADC core with built-in dual sample-and-hold (S/H)
- Simultaneous sampling or sequential sampling modes
- Analog input: 03 V
- Fast conversion time runs at 12.5 MHz, ADC clock, or 6.25 MSPS
- Sixteen result registers storing conversion values

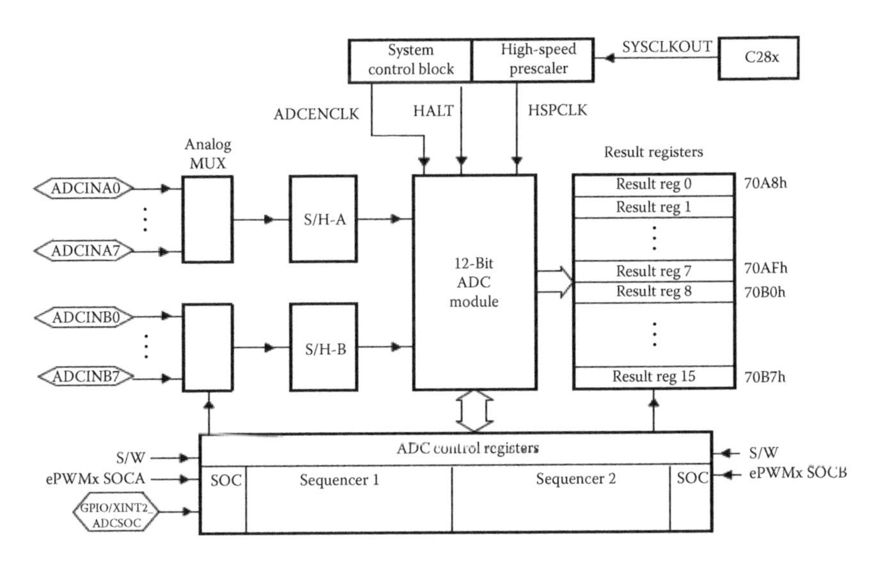

FIGURE 12.33
Block diagram of the ADC module. (Adapted from TMS320x2833x Analog-to-Digital Converter [ADC Module] Reference Guide [SPRU812A].)

The digital value of the input analog voltage is derived by

Digital value = 0, when input ≤ 0 V

Digital value

$$= 4096 \times \frac{\text{Input analog voltage} - \text{ADCLO}}{3} \quad \text{when } 0\,\text{V} < \text{input} < 3\,\text{V}$$

Digital value = 4096, when input ≥ 3 V

- Multiple trigger sources for the start-of-conversion (SOC) sequence
    - S/W—software immediate start
    - ePWM 1-6
    - GPIO XINT2

The significant features of the ADC module are shown above. For controlling the ADC module and gathering the ADC results, the following registers (summarized in Table 12.6 [6]) are normally involved.

The initializing settings for ADC sampling involve several steps: set ADC clock frequency; set maximum conversion channels; select the trigger source for the start of the conversion; and enable/disenable the ADC interrupt. A good example of ADC initializations is shown in Figure 12.34.

Reading ADC converted values is relatively simple, compared to ADC initialization. The syntax for obtaining the ADC values can be written as

```
Global Variable/Local Variable = AdcRegs.ADCRESULTx >> 4;
```

Users may prefer to use interrupts for fetching ADC values in terms of implementing a practical digital control system. Note that users should keep in mind that the interrupt flag bits must be set to 0 after the ISR is handled, and then the SEQ1/SEQ2 should be reset at the end of the procedure of handling the ADCINT callback function. Three-syntax is to be written at the end of the ISR for ADCINT, for example:

```
// Reinitialize for next ADC sequence
AdcRegs.ADCTRL2.bit.RST_SEQ1 = 1; // Reset SEQ1
AdcRegs.ADCST.bit.INT_SEQ1_CLR = 1; // Clear INT SEQ1 bit
PieCtrlRegs.PIEACK.all = PIEACK_GROUP1;// Acknowledge INT to PIE
```

In fact, Texas Instrument provides a convenient method for helping users quickly develop DSPs based on MATLAB/Simulink. Users can build control diagrams in the MATLAB/Simulink environment, throughout the Embedded Coder, where the user can find the Embedded Coder toolbox (shown in Figure 12.35) in the Simulink Library Browser.

TABLE 12.6

ADC Registers and Corresponding Functions

Register	Description
ADCCTRL2	ADC module control register2 The function bit/bits involve: Bit 15—Enable ePWM SOCB for starting sequencers Bit 14—Reset sequencer to sequencer to state CONV00 Bit 13—Start-of-conversion (SOC) trigger sources S/W, ePWM SOCA, ePWM SOCB, external pin Bit 11—Enable SEQ1 interrupt mode Bit 8—Enable ePWM SOCA for starting SEQ1 Bit 6—Reset sequencer to sequencer to state CONV08 Bit 3—Enable SEQ2 interrupt mode Bit 0—Enable ePWM SOCB for starting SEQ2
ADCMAXCONV	Maximum Conversion Channels Register The function bit/bits involve: Bit 3-0—Define maximum conversion for SEQ1 Bit 6-4—Define maximum conversion for SEQ2 Bit 3-0—Define maximum conversion for SEQ (cascaded mode)
ADCST	ADC Status and Flag Register The function bit/bits of this register involve: Bit 5—SEQ2 interrupt clear bit Bit 4—SEQ1 interrupt clear bit Bit 3—SEQ2 busy status bit Bit 2—SEQ1 busy status bit Bit 1—SEQ2 interrupt flag bit Bit 0—SEQ1 interrupt flag bit
ADCCHSELSEQx	ADC Input Channel Select Sequencing Control Registers The function bit/bits involve: Bit 15-12—CONV03 (e.g., 0x0000 ↔ ADCINA0) Bit 11-8—CONV02 Bit 7-4—CONV01 Bit 3-0—CONV00
ADCRESULTn	ADC Conversion Result Buffer Registers Bit 15-0 presents a Hex number, where this number is the converted value corresponding to the ADCINAx/ADCINBx

Source: Adapted from TMS320x2833x Analog-to-Digital Converter (ADC Module) Reference Guide (SPRU812A).

Users can build control diagrams for programming C2000 DSP by following these steps:

1. First step is to create a new MATLAB/Simulink model. The second step is to drag the *Target Preferences* to the new model, where the component, *Target Preferences* can be found in the list, *Embedded Targets/Embedded Coder*. Figure 12.36 shows the step 1.

```
#include "DSP2833x_Device.h"
#include "DSP2833x_GlobalPrototypes.h"
#include "rtwtypes.h"
#include "untitled1.h"
#include "untitled1_private.h"

void config_ADC_A(uint16_T maxConv, uint16_T adcChselSEQ1Reg, uint16_T
                  adcChselSEQ2Reg, uint16_T adcChselSEQ3Reg, uint16_T
                  adcChselSEQ4Reg)
{
  AdcRegs.ADCTRL1.bit.SUSMOD = 0x0;      // Emulation suspend ignored
  AdcRegs.ADCTRL1.bit.ACQ_PS = 4;        // Acquisition window size
  AdcRegs.ADCTRL1.bit.CPS = 1;           // Core clock pre-scaler
  AdcRegs.ADCTRL3.bit.ADCCLKPS = 3;      // Core clock divider
  AdcRegs.ADCREFSEL.bit.REF_SEL = 0 ;    // Set Reference Voltage
  AdcRegs.ADCOFFTRIM.bit.OFFSET_TRIM = 0;// Set Offset Error Correctino Value
  AdcRegs.ADCTRL1.bit.CONT_RUN = 0;      // 0:Start-Stop or continuous sequencer mode
  AdcRegs.ADCTRL3.bit.ADCBGRFDN = 0x3;   // Bandgap and reference powered up
  AdcRegs.ADCTRL3.bit.SMODE_SEL = 0 ;    // 1:Simultaneous, 0:Sequential sampling
  AdcRegs.ADCMAXCONV.bit.MAX_CONV1 = maxConv;// Number of conversions in CONV2 when using B module
  AdcRegs.ADCTRL1.bit.SEQ_CASC = 0;      // 1:Cascaded, 0:Dual sequencer mode
  AdcRegs.ADCCHSELSEQ1.all = adcChselSEQ1Reg;// Channels for conversion
  AdcRegs.ADCCHSELSEQ2.all = adcChselSEQ2Reg;// Channels for conversion
  AdcRegs.ADCTRL2.bit.INT_MOD_SEQ1 = 0;  //Interrupt will be generated at the end of every SEQ1
  AdcRegs.ADCTRL2.bit.INT_ENA_SEQ1 = 1;  //Enable SEQ1 interrupt
  AdcRegs.ADCTRL2.bit.EPWM_SOCA_SEQ1 = 1;// Enable ePWMxA SOC
  AdcRegs.ADCTRL2.bit.RST_SEQ1 = 0x1;    // Reset SEQ1
  AdcRegs.ADCTRL3.bit.ADCCLKPS = 3;
}
```

FIGURE 12.34
The subroutine for initializing ADC module.

2. Users need to click the *Target Preferences* shown in Figure 12.36 for setting information of the target DSP board, where the target DSP board is TI F28335 in our case. The proper configuration is shown in Figure 12.37.
3. Users can find ADC module in the list, *C28x3x/Texas Instruments C2000/Processors/Embedded Targets/Embedded Coder*. TI F28335 Simulink components are shown in Figure 12.38.
4. Users can paste the ADC component to the new model, and set the ADC module in MATLAB/Simulink environment. The procedure is shown Figure 12.39.

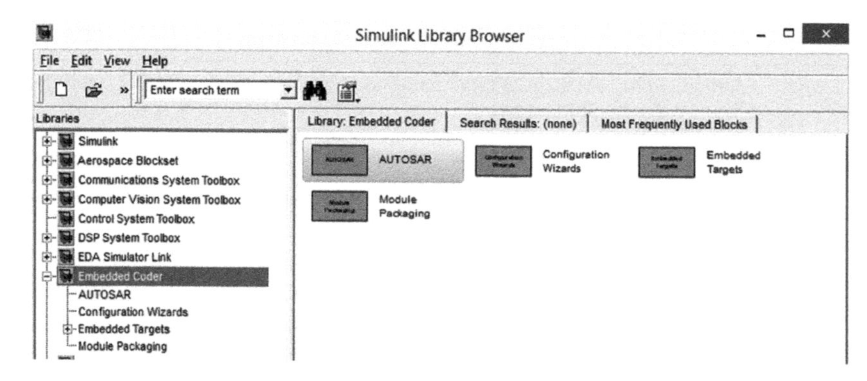

FIGURE 12.35
Embedded Coder Toolbox in Simulink Library Browser.

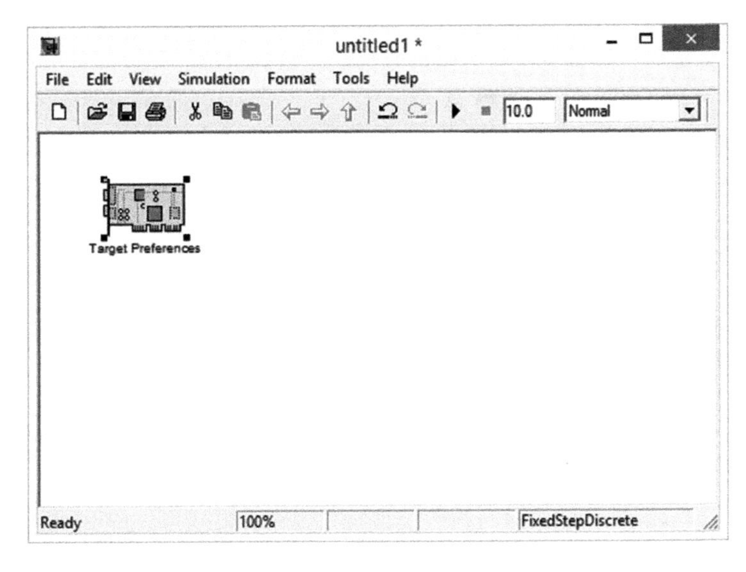

FIGURE 12.36
Step 1 for drawing a control block: new model.

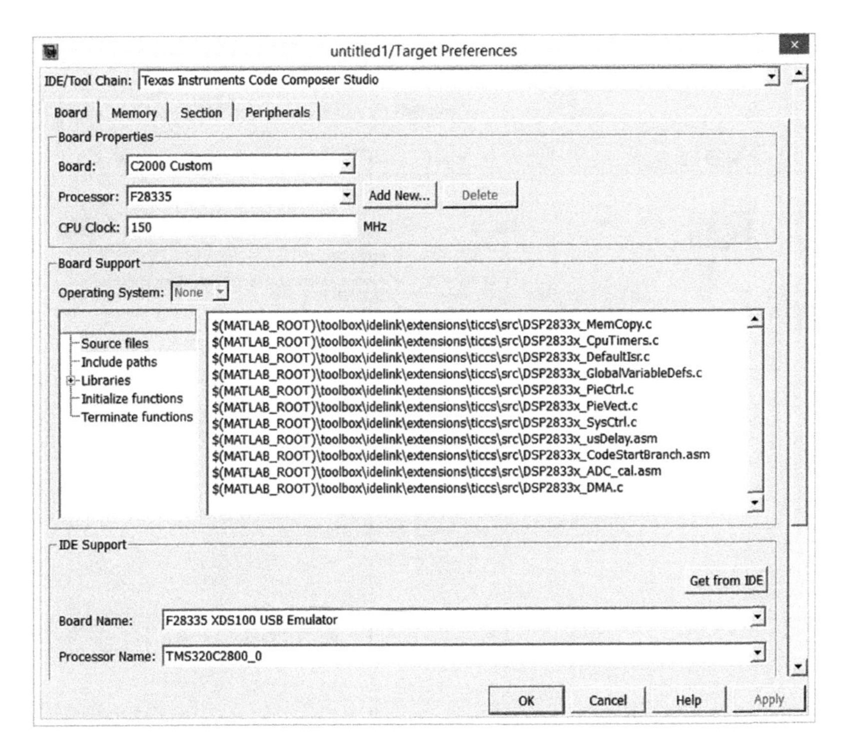

FIGURE 12.37
Configuration for setting *Target Preferences*.

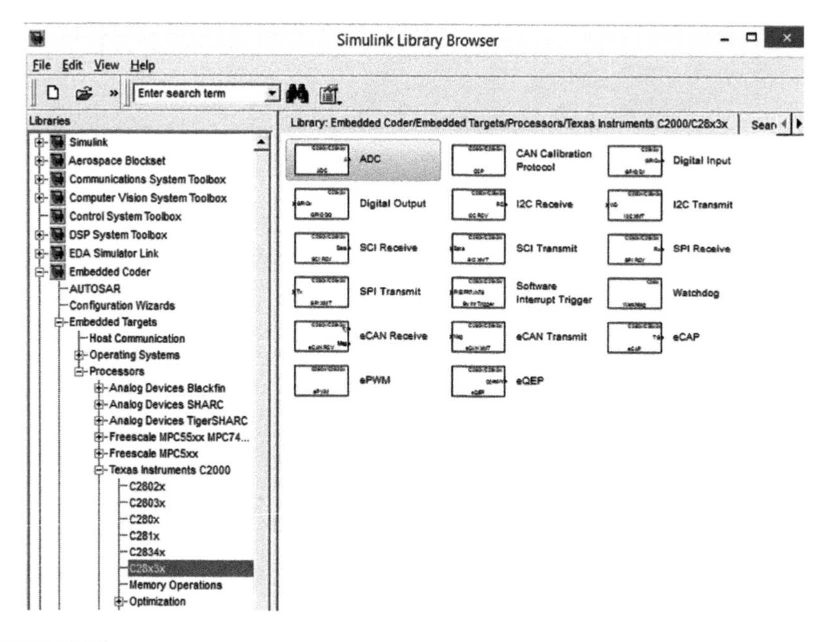

FIGURE 12.38
TI F28335 component list.

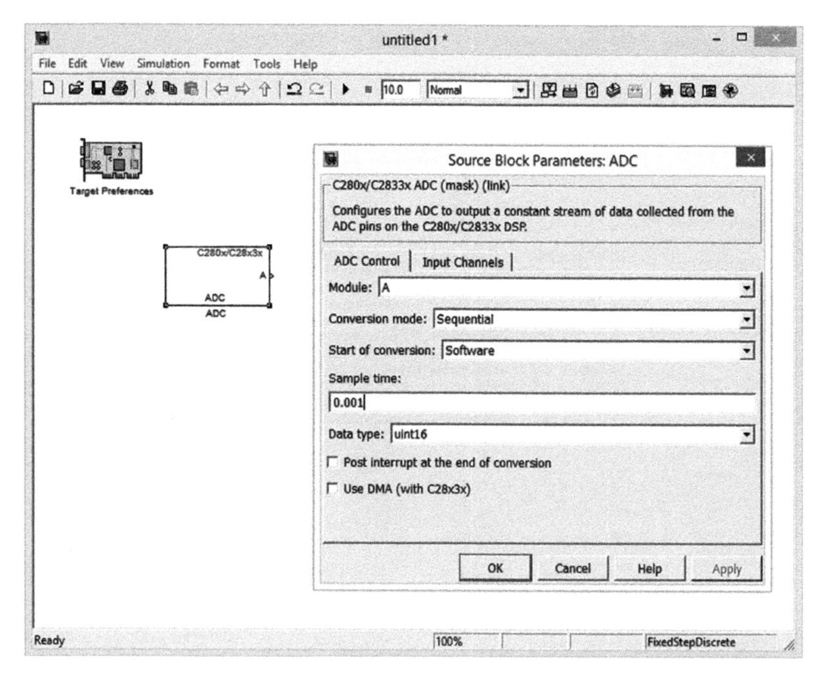

FIGURE 12.39
Configuration for TI F28335 ADC module.

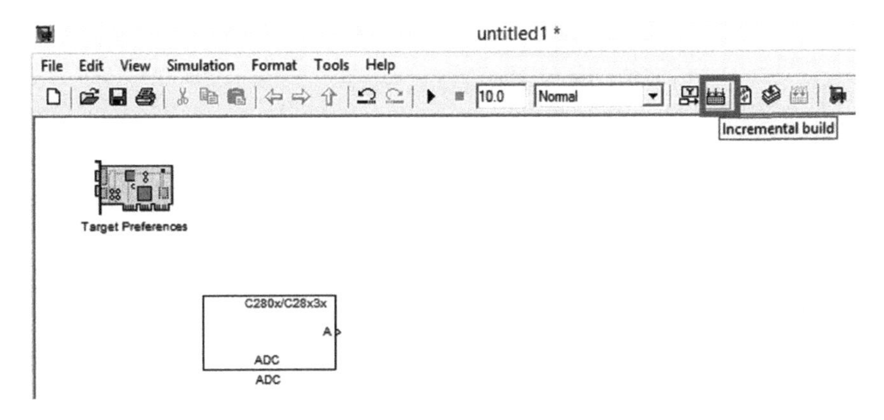

FIGURE 12.40
Download the control diagram to target DSP board.

5. For simply demonstrating operations of setting ADC modules, the author can use ADC module A in this example, where the ADCINA0–ADCINA7 channel could be sampled throughout this ADC Simulink block. Conversion mode is set to be sequential. The trigger source is selected as the software trigger source. The sampling time is set to 0.001, which means that ADC reading frequency is set to 1000 Hz. In fact, the *Sampling Time* in the ADC module is not the actual conversion frequency of the ADC module. The sampling time indicates that the CPU will access ADC result registers with the frequency corresponding to the sampling time. After users finish the configurations, click "OK" as shown in Figure 12.39. Users can convert the Simulink control blocks to an embedded code by clicking the "OK" button, incremental build highlighted by the rectangular box shown in Figure 12.40. On the condition that the control block is successfully converted to the embedded code, users can observe the embedded code in the CCStudio IDE v3.3. The converted embedded code is shown in Figure 12.41.

The highlighted project files are generated by MATLAB/Simulink. If the control diagram is designed properly, TI F28335 will fulfill desired functions. In this example, ADCINA0, ADCINA1, and ADCINA2 result registers will be accessed 1000 times per second. The monitoring window of ADC result registers is presented in Figure 12.42.

12.6 ePWM Generator

TI F28335 ePWM generators can generate high-resolution PWM signals varying from 10 kHz to 1 MHz. Users can set multiple parameters, for example,

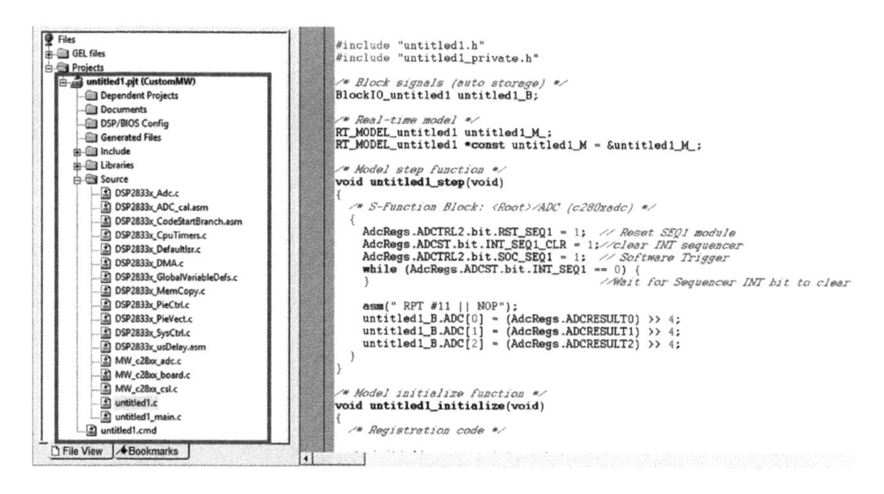

FIGURE 12.41
Embedded code converted by MATLAB/Simulink.

Name	Value	Type	Radix
ADCRESULT0	0xFFF0	int	hex
ADCRESULT1	0xEFA0	int	hex
ADCRESULT2	0xEB70	int	hex
ADCRESULT3	0x0000	int	hex

FIGURE 12.42
ADCRESULT registers: ADCINA0, ADCINA1, and ADCINA2.

switching frequency, duty-ratio, trigger-event, synchronous/asynchronous, phase shift, etc. In this section, fundamental ePWM registers and their functionalities will be briefly introduced.

12.6.1 Event-Trigger Selection Register

The main function of this register is to set the trigger mode/option for ADC start of conversion and interrupt of ePWM. The structure of the register is shown in Figure 12.43 [5].

15	14–12	11	10–8
SOCBEN	SOCBSEL	SOCAEN	SOCASEL
R/W-0	R/W-0	R/W-0	R/W-0

7–4	3	2–0
Reserved	INTEN	INTSEL
R-0	R/W-0	R/W-0

Legend: R/W = Read/Write; R = Read only; -n = value after reset

FIGURE 12.43
Event-trigger selection register. (Adapted from TMS320x 2833x, 2823x Enhanced Pulse Width Modulator [ePWM] Module [SPRUG04A].)

Writing 1 to SOCBEN/SOCAEN bit can enable the EPWMxSOCB/EPWMxSOCA so that the ADC Start of Conversion can be triggered by the EPWMx pulses, where the EPWM modules can generate such pulses referring to the selected EPWMxSOCA/EPWMxSOCB Options. Writing 1 to INTEN can enable the ePWM to interrupt.

12.6.2 Counter-Compare A Register

The value in the active CMPA register is continuously compared to the time-base counter (TBCTR). When the values are equal, the counter-compare module generates a "time-base counter equal to counter compare A" event. This event is sent to the action-qualifier for generating one or more actions as follows:

- Do nothing; the event is ignored
- Clear: Pull the EPWMxA/EPWMxB signal low
- Set: Pull the EPWMxA/EPWMxB signal high
- Toggle the EPWMxA/EPWMxB signal

In other words, changing the value storing in the CMPA can change the duty-ratio of the EWPM signals. The structure of the CMPA register is shown in Figure 12.44 [5].

12.6.3 Time-Base Period Register

The value of this register determines the period of the TBCTR. Varying the value of this register will change the frequency of the EPWM signals. The structure of the TBPRD is shown in Figure 12.45 [5].

Users can also configure ePWM module in the MATLAB/Simulink environment. The values of the above registers can be in decimal-format, or in Hex-format. The module can be found in the list, *C28x3x/Texas Instruments*

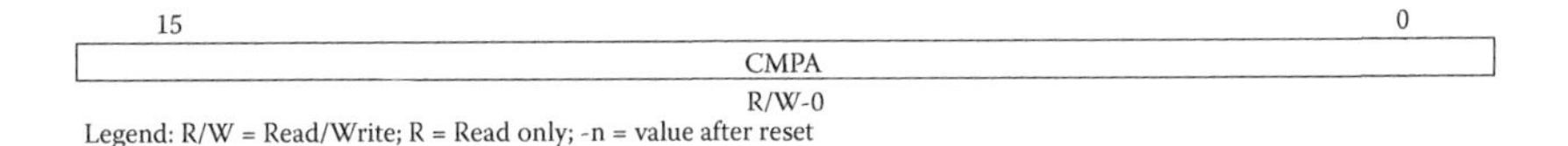

FIGURE 12.44
Counter-compare a register. (Adapted from TMS320F2833x, 2823x Digital Signal Controllers (DSCs) (SPRS439M).)

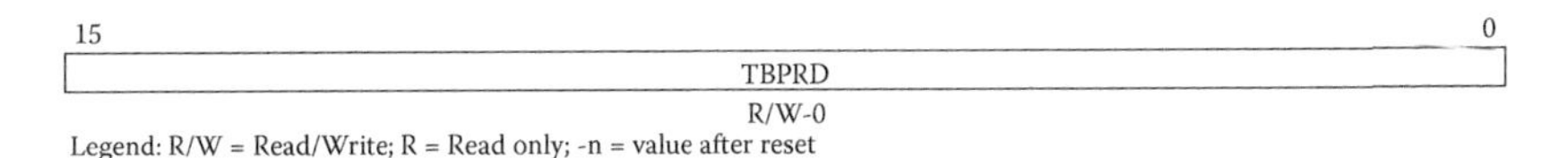

FIGURE 12.45
Time-base period register. (Adapted from TMS320x 2833x, 2823x Enhanced Pulse Width Modulator [ePWM] Module [SPRUG04A].)

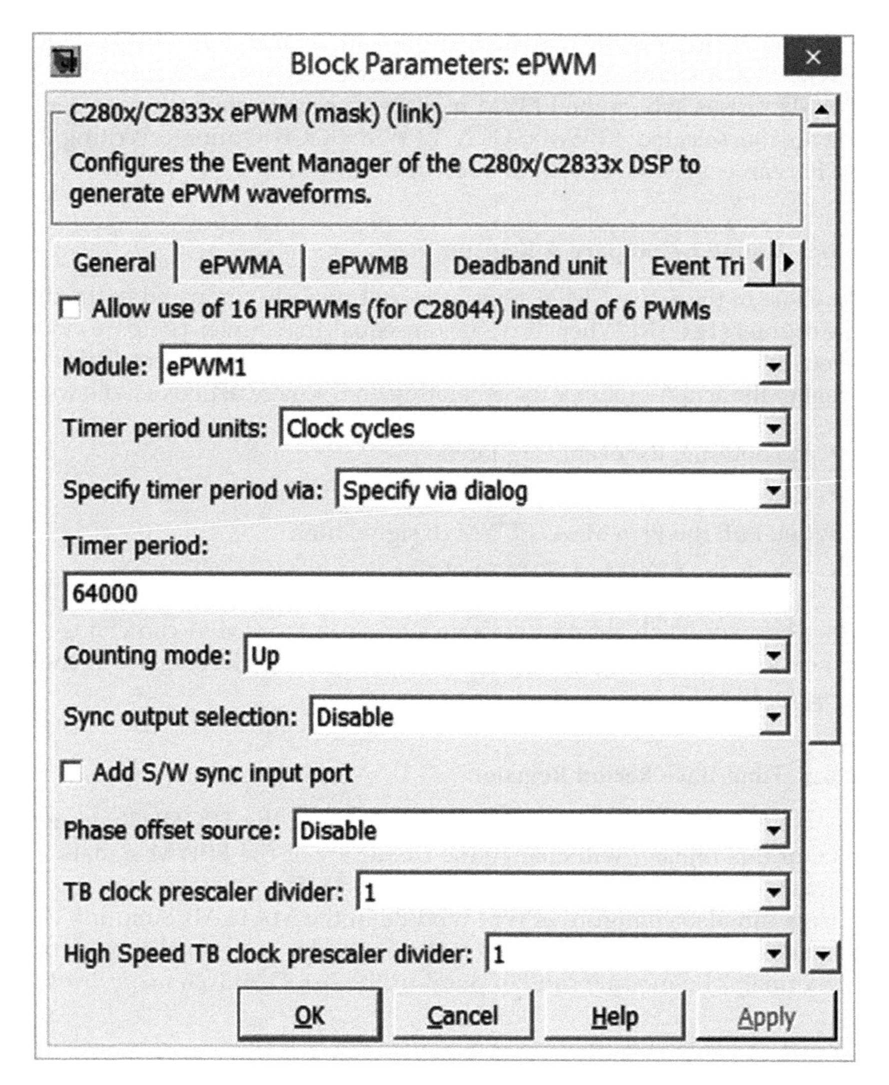

FIGURE 12.46
Configuration dialog of the ePWM module.

C2000/Processors/Embedded Targets/Embedded Coder. The configuration of the Simulink ePWM module is shown in Figure 12.46.

12.7 Utilizing DSP for Power Electronic Applications

C2000 series DSPs are widely used for power electronics applications because C2000 series DSPs have versatile advantages in terms of the wide-band sampling frequency, high-resolution PWM generators, configurable GPIO (general-purpose input/output) pins, low cost, and simplified programming environment. In this chapter, for an example, a DSP-controlled boost converter is provided for explaining how to use DSP control in a power electronics system, and what kinds of practical concerns related to DSP implementation should be taken into account.

12.7.1 Control of a Boost Converter

The nonisolated boost converter consists of the input power source, input inductor, switching device, current-direction regulating diode, output filtering capacitor, and output load. Depending on the type of the output load, the voltage and current regulation of the boost converter could be changed. In the case of the resistive output load, the controller for the boost circuit can be a two-cascaded controller (the outer-loop voltage controller and inner-loop current controller) or a direct voltage controller. The outer-loop voltage controller will be used for generating a current reference signal referring to the feedback signal which is the error between the voltage reference and the feedback voltage signal. And the inner-loop current controller will generate a PWM signal according to the error signal that is the difference between the reference current signal and feedback signal. The topology of the cascaded voltage–current control for a boost converter is shown in Figure 12.47. In the case of the direct voltage control for a boost converter system, the direct voltage controller can generate a duty-ratio signal for perturbing the pulse-width of the PWM signal sent to the switching device in the boost converter referring to the error signal. The direct voltage control for a boost converter is given by Figure 12.48.

The outer-loop controller shown in Figure 12.47 generally operates at a relatively lower frequency, compared to that of the inner-loop controller. This

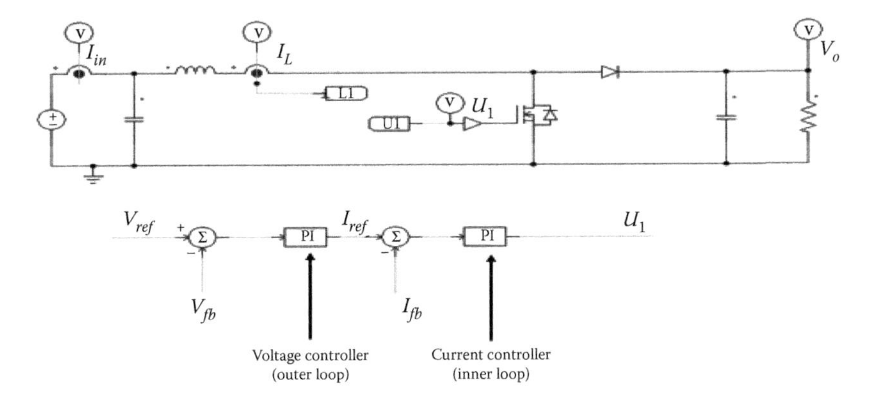

FIGURE 12.47
The cascaded control for a boost converter.

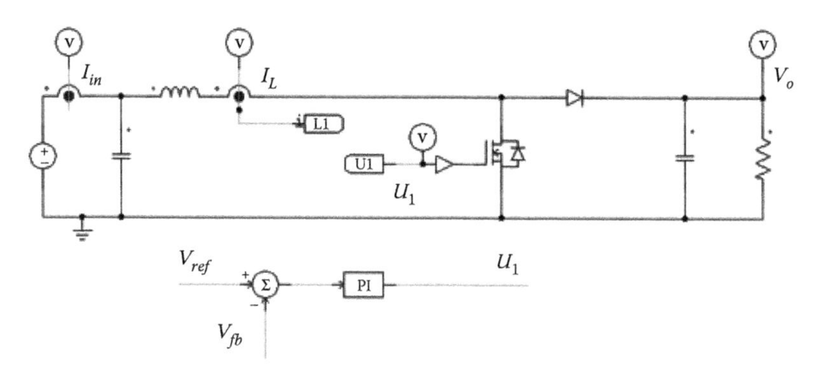

FIGURE 12.48
Direct voltage control for a boost converter.

is because each control period of the outer-loop voltage controller will hold a certain amount of value for the current reference. If the outer-loop voltage controller can hold the current reference value with enough time periods, the inner current controller will have enough time for regulating current dynamics to trace the current reference. Hence, the operating frequency of the outer-loop voltage controller will be one or several decades below that of the inner-loop controller. Two PI controllers present the voltage controller and current controller. In practice, the voltage controller and current controller may have different structures with various poles and zeros in terms of providing desired gain margin and phase margin.

In the case shown in Figure 12.48, the designer can implement one signal controller in the DSP, if the operating point of the system can be limited within a certain range, which means that the transfer function related to output voltage and duty-ratio will only vary within a small band. If so, the designer can derive a signal controller with a poles-zeros controller for regulating system dynamics for achieving the target output voltage–current.

In the following example of the DSP-controlled boost converter, the current reference signal is given, and the work is focussed on the inner-loop current controller. The inner-loop controller will adopt one of the nonlinear control algorithms, Sliding Mode Control. The currents flowing through key components of the boost converter are shown in Figure 12.49, assuming that the system operates at steady state.

The triangular current waveform shown in Figure 12.49 is caused by the switching pattern of the boost converter, that is indeed the nature of the boost converter. By applying a PWM signal into a boost converter, a periodical triangular current waveform can be observed throughout the input inductor, where the ramp rates of the rising edge and falling edge are determined by the system predesigned parameters such as the value of the input inductor, input voltage level, and the target output voltage level. The chattering around the current reference signal can be presented by Figure 12.50 [1].

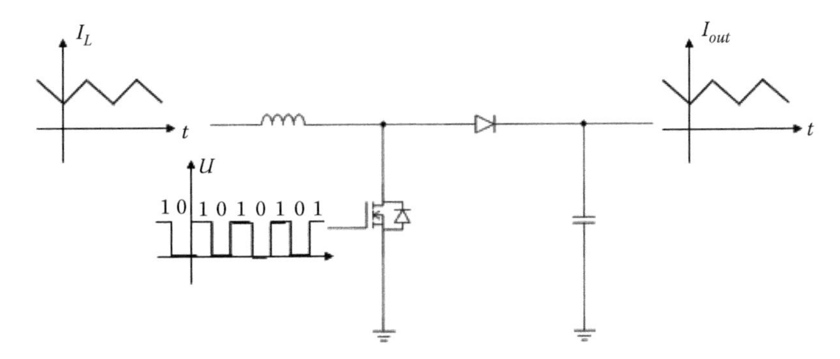

FIGURE 12.49
Current waveforms of key components in a boost converter.

The ramp rate of the rising edge is given by Equation 12.1, and the ramp rate of the falling edge is given by Equation 12.2. The switching period can be calculated through Equation 12.3.

$$\dot{S}_+(S>0) = -\frac{V_{in}}{L} + \frac{V_c}{L} \tag{12.1}$$

$$\dot{S}_-(S<0) = -\frac{V_{in}}{L} \tag{12.2}$$

$$T_s = \frac{2\Delta}{\dot{S}_+} - \frac{2\Delta}{\dot{S}_-} = -\frac{2\Delta(\dot{S}_+ - \dot{S}_-)}{\dot{S}_+ * \dot{S}_-} = \frac{2\Delta(V_c/L)}{V_{in}(V_c - V_{in})/L^2} = \frac{2\Delta(V_c)L}{V_{in}(V_c - V_{in})} \tag{12.3}$$

where

$$S = I_{ref} - I_L \tag{12.4}$$

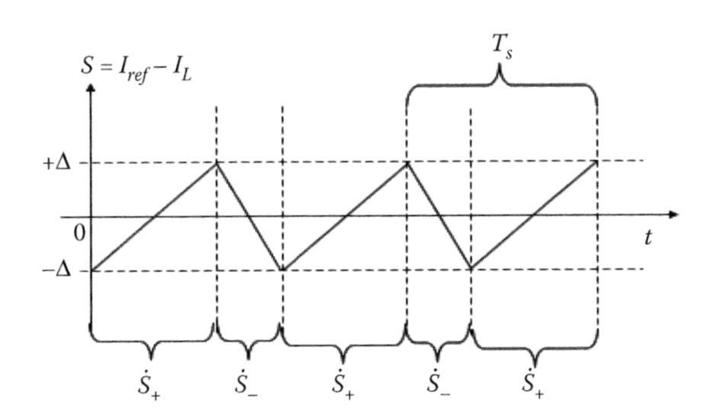

FIGURE 12.50
The chattering of the system dynamic regarding the sliding surface. (Adapted from Lee, H., Utkin, V., and Malinin, A., *International Journal of Control*, 82(9), 1720–1737, 2009.)

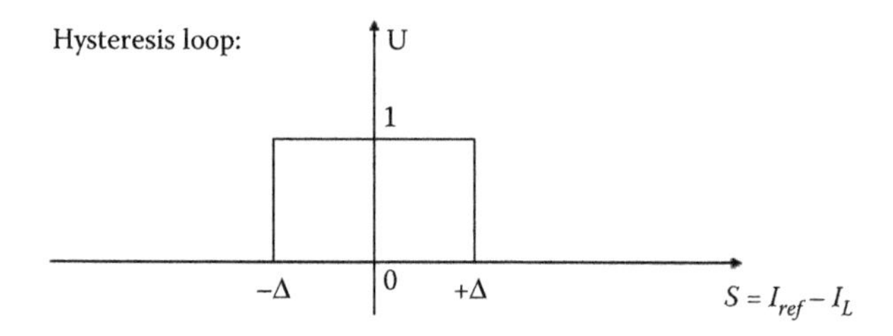

FIGURE 12.51
Hysteresis loop.

The chattering width, Δ in Equation 12.3 is the actual amplitude of the ripple component of the inductor current. According to Equations 12.1 through 12.4, readers can easily find the principles for controlling dynamics of the boost converter. Knowing that the inductor current will increase if the switching device of the system is turned on, which means the input sources of the power converter only charges the input inductor, as long as the switching device (MOSEFTs, IGBTs) are turned on; the ramp rate of the chattering shown in Figure 12.50 will be greater than zero. On the contrary, the switching-device is turned off; the input source associating with the input inductor will supply the output load, where the input inductor will serve as the secondary power sources discharging its energy in the inductor. Hence, the ramp rate of the chattering should be less than zero. If the power converter operates at continuous conduction mode (CCM), then chattering is periodical, that means the positive amplitude and negative amplitude, $|+\Delta|$ and $|-\Delta|$ are equal.

For fulfilling the control of the boost converter, we only need to set the bandwidth of the chattering and hysteresis loop in the DSP. The hysteresis plane can be illustrated by Figure 12.51, while the mechanism for implementing the hysteresis plane in DSP environment can be presented by Figure 12.52.

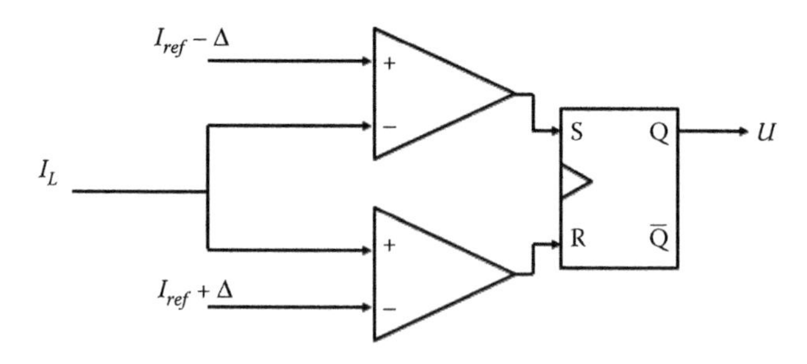

FIGURE 12.52
Flip-Flop for implementing hysteresis loop.

Before implementing control strategies on the DSP board, several considerations should be taken:

- *Amplitude of the chattering:* The ripple component of the inductor current is periodical if the system operates under CCM. Users normally need the periodical ripple component for regulating the system behaviors to guarantee the CCM of the system is the top consideration. To guarantee the systems operates under CCM, the average inductor current must be greater than or at least be equal to the amplitude of the chattering, in this case, the amplitude of the chattering is $|\Delta|$.
- *Frequency of the switching signal:* There are two main approaches to suppressing the amplitude of the ripple component of the inductor current for assisting the functionalities of the CCM. The first one is to increase the switching frequency, but a high-frequency switching signal may increase power loss, moreover it may introduce difficulties to the current sensing. The second one is to increase the size of the inductor, but the cost of the system will be increased and the system becomes bulky.
- *Sampling rate for the current sensing:* High sampling rate is good for capturing current information and reconstructing current waveforms in DSPs, while the high sampling rates are always faced with practical limitations. In this case, the sampling rate is not the natural sampling rate of the ADC channel. The sampling rate in this chapter is the rate of the CPU accessing the ADCRESULT registers. The CPU can normally fulfill high performance in terms of high-frequency ADC sampling and handling high-frequency ISRs. However, the rate of handling ISRs is mainly affected by the structure of the embedded code and numbers of the interrupt events. Knowing that each ISR contains multiple C codes (which consume plenty of time) and the next ISR can be only handled after the previous one is served, the frequency of the CPU accessing the ISRs is affected by the length of embedded code. For gathering sufficient information of the current waveform, the sampling rate for the DSP control should be at least 20 times greater than the fundamental frequency of the sensing signal.

12.7.2 MATLAB/Simulink Control Diagram for the DSP Implementation

Once deciding all required parameters of the boost converter, users can begin to design a control diagram in the MATLAB/Simulink environment with Embedded Coder. To fulfill the control strategy shown in Figure 12.52, the control diagram can be drawn as shown in Figure 12.53.

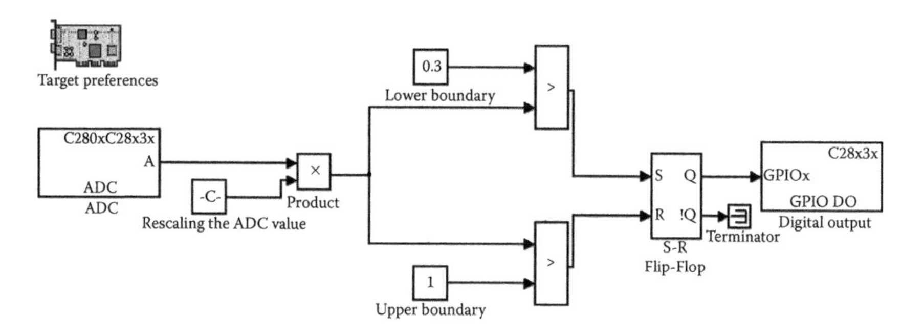

FIGURE 12.53
Control diagram for implementing the Flip-Flop in DSP board.

According to the different scaling factor of the current sensing circuit, the value of the ADCRESULTx register is needed to be rescaled for reconstructing accurate information of the sensed signal. To achieve the rescaling, a constant component and product component are used. The output value of the product component presents the correct instant value of the sensing signal. Two comparison components are used for building the Flip-Flop block. Each comparator will generate "1" as long as the condition is satisfied. The S-R Flip-Flop component will generate a binary value referring to the inner logic table. The GPIO component will pull up/down the corresponding GPIO pin on DSP board, according to its input binary value. This approach is a typical Sliding Mode Control.

There is another type of control diagram generally used for voltage regulation of the power electronics circuit. This control diagram presents the control strategy shown in Figure 12.54. In the case of the DSP implementation, the direct voltage controller is normally a digital PID controller or a digital poles-zeros controller. The RTDX component shown in Figure 12.54 can be used for

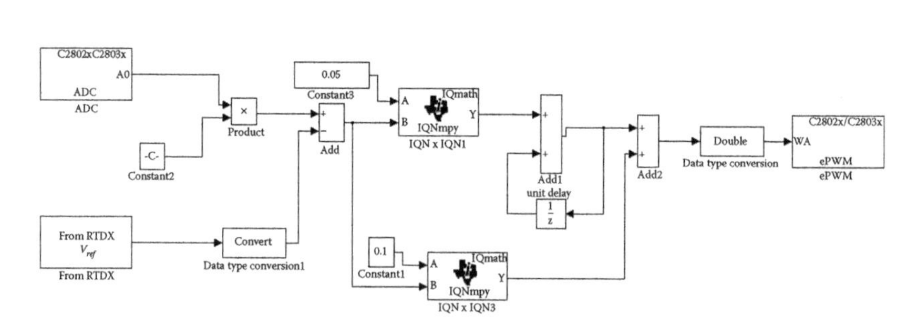

FIGURE 12.54
Digital PID controller.

real-time setting the output voltage reference. The *IQN*IQNx* components are used to fulfill product operation and format the output data type. The output of the digital PID controller, the value of duty-ratio of the generated PWM signal will be changed in the next switching period. Subsequently, the ePWM generate will regulate the system dynamics throughout manipulating pulse by pulse.

12.8 Conclusion

In this chapter, basic steps and elements for using DSPs are introduced. For DSP-controlled power electronics systems, several significant embedded modules such as ADC modules, PIE, and ePWM generators should be mastered by users. Significant control registers and data registers are explained in detail in this chapter for helping users quickly understand the operations of C2000 DSPs. Depending on practical limitations, users may need to gather more information of each embedded module in C2000 DSP in terms of creating mechanisms for a better inner cooperation. In cases of high-frequency applications, synchronous control and asynchronous control are generally involved in DSP designs. If so, users may need to be concerned with implementing the time of each embedded subroutines, besides setting the proper phase shift between synchronous clocks. For DSP-controlled power electronics systems, the desired analog S-domain controller is converted to a Z-domain digital controller so that the target phase margin will be varied by the time-delay always considered in the DSP implementation. To enhance the functionalities of the time-delay system and other control functionalities, tuning work may be necessary.

References

1. H. Lee, V. Utkin, and A. Malinin, Chattering reduction using multiphase sliding mode control, *International Journal of Control*, 82(9), 1720–1737, 2009.
2. TMS320F2833x, 2823x Digital Signal Controllers (DSCs) (SPRS439M).
3. TMS320x2833x, 2823x Serial Peripheral Interface (SPI) Reference Guide (SPRUEU3A).
4. TMS320x2833x, 2823x System Control and Interrupts (SPRUFB0D).
5. TMS320x 2833x, 2823x Enhanced Pulse Width Modulator (ePWM) Module (SPRUG04A).
6. TMS320x2833x Analog-to-Digital Converter (ADC Module) Reference Guide (SPRU812A).

Appendix A: Linear Control

A.1 Introduction

Since this book is aimed at automotive and power industry engineers, as well as researchers and graduate students in the areas of renewable energy, circuits, and control, it has been assumed that the reader has some basic knowledge of differential equations, the Laplace transform, transfer functions, poles and zeros, as well as matrix algebra. Furthermore, it would be desirable for the reader to have taken a course on continuous-time linear control systems. But to make this book accessible to a wider audience, we include a brief review of linear systems and control in this appendix. Furthermore, we describe certain aspects of nonlinear control theory that have been applied to obtain various results described in this book.

A.2 Linear Systems and Control

A.2.1 State Variables and State Equations

The *state variables* of a dynamic system are a *minimal* set of variables, usually denoted as $x_1(t), x_2(t),\ldots, x_n(t)$, that can describe the system's dynamic behavior completely in terms of n first-order linear differential equations

$$\begin{aligned}
\dot{x}_1(t) &= a_{11}x_1 + a_{12}x_2 + \cdots + a_{1n}x_n + b_{11}u_1 + \cdots + b_{1m}u_m \\
\dot{x}_2(t) &= a_{21}x_1 + a_{22}x_2 + \cdots + a_{2n}x_n + b_{21}u_1 + \cdots + b_{2m}u_m \\
&\vdots \\
\dot{x}_n(t) &= a_{n1}x_1 + a_{n2}x_2 + \cdots + a_{nn}x_n + b_{n1}u_1 + \cdots + b_{nm}u_m
\end{aligned} \tag{A.1}$$

where $u_i(t)$, $i = 1, 2,\ldots, m$, represent the system's (control) *inputs*, and all the a and b coefficients are (time-invariant) constants. These first-order linear differential equations are then known as the *state equations* of the dynamic system.

The system's *outputs* are those variables that are measured (using sensors), usually denoted by $y(t)$. If these linearly depend on the system's state variables and inputs, then, they can be mathematically described by the equations

$$y_1(t) = c_{11}x_1 + c_{12}x_2 + \cdots + c_{1n}x_n + d_{11}u_1 + \cdots + d_{1m}u_m$$
$$y_2(t) = c_{21}x_1 + c_{22}x_2 + \cdots + c_{2n}x_n + d_{21}u_1 + \cdots + d_{2m}u_m$$
$$\vdots \qquad \vdots$$
$$y_p(t) = c_{p1}x_1 + c_{p2}x_2 + \cdots + c_{pn}x_n + d_{p1}u_1 + \cdots + d_{pm}u_m \tag{A.2}$$

where $p \leq n$, and all the c and d coefficients are constants. Then, Equations A.1 and A.2 can be written much more compactly in vector–matrix form as

$$\dot{x}(t) = Ax(t) + Bu(t)$$
$$y(t) = Cx(t) + Du(t) \tag{A.3}$$

where $\dot{x}(t)$, $x(t)$, $u(t)$, and $y(t)$ are $n \times 1$, $n \times 1$, $m \times 1$, and $p \times 1$ vectors of the state derivatives, state variables, inputs, and outputs, respectively, while A, B, C, and D are $n \times n$, $n \times m$, $p \times n$, and $p \times m$ matrices, respectively.

EXAMPLE

State equations describing a mass–spring–damper system.

First, note that a mass–spring–damper system's dynamic behavior is completely described by the second-order linear differential equation

$$M\ddot{y}(t) + B\dot{y}(t) + Ky(t) = u(t) \tag{A.4}$$

where $y(t)$ represents the displacement of the mass M when it is subject to the force $u(t)$ as shown in Figure A.1; B and K are the system's damping and spring constants, respectively.

Let us define two new variables $x_1(t) = y(t)$ and $x_2(t) = \dot{y}(t)$ to be this system's state variables. Then, we can rewrite Equation A.4, after some simple algebra, equivalently as the following first-order linear differential

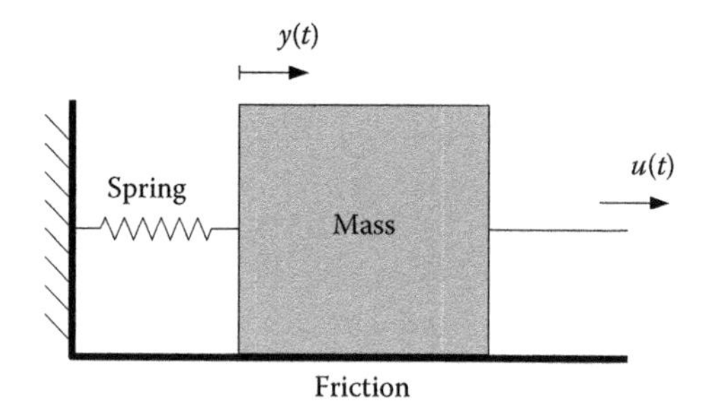

FIGURE A.1
Mass–spring–damper system.

equations, which represent a second-order state-equation description of this mass–spring–damper system.

$$\begin{aligned}\dot{x}_1(t) &= \dot{y}(t) = x_2(t)\\ \dot{x}_2(t) &= \ddot{y}(t) = -\frac{K}{M}x_1(t) - \frac{B}{M}x_2(t) + u(t)\end{aligned} \quad \text{(A.5)}$$

This can be expressed in vector–matrix form as

$$\begin{pmatrix}\dot{x}_1\\ \dot{x}_2\end{pmatrix} = \begin{pmatrix}0 & 1\\ -\dfrac{K}{M} & -\dfrac{B}{M}\end{pmatrix}\begin{pmatrix}x_1\\ x_2\end{pmatrix} + \begin{pmatrix}0\\ 1\end{pmatrix}u \quad \text{(A.6)}$$

A.2.2 Linear Approximation of a Nonlinear System

For a nonlinear dynamic system, it is also possible to define state variables and thereby obtain a set of state equations, although these will be nonlinear, for describing the dynamic behavior of that system. However, assuming that the system operates in a limited range, i.e., the variables do not change by much, about an operating or equilibrium point defined by the set of state variable and input values (x_0, u_0), we can approximate the system's behavior within that range (its small-signal behavior, using electrical engineering terminology) by a set of *linear* state equations. A standard procedure for doing so, via the mathematical concept of Taylor series expansion, can be found in many undergraduate control systems textbooks such as Reference 1. This same approximation technique has been applied to obtain the PEMFC linear model described in Section 3.5.1 of this book.

EXAMPLE

Linear approximation of a generic second-order nonlinear system.

Suppose that we are given a second-order dynamic system, with state variables x_1, x_2, and a single input u, which is described by the nonlinear state equations

$$\begin{aligned}\dot{x}_1(t) &= f_1(x_1, x_2, u)\\ \dot{x}_2(t) &= f_2(x_1, x_2, u)\end{aligned} \quad \text{(A.7)}$$

where f_1 and f_2 are some nonlinear functions. Denoting the system's operating point values to be (x_{10}, x_{20}, u_0), then, a linear approximation of Equation A.7 are the equations

$$\begin{aligned}\dot{x}_1(t) &= \left.\frac{df_1}{dx_1}\right|_{x_{10},x_{20},u_0} x_1 + \left.\frac{df_1}{dx_2}\right|_{x_{10},x_{20},u_0} x_2 + \left.\frac{df_1}{du}\right|_{x_{10},x_{20},u_0} u\\ \dot{x}_2(t) &= \left.\frac{df_2}{dx_1}\right|_{x_{10},x_{20},u_0} x_1 + \left.\frac{df_2}{dx_2}\right|_{x_{10},x_{20},u_0} x_2 + \left.\frac{df_2}{du}\right|_{x_{10},x_{20},u_0} u\end{aligned} \quad \text{(A.8)}$$

Note that the evaluation of the calculated partial derivatives at the operating point values will yield constant coefficients for x_1, x_2, and u in the equations.

A.2.3 Characteristic Equation, Characteristic Roots, and Eigenvalues of a Linear System

For the linear dynamic system described by the state equations A.3, its input–output *transfer function* can be expressed as

$$Y(s)/U(s) = C(sI - A)^{-1}B + D \tag{A.9}$$

where s is the Laplace transform's complex frequency variable and I represents the identity matrix. Furthermore, the system's *characteristic roots* are given by the roots of the matrix determinant of $sI-A$ (an nth-order polynomial of s) equated to 0, i.e.,

$$|sI - A| = 0 \tag{A.10}$$

These n roots are also the *poles* of the system (subject to exact cancelation of the system's zeros) and which be alternatively obtained by finding the *eigenvalues* of the system matrix A.

A.2.4 Linear-State Feedback Control

We consider a linear dynamic system described by the state equations A.3, where the control has the form

$$u(t) = -Kx(t) \tag{A.11}$$

where the state feedback gain K is a vector (or a matrix in the multiple-input case) with constant elements. This form of control indicates that

1. All state variables can be, and are, measured
2. The control is a linear combination of the state variables
3. The closed-loop control system consisting of Equations A.3 and A.11 can now be described by

$$\dot{x}(t) = (A - BK)x(t) \tag{A.12}$$

4. The characteristic equation of the closed-loop system is then given by $|sI-(A - BK)| = 0$

Now, if the pair of matrices (A, B) are such that the open-loop system (A.3) is completely state controllable [1], then, it is well known that the closed-loop system's characteristic roots can be "placed" anywhere in the complex s-plane (subject to the complex roots being in conjugate pairs), by appropriately choosing the elements of K. This state feedback gain design procedure is typically called a *pole-placement* design.

EXAMPLE

State feedback pole placement for a generic second-order system.

Suppose that we are given a dynamic system that is described by the state equation

$$\begin{pmatrix} \dot{x}_1 \\ \dot{x}_2 \end{pmatrix} = \begin{pmatrix} 0 & 1 \\ -a_0 & -a_1 \end{pmatrix} \begin{pmatrix} x_1 \\ x_2 \end{pmatrix} + \begin{pmatrix} 0 \\ 1 \end{pmatrix} u \tag{A.13}$$

This system's characteristic equation is then given by

$$s^2 + a_1 s + a_0 = 0 \tag{A.14}$$

with two eigenvalues (poles) that may be in undesirable locations, e.g., in the right-half s-plane, which corresponds to an unstable system. Since (A, B) is a completely controllable pair for this system, we can apply state feedback control to place the closed-loop system's poles at any other locations; but it is necessary to have the measured outputs $y(t) = x(t)$.

Then, the needed state feedback gain is given by $K = (k_0\ k_1)$, which means that the system under state feedback control is described by

$$\begin{pmatrix} \dot{x}_1 \\ \dot{x}_2 \end{pmatrix} = \begin{pmatrix} 0 & 1 \\ -[a_0 + k_0] & -[a_1 + k_1] \end{pmatrix} \begin{pmatrix} x_1 \\ x_2 \end{pmatrix} \tag{A.15}$$

This yields the characteristic equation

$$s^2 + (a_1 + k_1)s + (a_0 + k_0) = 0 \tag{A.16}$$

so that the elements of K can therefore be chosen appropriately to obtain the desired characteristic equation that corresponds to having the closed-loop system's poles at those locations that would yield the desired system stability and performance characteristics.

Reference

1. K. Ogata, *Modern Control Engineering*, Prentice-Hall, Upper Saddle River, New Jersey, 2002.

Appendix B: Nonlinear Control

In nature, almost all systems are nonlinear systems, for example, a fuel cell system, power grids, etc. Linear control is based on the linearized modeling of the system, which is accurate only at the operating point of linearization. To achieve a better control performance, nonlinear modeling and nonlinear control need to be considered. Nonlinear modeling of a fuel cell system is previously given in Chapter 3. This appendix provides a preliminary introduction to a nonlinear control technique, especially the exact linearization approach [1–3].

B.1 Nonlinear Coordinate Transformation and Diffeomorphism

For a given set of nonlinear equations

$$Z = \Phi(X) \tag{B.1}$$

where Z and X are vectors with $(n \times 1)$ dimension, Φ is a nonlinear vector function with $(n \times 1)$ dimension, and the set of a nonlinear transformation in Equation B.1 is called a nonlinear coordinate transformation, and the following two conditions are satisfied:

1. The inverse transformation of Φ exists, which means that Φ^{-1} exists
2. Both Φ and Φ^{-1} are smooth vector functions, that is, the function of each component of both Φ and Φ^{-1} has continuous partial derivatives of any orders

If the above two conditions are satisfied, the coordinate transformation $\Phi(X)$ is also called a diffeomorphism between two coordinate spaces.

From a geometric point of view, the coordinate transformation $Z = \Phi(X)$ and $X = \Phi^{-1}(Z)$ can be considered as a mapping between two spaces with the same dimension X and Z.

B.2 Local Diffeomorphism

The above two conditions may be satisfied only for a neighborhood of a specific point X^0 rather than all points in the space. A local diffeomorphism is then defined on a certain domain.

A property of a local diffeomorphism is given as follows:

Suppose $\Phi(X)$ is a smooth function defined on a certain subset S of space R^n. If the Jacobian matrix $\partial\Phi/\partial X$ at $X = X^0$ is nonsingular, then, $\Phi(X)$ is a local diffeomorphism in an open subset S^0 including X^0.

B.3 Coordinate Transformation of Nonlinear Control Systems

Given a nonlinear control system

$$\begin{cases} \dot{X} = f(X) + g(X)u \\ y = h(X) \end{cases} \tag{B.2}$$

where $X \in R^n$ and $u \in R$ are state variables and control variables, respectively, $y \in R$ is the output variable, f and g are nonlinear function vectors, and h is a nonlinear function.

Let us select a local diffeomorphism $Z = \Phi(X)$. Then, considering Equation B.2, we have

$$\dot{Z} = \frac{d\Phi}{dX} = \frac{\partial\Phi}{\partial X}\frac{dX}{dt} = \frac{\partial\Phi}{\partial X}\dot{X} = \frac{\partial\Phi}{\partial X}(f(X) + g(X)u)$$

And, considering $X = \Phi^{-1}(Z)$, therefore, we get

$$\begin{cases} \dot{Z} = \dfrac{\partial\Phi}{\partial X}[f(X) + g(X)u] = \dfrac{\partial\Phi}{\partial X}[f(\Phi^{-1}(Z)) + g(\Phi^{-1}(Z))u] \\ y = h(X) = h(\Phi^{-1}(Z)) \end{cases}$$

Define

$$\tilde{f}(Z) = \frac{\partial\Phi}{\partial X} f(X)|_{X=\Phi^{-1}(Z)}$$

$$\tilde{g}(Z) = \frac{\partial\Phi}{\partial X} g(X)|_{X=\Phi^{-1}(Z)}$$

$$\tilde{h}(Z) = h(X)|_{X=\Phi^{-1}(Z)}$$

Then, we have

$$\begin{cases} \dot{Z} = \tilde{f}(Z) + \tilde{g}(Z)u \\ y = \tilde{h}(Z) \end{cases} \tag{B.3}$$

Equation B.3 represents the transformed nonlinear system using the selected local diffeomorphism.

B.4 Affine Nonlinear Control Systems

A control system is called an affine nonlinear control system if it is nonlinear to a state variable vector but linear to a control variable vector. It has the following format:

$$\begin{cases} \dot{X} = f(X) + \sum_{i=1}^{m} g_i(X)u_i \\ Y = h(X) \end{cases}$$

where $X \in R^n$ and $u \in R$ are state variable vectors and control variable vectors, respectively, $y \in R$ is the output variable, f and g are nonlinear function vectors, and h is a nonlinear function.

Nonlinear function vectors $f(X)$ and $g(X)$ are also called vector fields of state space.

B.5 Derived Mapping of Vector Fields

Derived mapping of vector fields is very important for the exact linearization technique. Here is the definition.

For a given diffeomorphism, $Z = \Phi(X)$, and a vector field $f(X)$, if the Jacobian matrix of $\Phi(X)$ is $J_\Phi = \partial\Phi/\partial X$, then, the derived mapping of $f(X)$ under the mapping $\Phi(X)$ is defined as

$$\Phi_*(f) = J_\Phi(X)f(X)\big|_{X=\Phi^{-1}(Z)}$$

The derived mapping of a vector field $f(X)$ is a transformation that moves the vector field $f(X)$ from the state space to Z space based on the diffeomorphism $Z = \Phi(X)$.

B.6 Lie Derivative and Lie Bracket

Two essential geometric concepts of nonlinear systems are Lie derivative and Lie bracket, which are used in the application of the exact linearization technique.

B.6.1 Lie Derivative

Given a differentiable scalar function $\xi(X) = \xi(x_1, x_2, \ldots, x_n)$ and a vector field $f(X)$ of X, a new scalar function, denoted by $L_f\xi(X)$, is obtained by the following operation:

$$Lf\xi(X) = \frac{\partial \xi(X)}{\partial X} f(X) = \sum_{i=1}^{n} \frac{\partial \xi(X)}{\partial x_i} f_i(X)$$

and called the Lie derivative of function $\xi(X)$ along the vector field $f(X)$.

B.6.2 Lie Bracket

Suppose there are two vector fields $f(X)$ and $g(X)$. The Lie bracket of $g(X)$ along $f(X)$ is defined by the following operation:

$$[f, g] = ad_f g = \frac{\partial g}{\partial X} f - \frac{\partial f}{\partial X} g$$

Since Lie bracket of $g(X)$ is a new vector field, it can be used to calculate the Lie bracket along $f(X)$ once more.

$$ad_f^{\,2} g(X) = [f, [f, g]](X)$$

$$\ldots$$

$$ad_f^{\,k} g(X) = [f, ad_f^{\,k-1} g](X)$$

The main operational rules of Lie bracket are

1. Lie bracket is skew symmetric

$$[f, g] = -[g, f]$$

2. If λ_1 and λ_2 are two real numbers, then

$$[f, \lambda_1 g_1 + \lambda_2 g_2](X) = \lambda_1 [f, g_1](X) + \lambda_2 [f, g_2](X)$$

3. If $\gamma(X)$ is another vector field, then

$$[f,[g,\gamma]]+[\gamma,[f,g]]+[g,[\gamma,f]]=0$$

4. If $f(X)$ and $g(X)$ are vector fields, and $\xi(X)$ is a scalar function, then, the Lie derivative of $\xi(X)$ along the vector field $[f(X), g(X)]$ is

$$L_{[f,g]}\xi(X)=L_f L_g \xi(X)-L_g L_f \xi(X)$$

B.7 Involutivity of Vector Field Sets

Suppose there are k vector fields $g_1(X),\ldots, g_n(X)$, all with an n-dimension, let us form the matrix

$$G=[g_1(X)\; g_2(X)\cdots g_k(X)$$

If the matrix G has rank k at $X = X^0$ and the following augmented matrix

$$[g_1(X)\; g_2(X)\cdots g_k(X)\,[g_i, g_j]$$

has the same rank k at $X = X^0$ for any i and j where $1 \le i, j \le \mathrm{k}$, then, the vector field set $\{g_1, g_2,\ldots, g_k\}$ is called an involutive one or we say that it has the property of involutivity.

B.8 Relative Degree of a Control System

Suppose there is an SISO nonlinear control system

$$\begin{cases} \dot{X}=f((X)+g(X)u) \\ \quad Y=h(X) \end{cases}$$

where $X \in R^n$, $u \in R$, $y \in R$, $f(X)$, and $g(X)$ are vector field items.

The above is said to have a relative degree γ in a neighborhood Ω of $X = X^0$, if the following two conditions are satisfied:

1. The Lie derivative of the function $L_f^{\,k} h(X)$ along g equals zero in Ω, i.e.,

$$L_g L_f^{\,k} h(X)=0,\quad k<\gamma-1,\quad \forall x\in\Omega$$

2. The Lie derivative of the function $L^{\gamma-1}{}_f h(X)$ along the vector field $g(X)$ is not equal to zero in Ω, i.e.,

$$L_g L_f^{\gamma-1} h(X) \neq 0$$

B.9 Exact Linearization Control

The exact linearization control design principle is to exactly linearize a nonlinear control system into a controllable linear system, and to get the state nonlinear feedback control law based on the linear control system. Converting the obtained control law back into the original nonlinear space results in the nonlinear control design for the original control system.

B.9.1 Conditions of the Exact Linearization

To introduce the conditions of the exact linearization, we need to introduce a theorem.

Frobenius Theorem

Consider the following partial differential equations set:

$$\frac{\partial h(X)}{\partial X}\left[Y_1(X)\; Y_2(X) \cdots Y_k(X)\right] = 0 \tag{B.4}$$

where $\partial h(X)/\partial X$ is the gradient vector of $h(X)$; $Y_1(X)$, $Y_2(X)$,..., $Y_k(X)$ are the n-dimensional vector fields defined on X space.

Suppose the matrix $[Y_1(X)\; Y_2(X) \cdots Y_k(X)]$ has rank k at $X = X^0$. If and only if the augmented matrix $[Y_1(X)\; Y_2(X) \cdots Y_k(X)\; [Y_i, Y_j]]$ still has rank k for all X in a neighborhood of X^0, $n-k$ scalar functions must exist, defined in a neighborhood Ω of X^0, which are the solutions of Equation B.4, such that the Jacobian matrix $\partial h(X)/\partial X$ has rank $n-k$ at $X = X^0$. ■

The conditions proposed in Frobenius theorem are actually the conditions of involutivity of a vector field $\{Y_1, Y_2, \ldots, Y_k\}$.

B.9.2 Nonlinear Control Design by Exact Linearization

Consider a SISO nth-order nonlinear control system as follows:

$$\begin{cases} \dot{X} = f(X) + g(X)u \\ y = h(X) \end{cases} \tag{B.5}$$

If we assume that the relative degree of the system $r = n$, then, the system could be transformed into a controllable linear system

$$\dot{Z} = AZ + Bv \tag{B.6}$$

where the matrices A and B are Brunovsky normal form, and Z and v are state variable vectors and control variables, respectively.

This linearization can be generated by using the nonlinear-state feedback

$$u = -\frac{\alpha(X)}{\beta(X)} + \frac{1}{\beta(X)}v \tag{B.7}$$

where

$$\begin{cases} \alpha(X) = L_f^{\,n} h(X) \\ \beta(X) = L_g L_f^{\,n-1} h(X) \end{cases}$$

and the diffeomorphism coordinate transformation is

$$Z = \Phi(X) = \begin{bmatrix} h(X) \\ L_f h(X) \\ \vdots \\ L_f^{\,n-1} h(X) \end{bmatrix} \tag{B.8}$$

The optimal control of the linear system in Equation B.6 is

$$v^* = -B^T P^* Z(t)$$

where P^* is the solution of the Riccati matrix equation.

Considering the coordinate transformation $Z = \Phi(X)$, we can obtain the following nonlinear control law:

$$u = \frac{-(L_f^{\,n} h(X) + k_n^* L_f^{\,n-1} h(X) + \cdots + k_2^* L_f h(X) + k_1^* h(X))}{L_g L_f^{\,n-1} h(X)}$$

The Brunovsky normal form is given as follows:

$$\dot{Z} = AZ + Bv$$

where

$$Z = \begin{bmatrix} z_1 \\ z_2 \\ \vdots \\ z_n \end{bmatrix}$$

$$A = \begin{bmatrix} 0 & 1 & 0 & \cdots & 0 & 0 \\ 0 & 0 & 1 & \cdots & 0 & 0 \\ \vdots & \vdots & \vdots & \vdots & \vdots & \vdots \\ 0 & 0 & 0 & \cdots & 0 & 1 \\ 0 & 0 & 0 & 0 & 0 & 0 \end{bmatrix}$$

$$B = \begin{bmatrix} 0 \\ 0 \\ \vdots \\ 0 \\ 1 \end{bmatrix}$$

References

1. A. Isidori, *Nonlinear Control System: An Introduction* (3rd Edition), Springer-Verlag, New York, 1995.
2. D. Cheng, T.J. Tarn, and A. Isidori, Global linearization of nonlinear system via feedback, *IEEE Transactions on AC*, 30(8), 808–811, 1985.
3. Q. Lu, Y. Sun, and S. Mei, *Nonlinear Control Systems and Power System Dynamics,* Kluwer Academic Publishers, Boston, Massachusetts, 2001.

Appendix C: Induction Machine Modeling and Vector Control for Fuel Cell Vehicle Applications

C.1 Voltage Equations of the Induction Machine

The idealized induction machine is shown in Figure C.1.

If the winding of the stator is sinusoidally distributed, then, stator and rotor voltage equations are expressed as follows [1]:

$$\begin{aligned} V_{abcs} &= R_s i_{abcs} + p\bar{\lambda}_{abcs} \\ V_{abcr} &= R_r i_{abcr} + p\bar{\lambda}_{abcr} \end{aligned} \tag{C.1}$$

where $p = d/dt$, $V_{abcs} = [V_{as}\ \ V_{bs}\ \ V_{cs}]^T$, $i_{abcs} = [i_{as}\ \ i_{bs}\ \ i_{cs}]^T$, $\bar{\lambda}_{abcs} = [\lambda_{as}\ \ \lambda_{bs}\ \ \lambda_{cs}]^T$, $V_{abcr} = [V_{ar}\ \ V_{br}\ \ V_{cr}]^T$, $i_{abcr} = [i_{ar}\ \ i_{br}\ \ i_{cr}]^T$, and $\lambda_{abcs} = [\lambda_{as}\ \ \lambda_{bs}\ \ \lambda_{cs}]^T$

The s and r subscripts represent the equations in terms of stator and rotor windings.

The flux linkage equations are given

$$\begin{bmatrix} \bar{\lambda}_{abcs} \\ \bar{\lambda}_{abcr} \end{bmatrix} = \begin{bmatrix} L_s & L_{sr} \\ L_{sr}^T & L_r \end{bmatrix} \cdot \begin{bmatrix} i_{abcs} \\ i_{abcr} \end{bmatrix} \tag{C.2}$$

where

$$L_s = \begin{bmatrix} L_{ls} + L_{ms} & -\frac{1}{2}L_{ms} & -\frac{1}{2}L_{ms} \\ -\frac{1}{2}L_{ms} & L_{ls} + L_{ms} & -\frac{1}{2}L_{ms} \\ -\frac{1}{2}L_{ms} & -\frac{1}{2}L_{ms} & L_{ls} + L_{ms} \end{bmatrix}; \ L_r = \begin{bmatrix} L_{lr} + L_{mr} & -\frac{1}{2}L_{mr} & -\frac{1}{2}L_{mr} \\ -\frac{1}{2}L_{mr} & L_{ls} + L_{mr} & -\frac{1}{2}L_{mr} \\ -\frac{1}{2}L_{mr} & -\frac{1}{2}L_{mr} & L_{ls} + L_{mr} \end{bmatrix} \text{ and}$$

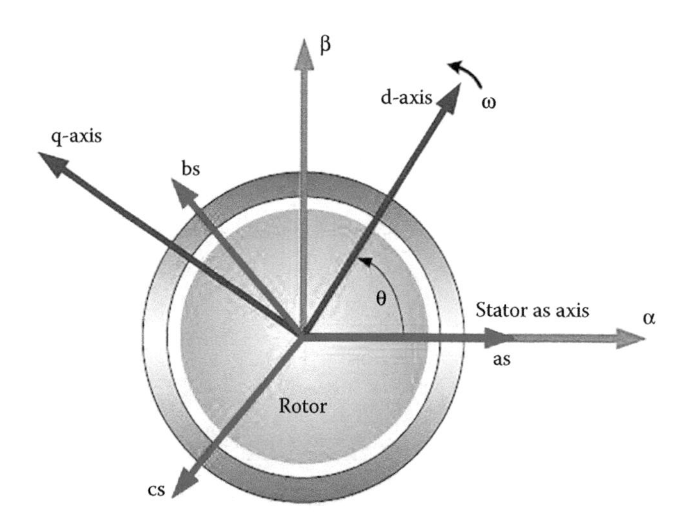

FIGURE C.1
Idealized induction machine. (Adapted from J. Liu, Modeling, analysis and design of integrated starter generator system based on field oriented controlled induction machines, PhD thesis, Ohio State University, Ohio, 2005.)

$$L_{sr} = L_{sr}\begin{bmatrix} \cos\theta_r & \cos\left(\theta_r + \frac{2\pi}{3}\right) & \cos\left(\theta_r - \frac{2\pi}{3}\right) \\ \cos\left(\theta_r - \frac{2\pi}{3}\right) & \cos\theta_r & \cos\left(\theta_r + \frac{2\pi}{3}\right) \\ \cos\left(\theta_r + \frac{2\pi}{3}\right) & \cos\left(\theta_r - \frac{2\pi}{3}\right) & \cos\theta_r \end{bmatrix}$$

where L_{ms} is the stator-magnetizing inductance; L_{mr} the rotor-magnetizing inductance; L_{ls} the stator leakage inductance; L_{lr} the rotor leakage inductance; and L_{sr} is the mutual inductance between stator and rotor windings.

C.2 Voltage Equations on the Stationary Reference Frame

Using reference frame theory (Appendix D), we can transform the induction machine variables (voltage, current, and linkage) to another reference frame as long as the condition that $f_a + f_b + f_c = 0$ (f_a, f_b, and f_c are each phase of v, i, and λ components) is satisfied. Using Park transformation, these three phases can be changed on the stationary reference frame

$$f_s^s = f_{ds}^s + jf_{qs}^s = \frac{2}{3}(f_a + f_b e^{j2\pi/3} + f_b e^{j4\pi/3}) \tag{C.3}$$

or equivalently

$$\begin{bmatrix} f_{ds}^s \\ f_{qs}^s \end{bmatrix} = \frac{2}{3}\begin{bmatrix} 1 & -\frac{1}{2} & -\frac{1}{2} \\ 0 & \frac{\sqrt{3}}{2} & -\frac{\sqrt{3}}{2} \end{bmatrix}\begin{bmatrix} f_a \\ f_b \\ f_c \end{bmatrix} \tag{C.4}$$

Simply, we can write stator and rotor voltage equations as

$$\text{Stator voltage equation: } v_s^s = R_s i_s^s + p\lambda_s^s \tag{C.5}$$

$$\text{Rotor voltage equation: } v_r^r = R_r i_r^r + p\lambda_r^r \tag{C.6}$$

Assuming that there is a relationship between the stationary reference frame and the synchronous reference frame

$$f_r^s = e^{j\theta_r} \cdot f_r^r \tag{C.7}$$

where $\theta_r = \omega_r t$ is the phase angle difference between the stationary reference frame and the synchronous frame.

$$v_r^r = R_r i_r^r + p\lambda_r^r \tag{C.8}$$

$$\begin{aligned} v_r^s \cdot e^{-j\theta_r} &= R_r \cdot i_r^s \cdot e^{-j\theta_r} + p\left(\lambda_r^s \cdot e^{-j\theta_r}\right) \\ &= R_r \cdot i_r^s \cdot e^{-j\theta_r} + p\left(\lambda_r^s \cdot e^{-j\theta_r}\right) \\ &= R_r \cdot i_r^s \cdot e^{-j\theta_r} + e^{-j\theta_r} \cdot p\lambda_r^s - p(e^{-\theta_r}) \cdot \lambda_r^s \\ &= R_r \cdot i_r^s \cdot e^{-j\theta_r} + e^{-j\theta_r} \cdot p\lambda_r^s - \omega_r \cdot e^{-\theta_r} \cdot \lambda_r^s \\ \Rightarrow v_r^s &= R_r \cdot i_r^s + (p - j\omega_r)\lambda_r^s \end{aligned} \tag{C.9}$$

So, the voltage equations on the stationary frame are

$$\text{Stator voltage equation: } v_s^s = R_s i_s^s + p\lambda_s^s \tag{C.10}$$

$$\text{Rotor voltage equation: } v_r^s = R_r i_r^s + (p - j\omega)\lambda_r^s \tag{C.11}$$

C.3 Voltage Equations on the Synchronous Reference Frame

Where

$\lambda_s^s = L_s i_s^s + L_m i_r^s$ (L_s: stator self-inductance and L_m: mutual inductance)

$$\lambda_r^s = L_r i_r^s + L_m i_s^s \ (L_s\text{: rotor self-inductance}) \tag{C.12}$$

By substituting Equation C.12 into Equations C.10 and C.11, the dq model voltage equations on the stationary frame can be derived.

$$\begin{bmatrix} v_{ds} \\ v_{qs} \\ v_{dr} \\ v_{qr} \end{bmatrix} = \begin{bmatrix} R_s + L_s p & 0 & L_m p & 0 \\ 0 & R_s + L_s p & 0 & L_m p \\ L_m p & \omega_r L_m & R_r + L_r p & \omega_r L_r \\ -\omega_r L_m & L_m p & -\omega_r L_r & R_r + L_r p \end{bmatrix} \begin{bmatrix} i_{ds} \\ i_{qs} \\ i_{dr} \\ i_{qr} \end{bmatrix} \tag{C.13}$$

The voltage equations on the stationary frame can be transferred on the synchronous frame by using the relationship given below

$$\begin{aligned} f_s^e &= e^{-j\theta_e} \cdot f_s^s \\ f_r^e &= e^{-j\theta_e} \cdot f_r^s \end{aligned} \tag{C.14}$$

where θ_e is the synchronous angular velocity.

The stator voltage equation becomes

$$\begin{aligned} v_s^s &= R_s i_s^s + p\lambda_s^s \\ e^{j\theta_e} \cdot v_s^e &= R_s i_s^e \cdot e^{j\theta_e} + p\left(e^{j\theta_e} \cdot \lambda_s^e\right) \\ &= R_s i_s^e \cdot e^{j\theta_e} + j\omega_e \cdot e^{j\theta_e} \cdot \lambda_s^e + e^{j\theta_e} \cdot p\lambda_s^e \\ \Rightarrow v_s^e &= R_s i_s^e + (p + j\omega_e)\lambda_s^e \end{aligned} \tag{C.15}$$

The rotor voltage equation becomes

$$\begin{aligned} v_r^s &= R_r i_r^s + (p - j\omega_r)\lambda_r^s \\ e^{j\theta_e} \cdot v_r^e &= R_r i_r^e \cdot e^{j\theta_e} + (p - j\omega_r)\left(\lambda_r^e \cdot e^{j\theta_e}\right) \\ &= R_r i_r^e \cdot e^{j\theta_e} + (p + j\omega_e - j\omega_r) \cdot \lambda_r^e \cdot e^{j\theta_e} \\ \Rightarrow v_r^e &= R_r i_r^e + (p + j(\omega_e - \omega_r)) \cdot \lambda_r^e \\ &= R_r i_r^e + (p + j\omega_{sl}) \cdot \lambda_r^e \end{aligned} \tag{C.16}$$

where the slip speed is defined as

$$\omega_{sl} = (\omega_e - \omega_r) \tag{C.17}$$

And the linkage flux equations can be expressed as

$$\begin{aligned} \lambda_s^e &= L_s i_s^e + L_m i_r^e \\ \lambda_r^e &= L_r i_r^e + L_m i_s^e \end{aligned} \tag{C.18}$$

Using Equations C.16 and C.17, the voltage equation can be described using current, resistance, reactance, and speeds.

$$\begin{bmatrix} v_{ds}^e \\ v_{qs}^e \\ v_{dr}^e \\ v_{qr}^e \end{bmatrix} = \begin{bmatrix} R_s + L_s p & -\omega_e L_s & L_m p & -\omega_e L_m \\ \omega_e L_m & R_s + L_s p & \omega_e L_m & L_m p \\ L_m p & -\omega_{sl} L_m & R_r + L_r p & -\omega_{sl} L_r \\ \omega_{sl} L_m & L_m p & \omega_{sl} L_r & R_r + L_r p \end{bmatrix} \begin{bmatrix} i_{ds}^e \\ i_{qs}^e \\ i_{dr}^e \\ i_{qr}^e \end{bmatrix} \tag{C.19}$$

C.4 *d–q* Equivalent Circuit

Figure C.2 shows the equivalent circuits in an arbitrary reference frame. According to the definition of the rotor speed ω on the reference frame, we can obtain the dynamic *d–q* model in a difference frame. When the rotating speed of the reference frame is zero where $\omega = 0$, the dynamic *d–q* model in the stationary reference frame can be obtained. When the rotating speed of the reference frame is synchronous speed where $\omega = \omega_e$, the dynamic *d–q* model in the synchronous rotating reference frame is obtained.

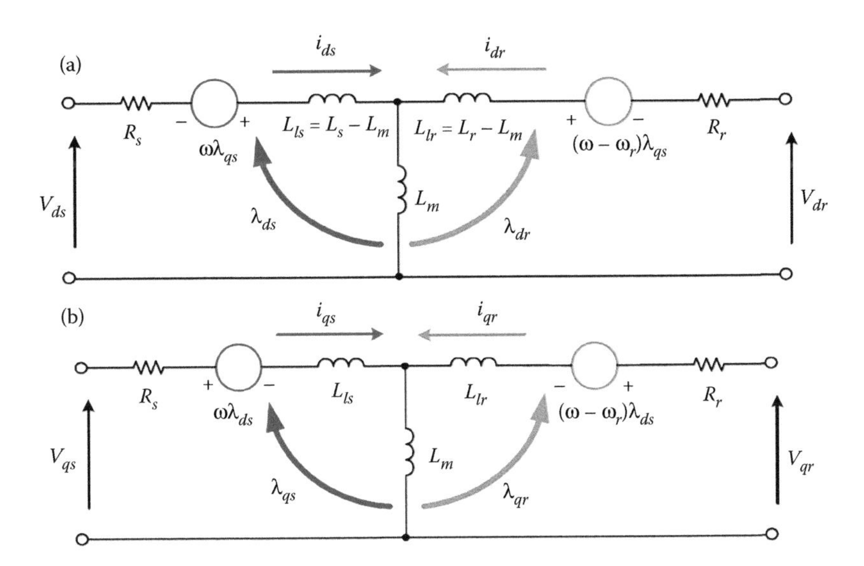

FIGURE C.2
Dynamic *d–q* equivalent circuit. (a) d-axis equivalent circuit. (b) q-axis equivalent circuit. (Adapted from J. Liu, Modeling, analysis and design of integrated starter generator system based on field oriented controlled induction machines, PhD thesis, Ohio State University, Ohio, 2005.)

C.5 Dynamic Induction Machine Model in the State-Space Form

To obtain the state-space equation about the induction machine, let us consider Equation C.20.

$$\begin{bmatrix} i_{ds}^e \\ i_{qs}^e \\ \lambda_{dr}^e \\ \lambda_{qr}^e \end{bmatrix} = \begin{bmatrix} 1 & 0 & 0 & 0 \\ 0 & 1 & 0 & 0 \\ L_m & 0 & L_r & 0 \\ 0 & L_m & 0 & L_r \end{bmatrix} \begin{bmatrix} i_{ds}^e \\ i_{qs}^e \\ i_{dr}^e \\ i_{qr}^e \end{bmatrix} \tag{C.20}$$

By inversing Equation C.20, the following equation is obtained:

$$\begin{bmatrix} i_{ds}^e \\ i_{qs}^e \\ i_{dr}^e \\ i_{qr}^e \end{bmatrix} = \begin{bmatrix} 1 & 0 & 0 & 0 \\ 0 & 1 & 0 & 0 \\ \frac{-L_m}{L_r} & 0 & \frac{1}{L_r} & 0 \\ 0 & \frac{-L_m}{L_r} & 0 & \frac{1}{L_r} \end{bmatrix} \begin{bmatrix} i_{ds}^e \\ i_{qs}^e \\ \lambda_{dr}^e \\ \lambda_{qr}^e \end{bmatrix} \tag{C.21}$$

By substituting Equation C.21 into Equation C.18, a new voltage equation is written as follows:

$$\begin{bmatrix} v_{ds}^e \\ v_{qs}^e \\ v_{dr}^e \\ v_{qr}^e \end{bmatrix} = \begin{bmatrix} R_s & -\sigma\omega_e L_s & 0 & -\frac{\omega_e L_m}{L_s} \\ -\sigma\omega_e L_s & R_s & \frac{\omega_e L_m}{L_s} & 0 \\ -\frac{L_m R_r}{L_r} & 0 & \frac{R_r}{L_r} & -\omega_{sl} \\ 0 & -\frac{L_m R_r}{L_r} & \omega_{sl} & \frac{R_r}{L_r} \end{bmatrix} \begin{bmatrix} i_{ds}^e \\ i_{qs}^e \\ \lambda_{dr}^e \\ \lambda_{qr}^e \end{bmatrix} + \begin{bmatrix} \sigma L_s & 0 & \frac{L_m}{L_r} & 0 \\ 0 & \sigma L_s & 0 & \frac{L_m}{L_r} \\ 0 & 0 & 1 & 0 \\ 0 & 0 & 0 & 1 \end{bmatrix} p \begin{bmatrix} i_{ds}^e \\ i_{qs}^e \\ \lambda_{dr}^e \\ \lambda_{qr}^e \end{bmatrix} \tag{C.22}$$

where $\sigma = 1 - \left(L_m^2 / L_s L_r\right)$

By defining

$$A = \begin{bmatrix} R_s & -\sigma\omega_e L_s & 0 & -\frac{\omega_e L_m}{L_s} \\ -\sigma\omega_e L_s & R_s & \frac{\omega_e L_m}{L_s} & 0 \\ -\frac{L_m R_r}{L_r} & 0 & \frac{R_r}{L_r} & -\omega_{sl} \\ 0 & -\frac{L_m R_r}{L_r} & \omega_{sl} & \frac{R_r}{L_r} \end{bmatrix}; \quad B = \begin{bmatrix} \sigma L_s & 0 & \frac{L_m}{L_r} & 0 \\ 0 & \sigma L_s & 0 & \frac{L_m}{L_r} \\ 0 & 0 & 1 & 0 \\ 0 & 0 & 0 & 1 \end{bmatrix}$$

The state-space equation can be obtained using the above definition

$$\frac{d}{dt}\begin{bmatrix} i_{ds}^e \\ i_{qs}^e \\ \lambda_{dr}^e \\ \lambda_{qr}^e \end{bmatrix} = -B^{-1}A\begin{bmatrix} i_{ds}^e \\ i_{qs}^e \\ \lambda_{dr}^e \\ \lambda_{qr}^e \end{bmatrix} + B^{-1}\begin{bmatrix} v_{ds}^e \\ v_{qs}^e \\ v_{dr}^e \\ v_{qr}^e \end{bmatrix} \tag{C.23}$$

$$\frac{d}{dt}\begin{bmatrix} i_{ds}^e \\ i_{qs}^e \\ \lambda_{dr}^e \\ \lambda_{qr}^e \end{bmatrix} = \begin{bmatrix} -\left(\frac{R_s}{\sigma L_s} + \frac{L_m^2}{\sigma L_s T_r L_r}\right) & \omega_e & \frac{K}{T_r} & \omega_r K \\ -\omega_e & -\left(\frac{R_s}{\sigma L_s} + \frac{L_m^2}{\sigma L_s T_r L_r}\right) & -\omega_r K & \frac{K}{T_r} \\ \frac{L_m}{T_r} & 0 & -\frac{1}{T_r} & \omega_{sl} \\ 0 & \frac{L_m}{T_r} & -\omega_{sl} & -\frac{1}{T_r} \end{bmatrix}\begin{bmatrix} i_{ds}^e \\ i_{qs}^e \\ \lambda_{dr}^e \\ \lambda_{qr}^e \end{bmatrix} + \begin{bmatrix} \frac{1}{\sigma L_s} & 0 \\ 0 & \frac{1}{\sigma L_s} \\ 0 & 0 \\ 0 & 0 \end{bmatrix}\begin{bmatrix} v_{ds}^e \\ v_{qs}^e \end{bmatrix} \tag{C.24}$$

where

$K = (L_m/\sigma L_s L_r)$ and $T_r = (L_r/R_r)$

C.6 Torque Equation

To derive the torque equation, let us consider the following voltage equation:

$$v = Ri + Lpi + G_m i \tag{C.25}$$

where

$$R = \begin{bmatrix} R_s & 0 & 0 & 0 \\ 0 & R_s & 0 & 0 \\ 0 & 0 & R_r & 0 \\ 0 & 0 & 0 & R_r \end{bmatrix}, L = \begin{bmatrix} L_s & 0 & L_m & 0 \\ 0 & L_s & 0 & L_m \\ L_m & 0 & L_r & 0 \\ 0 & L_m & 0 & L_r \end{bmatrix}, v = \begin{bmatrix} v_{ds}^e \\ v_{qs}^e \\ 0 \\ 0 \end{bmatrix}$$

$$G_m = \begin{bmatrix} 0 & -\omega_e L_s & 0 & -\omega_e L_s \\ \omega_e L_s & 0 & \omega_e L_m & 0 \\ 0 & -(\omega_e - \omega_r) L_m & 0 & -(\omega_e - \omega_r) L_r \\ (\omega_e - \omega_r) L_m & 0 & (\omega_e - \omega_r) L_r & 0 \end{bmatrix}, i = \begin{bmatrix} i_{ds}^e \\ i_{qs}^e \\ i_{dr}^e \\ i_{qr}^e \end{bmatrix}$$

If the voltage equation is multiplied by $1.5i^T$, the power equation can be obtained.

$$P = 1.5i^T v = 1.5(i^T Ri + i^T Lpi + i^T G_m i) \tag{C.26}$$

In Equation C.26, $1.5i^T v$ is the motor input power and $1.5i^T Ri$ is the copper loss. The item $1.5i^T Lpi$ represents the rate of exchange of the magnetic field energy between windings and $1.5i^T G_m i$ represents the rate of energy converted into mechanical work. Hence, the mechanical power can be defined as

$$\begin{aligned} P_m &= 1.5i^T G_m i \\ &= 1.5\left(i^T \omega_r G_e i + i^T \omega_r G_r i\right) \\ &= 1.5i^T \omega_r G_r i \end{aligned} \tag{C.27}$$

where

$$G_e = \begin{bmatrix} 0 & -L_s & 0 & -L_m \\ L_s & 0 & L_m & 0 \\ 0 & -L_m & 0 & -L_r \\ -L_m & 0 & L_r & 0 \end{bmatrix}, G_r = \begin{bmatrix} 0 & 0 & 0 & 0 \\ 0 & 0 & 0 & 0 \\ 0 & -L_m & 0 & L_r \\ L_m & 0 & L_r & 0 \end{bmatrix}$$

Since G_e is skew symmetric $\left(G_e^T = -G_e\right)$, $i^T\omega_r G_e i$ is not considered when we calculate the mechanical power P_m.

The torque is obtained by dividing the power by the mechanical speed

$$\begin{aligned} T_e = \frac{P_m}{\omega_m} &= 1.5 \cdot \frac{P}{2} \cdot i^T G_r i \\ &= 1.5 \cdot \frac{P}{2} \cdot L_m \left(i_{qs}^e i_{dr}^e - i_{ds}^e i_{qr}^e\right) \\ &= 1.5 \cdot \frac{P}{2} \cdot \frac{L_m}{L_r} \left(i_{qs}^e \lambda_{dr}^e - i_{ds}^e \lambda_{qr}^e\right) \end{aligned} \quad (C.28)$$

where P is the pole of the machine.

The power in the dq frame is

$$\begin{aligned} P_{DQ} &= v_{ds}^S i_{ds}^s + v_{qs}^s i_{qs}^s \\ &= \begin{bmatrix} v_{ds}^s & v_{qs}^s \end{bmatrix} \begin{bmatrix} i_{qs}^s \\ i_{qs}^s \end{bmatrix} \\ &= \frac{2}{3} [v_a \quad v_b \quad v_c] \begin{bmatrix} i_a \\ i_b \\ i_c \end{bmatrix} \\ &= \frac{2}{3} (v_a i_a + v_b i_b + v_c i_c) \\ &= \frac{2}{3} P_{abc} \end{aligned}$$

Then we get

$$P_{abc} = \frac{3}{2} P_{DQ} \quad (C.29)$$

C.7 Slip Calculation

The fourth row in Equation C.22 is

$$p\lambda_{qr}^e = -\frac{1}{T_r} \lambda_{qr}^e + \frac{L_m}{T_r} i_{qs}^e - \omega_{sl} \lambda_{qr}^e \quad (C.30)$$

For the field orientation control or vector control, I_{ds} is aligned with the rotor flux and I_{qs} is responsible for torque production. I_{ds} is perpendicular to I_{qs} so that the flux and the torque can be controlled separately like a DC machine. Since the rotor flux linkage is only aligned with the d-axis, the vector control condition is that $\lambda_{qr}^e = 0$. In Equation C.24, where $p\lambda_{qr}^e = 0$, Equation C.25 can be modified as

$$\begin{aligned} \omega_{sl}\lambda_{dr}^e &= \frac{L_m}{T_r} i_{qs}^e \\ \Rightarrow \omega_{sl} &= \frac{L_m}{T_r}\frac{i_{qs}^e}{\lambda_{dr}^e} \end{aligned} \tag{C.31}$$

C.8 Calculation of the Rotor Flux λ_{dr}^e on d-Axis

The third row in Equation C.22 is

$$p\lambda_{dr}^e = -\frac{1}{T_r}\lambda_{dr}^e + \frac{L_m}{T_r} i_{ds}^e + \omega_{sl}\lambda_{qr}^e \tag{C.32}$$

Considering the vector control condition $\left(\lambda_{qr}^e = 0\right)$, Equation C.32 can be rewritten as

$$\begin{aligned} p\lambda_{dr}^e + \frac{1}{T_r}\lambda_{dr}^e &= \frac{L_m}{T_r} i_{ds}^e \\ \Rightarrow \lambda_{dr}^e &= \frac{L_m}{1+T_r p} i_{ds}^e \end{aligned} \tag{C.33}$$

Using the steady-state condition, Equation C.33 can be modified as

$$i_{ds}^e = \frac{\lambda_{dr}^e}{L_m} \tag{C.34}$$

which means that i_{ds}^e controls the rotor flux linkage.

C.9 Calculation of the Torque

Since $\lambda_{qr}^e = 0$, the torque equation (C.28) becomes

$$T_e = 1.5 \cdot \frac{P}{2} \cdot \frac{L_m}{L_r} i_{qs}^e \lambda_{dr}^e \tag{C.35}$$

According to the vector control condition that the rotor flux is constant, i_{qs}^e controls the electromagnetic torque.

C.10 DQ Decoupling Control and Back EMF Compensation Control

From the first and second rows of Equation C.24, we can derive the voltage equations as follows, which have coupling terms due to the vector control condition, $\lambda_{qr}^e = 0$:

$$\begin{aligned} v_{ds}^e &= (R_s + \sigma L_s p) i_{ds}^e - \sigma L_s \omega_e i_{qs}^e \\ v_{qs}^e &= (R_s + \sigma L_s p) i_{qs}^e + \sigma L_s \omega_e i_{ds}^e + \omega_e \frac{L_m}{L_r} \lambda_{dr}^e \end{aligned} \tag{C.36}$$

The voltage equations (C.36) have coupling terms $\left(\sigma L_s \omega_e i_{qs}^e \text{ and } \sigma L_s \omega_e i_{ds}^e\right)$ and back EMF (electromagnetic force) term $(\omega_e (L_m/L_r)\lambda)$. These coupling and EMF terms must be controlled by using decoupling control and the back EMF compensation control. Otherwise, this vector control cannot control i_q and i_d separately due to these coupling terms.

The new decoupling voltage equations are

$$\begin{aligned} \bar{v}_{ds}^e &= v_{ds}^e + \sigma L_s \omega_e i_{qs}^e \\ \bar{v}_{qs}^e &= v_{qs}^e - \sigma L_s \omega_e i_{ds}^e \end{aligned} \tag{C.37}$$

By using Equation C.37, Equation C.38 can be derived

$$\begin{aligned} \bar{v}_{ds}^e &= (R_s + \sigma L_s p) i_{ds}^e \\ \bar{v}_{qs}^e &= (R_s + \sigma L_s p) i_{qs}^e + \omega_e \frac{L_m}{L_r} \lambda_{dr}^e \end{aligned} \tag{C.38}$$

To compensate for the back EMF term $\omega_e(L_m/L_r)\lambda$, we add $-\omega_e(L_m/L_r)\lambda$ to $\bar{v}_{qs}^e$. So, the new compensation voltage becomes

$$\tilde{v}_{qs}^{e} = \bar{v}_{qs}^{e} - \omega_e \frac{L_m}{L_r} \lambda_{dr}^{e} \tag{C.39}$$

Finally, new current equations, decoupling the coupling terms and compensating for the back EMF, can be written as

$$\begin{aligned} \frac{d}{dt} i_{ds}^{e} &= -\frac{R_s}{\sigma L_s} i_{ds}^{e} + \frac{1}{\sigma L_s} \bar{v}_{ds}^{e} \\ \frac{d}{dt} i_{qs}^{e} &= -\frac{R_s}{\sigma L_s} i_{qs}^{e} + \frac{1}{\sigma L_s} \bar{v}_{qs}^{e} \end{aligned} \tag{C.40}$$

Reference

1. B.K. Bose, *Modern Power Electronics and AC Drives*, Prentice-Hall PTR, Upper Saddle River, New Jersey, 2002.

Appendix D: Coordinate Transformation

D.1 *abc* Phase to *d–q*-Reference Frame Transformation

$$f_{qd0} = \begin{bmatrix} f_q \\ f_d \\ f_o \end{bmatrix} = Tf_{abc} = T \begin{bmatrix} f_a \\ f_b \\ f_c \end{bmatrix}$$

$$\begin{bmatrix} f_a \\ f_b \\ f_c \end{bmatrix} = T^{-1} \begin{bmatrix} f_q \\ f_d \\ f_o \end{bmatrix} \tag{D.1}$$

where

$$T = \frac{2}{3} \begin{bmatrix} \cos\theta & \cos(\theta-\gamma) & \cos(\theta+\gamma) \\ \sin\theta & \sin(\theta-\gamma) & \sin(\theta+\gamma) \\ \frac{1}{2} & \frac{1}{2} & \frac{1}{2} \end{bmatrix}; \quad \gamma = \frac{2\pi}{3}$$

$$T^{-1} = \begin{bmatrix} \cos\theta & \sin\theta & 1 \\ \cos(\theta-\gamma) & \sin(\theta-\gamma) & 1 \\ \cos(\theta+\gamma) & \sin(\theta+\gamma) & 1 \end{bmatrix}; \quad \theta: \text{arbitrary angle}$$

D.2 Synchronous Reference Frame

Assuming that q-axis is rotating with synchronous speed where $\omega_e = 2\pi f_e$, the phase angle difference between the q-axis and a-axis is

$$\theta = \theta_e = \omega_e t \tag{D.2}$$

The phase angle difference is shown in Figure D.1.

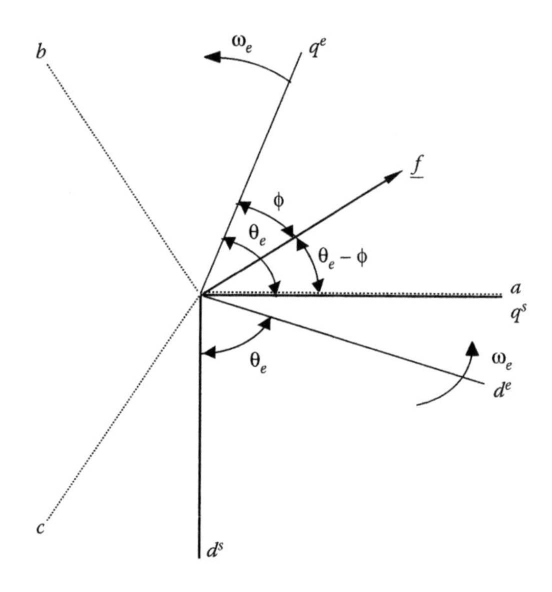

FIGURE D.1
Space vectors between the abc-axis and dq-axis. (Adapted from W. Na, A study on the output voltage control strategies of 3-phase PWM inverter for an uninterruptible power supply, Korea Master thesis, Kwangwoon University, Seoul, 1997.)

Using Equation D.1, the synchronous reference frame on the q-axis and d-axis is

$$\begin{aligned} f_q^e &= \frac{2}{3}[(f_a + f_b\cos\gamma + f_c\cos\gamma)\cos\theta_e + (f_b\sin\gamma - f_c\sin\gamma)\sin\theta_e)] \\ &= \frac{2}{3}\left(f_a - \frac{1}{2}f_b - \frac{1}{2}f_c\right)\cos\theta_e + \frac{1}{\sqrt{3}}(f_b - f_c)\sin\theta_e \end{aligned} \tag{D.3}$$

$$\begin{aligned} f_d^e &= \frac{2}{3}[(f_a + f_b\cos\gamma + f_c\cos\gamma)\sin\theta_e + (-f_b\sin\gamma + f_c\sin\gamma)\cos\theta_e)] \\ &= \frac{2}{3}\left(f_a - \frac{1}{2}f_b - \frac{1}{2}f_c\right)\sin\theta_e - \frac{1}{\sqrt{3}}(f_b - f_c)\cos\theta_e \end{aligned} \tag{D.4}$$

As long as the condition $f_a + f_b + f_c = 0$ is satisfied, Equations D.3 and D.4 can be converted into Equations D.5 and D.6, respectively, with two variables.

$$\begin{aligned} f_q^e &= f_a\cos\theta_e + \frac{1}{\sqrt{3}}(f_b - f_c)\sin\theta_e \\ &= f_a\cos\theta_e + \frac{1}{\sqrt{3}}(f_a + 2f_b)\sin\theta_e \end{aligned} \tag{D.5}$$

$$
\begin{aligned}
f_d^e &= f_a \sin\theta_e - \frac{1}{\sqrt{3}}(f_b - f_c)\cos\theta_e \\
&= f_a \sin\theta_e - \frac{1}{\sqrt{3}}(f_a + 2f_b)\sin\theta_e
\end{aligned}
\tag{D.6}
$$

D.3 Stationary Reference Frame

Assuming that the q-axis is aligned with the a-axis ($\theta = 0$), the synchronous frame on the q-axis and d-axis is

$$
\begin{aligned}
f_q^e &= \frac{2}{3}(f_a + f_b \cos\gamma + f_c \cos\gamma) \\
&= \frac{2}{3}\left(f_a - \frac{1}{2}f_b - \frac{1}{2}f_c\right)
\end{aligned}
\tag{D.7}
$$

$$
\begin{aligned}
f_d^e &= \frac{2}{3}(-f_b \sin\gamma + f_c \sin\gamma) \\
&= -\frac{1}{\sqrt{3}}(f_b - f_c)
\end{aligned}
\tag{D.8}
$$

For the condition $f_a + f_b + f_c = 0$, we obtain

$$f_q^s = f_a \tag{D.9}$$

$$f_d^s = -\frac{1}{\sqrt{3}}(f_b - f_c) = -\frac{1}{\sqrt{3}}(f_a + 2f_b) \tag{D.10}$$

D.4 Transformation between a Synchronous Frame and a Stationary Frame

Using Equation D.11, the stationary reference frame can be converted into the synchronous reference frame.

$$\begin{bmatrix} f_q^e \\ f_d^e \end{bmatrix} = \begin{bmatrix} \cos\theta & -\sin\theta \\ \sin\theta & \cos\theta \end{bmatrix} \begin{bmatrix} f_q^s \\ f_d^s \end{bmatrix} \tag{D.11}$$

Using Equation D.12, the stationary reference frame can be converted into the synchronous reference frame.

$$\begin{bmatrix} f_q^s \\ f_d^s \end{bmatrix} = \begin{bmatrix} \cos\theta & \sin\theta \\ -\sin\theta & \cos\theta \end{bmatrix} \begin{bmatrix} f_q^e \\ f_d^e \end{bmatrix} \tag{D.12}$$

Appendix E: Space Vector PWM

As mentioned in Chapter 7, SVPWM provides the maximum sinusoidal phase voltage $V_{max} = 0.57735\ V_{dc}$, which is 15.5% higher than the sinusoidal PWM [1]. Before we discuss SVPWM, the concept of the space voltage vectors in the inverter must be understood first with the help of the following voltage equations.

The voltages in Figure E.1 can be written as

$$\begin{aligned} v_{Aa} &= L_f \frac{d}{dt} i_{Aa} + v_{Ca} \\ v_{Ab} &= L_f \frac{d}{dt} i_{Ab} + v_{Cb} \\ v_{Ac} &= L_f \frac{d}{dt} i_{Ac} + v_{Cc} \end{aligned} \tag{E.1}$$

where L_f is inductance of the LC filter, and other voltages and currents can be found in Figure C.1.

The voltages in the stationary reference frame have the same form as those in Equation E.1:

$$\begin{aligned} v_{Aq}^s &= L_f \frac{d}{dt} i_{Aq}^s + v_{Cq}^s \\ v_{Ad}^s &= L_f \frac{d}{dt} i_{Ad}^s + v_{Cd}^s \end{aligned} \tag{E.2}$$

The space vector form of Equation E.2 is

$$V_{Aq}^s = L_f \frac{d}{dt} I_{Aq}^s + V_{Cq}^s \tag{E.3}$$

The right-hand side of Equation E.3, the capacitor voltage V_{Cq}^s in space vector form, is

$$V_C^s = V^* e^{j\theta} \tag{E.4}$$

where θ is an arbitrary phase angle in the space vector.

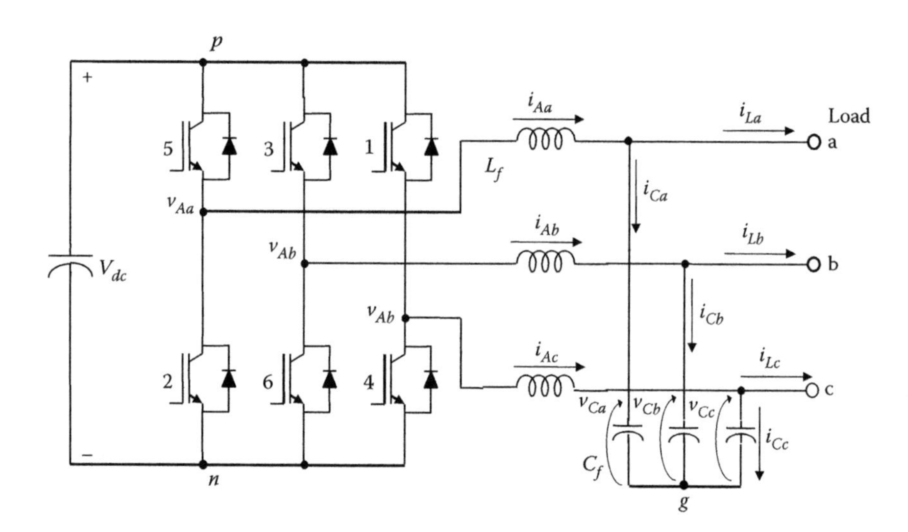

FIGURE E.1
Three-phase inverter with an LC filter. (Adapted from W. Na, A study on the output voltage control strategies of 3-phase PWM inverter for an uninterruptible power supply, Korea Master thesis, Kwangwoon Univerisity, Seoul, 1997.)

To obtain space voltage vectors in the inverter system given in Figure E.1, the switching function of each leg in the inverter can be defined as

$$S_a, S_b, S_c = \begin{cases} 0, \text{Negative leg is ON} \\ 1, \text{Positive leg is ON} \end{cases} \tag{E.5}$$

From the neutral ground point g, using the switching functions (E.5), the new inverter output voltages are

$$\begin{aligned} v_{Aa} &= S_a V_{dc} + v_{ng} \\ v_{Ab} &= S_b V_{dc} + v_{ng} \\ v_{Ac} &= S_c V_{dc} + v_{ng} \end{aligned} \tag{E.6}$$

where v_{ng} is the potential difference between the negative point n and the neutral ground g, and V_{dc} is the DC link input voltage of the inverter.

The new voltage equations of Equation E.6 in the stationary reference frame are

$$\begin{aligned} v_{Aq}^s &= V_{dc} S_q \\ v_{Aq}^s &= V_{dc} S_d \end{aligned} \tag{E.7}$$

According to the coordinate transformations given in Appendix D,

$$S_q = \frac{2}{3}\left(S_a - \frac{1}{2}S_b - \frac{1}{2}S_c\right)$$
$$S_d = \frac{1}{\sqrt{3}}(S_b - S_c) \tag{E.8}$$

and according to the switching function S_a, S_b, and S_c, the eight different states should exist. The detailed changes of the switching function are displayed in Table E.1 and the calculation of the switching time is explained here.

The SVPWM technique is a special technique for determining the switching-state sequence of the inverter. The inverter has eight possible switching states. As shown in Figure E.2, these states are mapped on the d–q axis. There are six valid voltage space vectors (V_1–V_6) and two null- or zero-voltage vectors (V_0, V_7), shown in Table E.1. A null- or zero-voltage vector occurs when all the upper switches are on simultaneously and the other bottom switches are off, or all the upper switches are off and the other bottom switches are on. This case leads to shorten the load and therefore, zero-voltage output is generated. For example, assuming that the reference voltage vector V^* is in area I, to generate the PWM output voltage, the adjacent voltage vectors V_1 and V_2 need to be used [2]. The switching cycle can be calculated by using Equation E.9 [3].

TABLE E.1

Changes of the Switching Function According to Switching States

Voltage Vector	Conducting Switches			S_a	S_b	S_c	S_q	S_d
V_0	2	4	6	0	0	0	0	0
V_1	6	1	2	1	0	0	$\frac{2}{3}$	0
V_2	1	2	3	1	1	0	$\frac{1}{3}$	$-\frac{1}{\sqrt{3}}$
V_3	2	3	4	0	1	0	$-\frac{1}{3}$	$-\frac{1}{\sqrt{3}}$
V_4	3	4	5	0	1	1	$-\frac{2}{3}$	0
V_5	4	5	6	0	0	1	$-\frac{1}{3}$	$\frac{1}{\sqrt{3}}$
V_6	5	6	1	1	0	1	$\frac{1}{3}$	$\frac{1}{\sqrt{3}}$
V_7	1	3	5	1	1	1	0	0

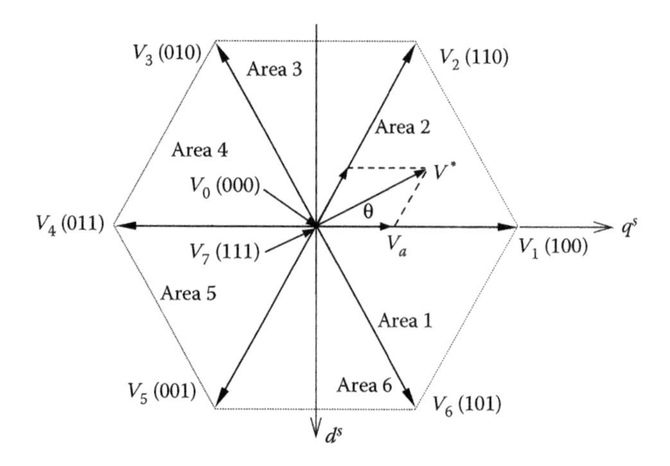

FIGURE E.2
Space vector diagram.

$$\int_0^{T_c} V^* dt = \int_0^{V_a} V_1 dt + \int_{V_a}^{V_a+V_b} V_2 dt + \int_{V_a+V_b}^{T_s} (V_0 \text{ or } V_7) dt \tag{E.9}$$

or

$$V^* \cdot T_c = V_1 \cdot t_a + V_2 \cdot t_b + (V_0 \text{ or } V_7) \cdot T_0 \tag{E.10}$$

where

$$T_a = \frac{V_a}{V_1} T_c$$
$$T_b = \frac{V_b}{V_2} T_c \tag{E.11}$$
$$T_0 = T_c - (T_a + T_b)$$

To minimize the switching frequency of each inverter leg, the switching pattern is arranged in a way shown in Figure E.3. By doing so, the symmetrical pulse pattern for two consecutive T_c intervals is created.

In Figure E.3, $T_s = 2T_c = 1/f_s$ (f_s = switching frequency) is the sampling frequency [2], and as shown in Figure E.3, during the first-half period, the switching sequence has to be (000 → 100 → 110 → 111) and in the following second period, the switching sequence has to be reverse (111 → 110 → 100 → 000). The zero- and null-voltage vectors (V_0, V_7) need to be replaced efficiently such that minimal output harmonics and torque ripple can be produced [3].

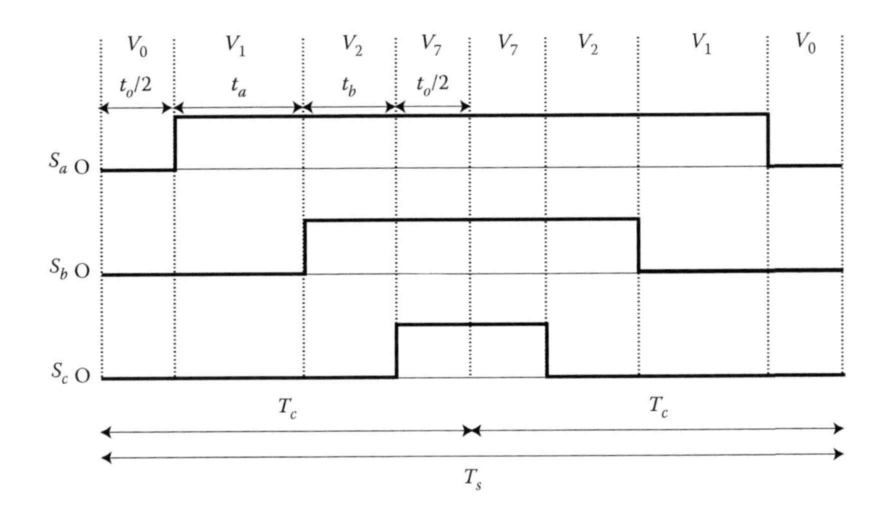

FIGURE E.3
Symmetrical pulse pattern for the three-phase inverter. (Adapted from B.K. Bose, *Modern Power Electronics and AC Drives*, Prentice-Hall PTR, Upper Saddle River, New Jersey, 2002.)

References

1. R. Valentine, *Motor Control Electronics Handbook*, McGraw-Hill, New York, 1998.
2. B.K. Bose, *Modern Power Electronics and AC Drives*, Prentice-Hall PTR, Upper Saddle River, New Jersey, 2002.
3. H.W.V.D. Broeck, H.C. Skudelny, and G.V. Stanke, Analysis and realization of a pulse modulator based on voltage space vectors, *IEEE Transactions on Industry Applications*, 24(1), 142–150, 1988.
4. W. Na, A study on the output voltage control strategies of 3-phase PWM inverter for an uninterruptible power supply, Korea Master thesis, Kwangwoon University, Seoul, 1997.

Index

F

K

L

O

R

S

T

V

W

Z